Treatise on Concrete

german concrete
the science of cement from trass to portland
1819-1877

Treatise on Concrete
Directed by Roberto Gargiani

salvatore aprea

german concrete the science of cement from trass to portland 1819-1877

EPFL Press

A Swiss academic publisher
distributed by Routledge

EPFL Press

The EPFL Press is the English language imprint of the Foundation of the Presses polytechniques et universitaires romandes (PPUR). The PPUR publishes mainly works of teaching and research of the Ecole polytechnique fédérale de Lausanne (EPFL), of universities and other institutions of higher education.

Presses polytechniques
et universitaires romandes
EPFL – Rolex Learning Center
Post office box 119
CH-1015 Lausanne, Switzerland
E-mail: ppur@epfl.ch
Phone: +41 (0)21 693 2130
Fax: +41 (0)21 693 4027

www.epflpress.org

ISBN 978-2-940222-96-4 (EPFL Press)
Printed in Italy

Routledge
Taylor & Francis Group

Taylor & Francis Group Ltd
2 Park Square, Milton Park
Abingdon, Oxford, OX14 4RN, UK

Routledge is an imprint of Taylor & Francis Group, An informa business.

Simultaneously published in the USA and Canada by Routledge, 711 Third Avenue, New York 10017

www.routledge.com

Library of Congress Cataloging-in-Publication Data
A catalog record for this book is available from the Library of Congress.

The graphical layout of the book and the cover design are by Paola Ranzini Pallavicini of Studio Pagina Milan, Italy. The photolithographic services were provided by Walter Bassani, Milan Italy.

ÉCOLE POLYTECHNIQUE
FÉDÉRALE DE LAUSANNE

Series Editors for the publication program in Architecture are Professors Jacques Lucan and Luca Ortelli.

The publisher and author express their thanks to the Ecole polytechnique fédérale de Lausanne (EPFL) for its generous support towards the translation of this book.

The author is grateful to the Institute of Architecture of the EPFL, who provided additional support.

Treatise on Concrete

Directed by Roberto Gargiani

The volumes of the *Treatise on Concrete* intend to offer a new and documented vision of the evolution of construction techniques and of the creative potential and formal expressions of concrete, from antiquity to the present time. Each book in the series will be based on specific archival research; the analysis of scientific, technical and iconographic original sources; and the investigation into documents made available by companies, engineers, architects and artists.
The full spectrum of the use of concrete will be covered, including the building of foundations, ports, moles, bridges, vaults, domes, fortresses, bunkers, houses and monuments, as well as the manufacture of artificial stones and sculptures. In addition, the volumes will address the theoretical issues of *béton brut*, Brutalism and the true nature of matter.
All volumes focus on the techniques and materials for the manufacture of formwork, from wood to fabric; on the meaning of the imprint; on the surface processing of concrete by hand or with machine; and on the composition of the mixture.
The series of books aims to provide the scientific community with an unique and updated reference on the historical and contemporary uses of concrete, which will open a new chapter in the knowledge of this crucial material for architectural construction, as well as for masterpieces of engineering and art.

Volumes published:
Concrete, From Archeology to Invention, 1700–1769
Louis I. Kahn, Exposed Concrete and Hollow Stones, 1949-1959
Louis I. Kahn, Towards the Zero Degree of Concrete, 1960-1974
The Rhetoric of Pier Luigi Nervi, Concrete and Ferrocement Forms

Volumes in preparation:
Tadao Ando, Concrete Spaces
Mnemosyne in Concrete
Research on the Roman Cement, 1770-1779
Concrete Sculpture, from 1950 to Today: Visions of a Material
Contemporary Ornament in Concrete
Bared Concrete of Natural and Artificial Pozzolanas, 1780-1817
Formless: Beyond Formwork
The Concrete of Perret and Pupils

Table of Contents

S

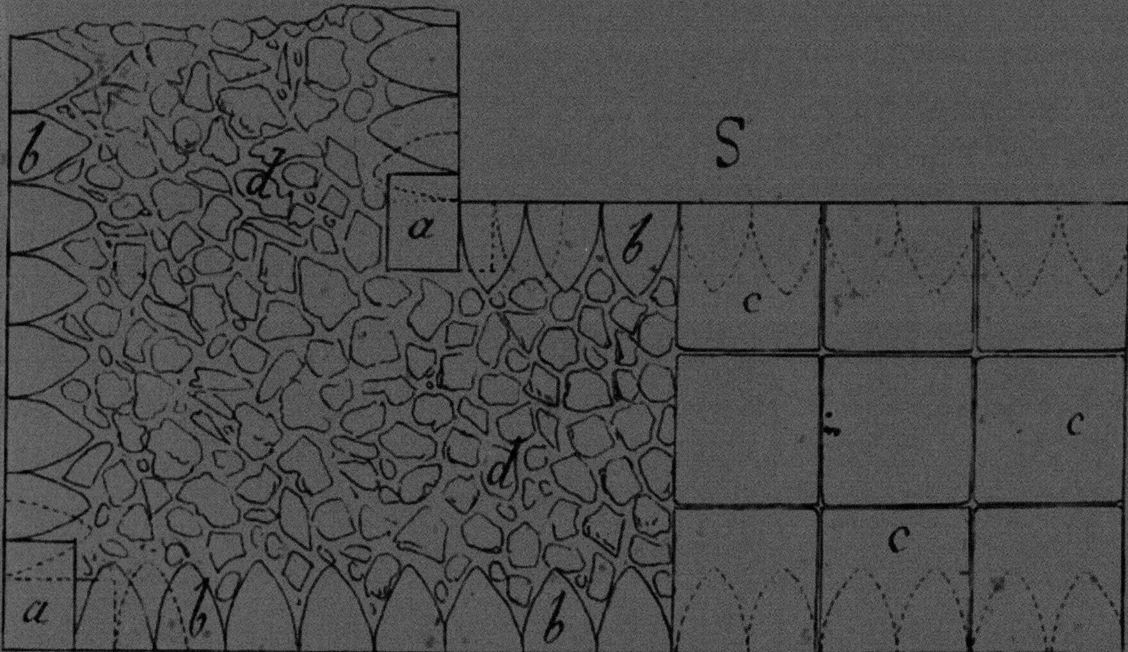

V

T

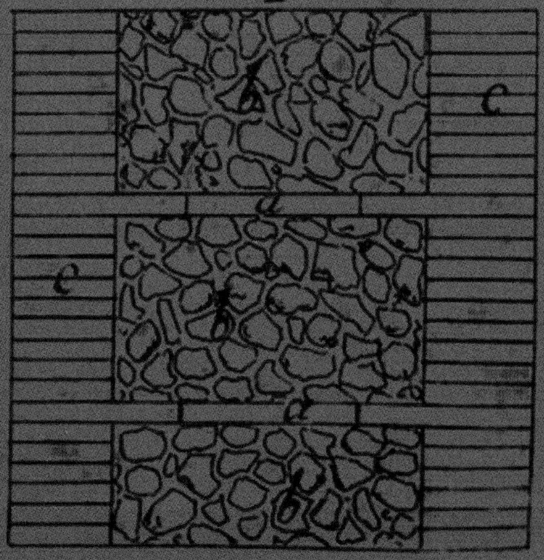

chapter one
lime, trass, mortar, concrete and artificial stones at the turn of the 19th century

Premises for the development of concrete

Around the turn of the 19th century, the culture of the Enlightenment, alongside the emerging needs for modernisation and industrialisation, give rise to a particular interest in technical aspects of construction. Originating in the scientific, mathematical and technical approach to architecture that several theoreticians of the Enlightenment promoted, this becomes stronger in the late 18th century through the unification and modernisation processes by which Germany is evolving from being an aggregate of several backward states and small political entities still based around medieval forms of administration and an agriculture-based economy into a united, modern and industrialised country.[1]

As it turns out, the construction of a national, cultural identity is linked to, inter alia, the ascription of a specific character to the German art of building. One peculiar aspect of this character is a prevailing interest in technical matters – considered a kind of German distinctiveness, as opposed to the otherwise dominant aesthetical argumentations concerning ancient Italian architectural styles. The *Ober-Bau-Department* (Building Department) of Prussia – the state that will lead the German unification process – strongly promotes this German way of approaching architecture from a technical perspective, as is evidenced in the preface to the first issue of the "Sammlung nützlicher Aufsätze und Nachrichten, die Baukunst betreffend," a periodical edited by the *Ober-Bau-Department* from 1797 to 1806.[2] The prevailing artistic issues concerning Italian architecture are criticized, notably targeting those treatises

[1] The reference goes to: Nicolaus Goldmann (1611–1665), Friedrich August Wolf (1759–1824), Leonhard Christoph Sturm (1669–1719), Johann Friedrich Penther (1693– 1749) and Lorenz Johann Daniel Suckow (1722–1801). See Mitglieder des König. Preuss. Ober–Bau–Departements, *Vorrede*, in, "Sammlung nützlicher Aufsätze und Nachrichten, die Baukunst betreffend," I, 1797, (pp. III–IX), p. IV.

[2] The Prussian *Ober-Bau-Department* is then changed into *Ober-Bau-Deputation* in 1804. See Katja Laude, *Das Oberbaudepartement und die Untertanenbauten Möglichkeiten und Grenzen obrigkeitlicher Einflussnahme*, in, "Brandemburgische Denkmalpflege," XVIII, 2009, pp. 61–75).

that focus on architectural orders, "as if raising and decorating a column were of major importance, whereas all other features that the art of building provides to humans, in each of their undertakings, were irrelevant."[3] Instead, the German technical attitude is described as the origin of architectural improvements, in opposition to the impressive constructions of the past in celebrated ancient lands that remained in ruins. However, this does not involve a rejection of the ancient art of building. Quite the contrary, ancient constructions inspire interest for both their aesthetical and technical aspects. As such, technical knowledge and building practice are considered on par with art and theory. A passage from the aforementioned preface refers directly to this. "We could be further developed if art and knowledge had sooner and deeper joined together in architecture. If combined, they can achieve a lot, but not if each of them goes its own way. Artists risk weirdness without a scientific guide, while savants lose their capability to reflect when they are robbed of artistic vision and feelings."[4]

At the same time as these theoretical argumentations are developed, early policies for modernisation and industrialisation suggest a pressing need to improve construction standards. The conditions of buildings, roads, and even entire villages, towns and landscape are to be developed, for which the availability of solid and durable building materials is crucial.[5] This appears particularly critical in the domain of hydraulic engineering, since, at the turn of the 19th century, many works are still made of timber and perishable materials such as bundle, ropes and clay, while the development of manufacturing and trade requires the construction of efficient roads, railways, waterways and ports provided with solid and durable locks, weirs, bridges and embankments.

The use of masonry, which already has a long tradition in the German construction culture, is encouraged in place of timber and perishable materials.[6] Consequently, increasing attention is paid to the suitable materials for producing mortar, since, as Johann Albert Eytelwein (1765–1849), a main member of the *Ober-Bau-Department* and of the Berlin *Bauakademie*, asserts,

[3] "Als ob die Aufstellung und Verzierung einer Säule die Hauptsache, das Übrige aber was die Baukunst dem Menschen fast bey allen seinen Unternehmungen leistet, nur Nebensache wäre" (Mitglieder des König. Preuss. Ober–Bau–Departements, *Vorrede*, cit., pp. III–IV).

[4] "Weiter würden wir wahrscheinlich schon seyn, wenn Kunst und Gelehrsamkeit sich früher und inniger für die Architektin: vereiniget hätten. Gemeinschaftlich können beyde viel leisten, aber wenn jede ihren eigenen Gang gehet, so ist dieses nicht möglich. Ohne wissenschaftliche Leitung fällt der Künstler in das Gewagte und Abentheuerliche; ohne Künstlerauge und Kunstgefühl entgehet dem Gelehrten der Stoff zur Betrachtung" (*ibid.*, pp. IV–V).

[5] The publishing of several periodicals about architecture, building, gardens and even taste, since about the 1790s, is further evidence of the general desire to improve living conditions; the main ones are: "Allgemeine Magazin für die Bürgerliche Baukunst" (1789–96), "Magazin für freunde des guten Geschmacks" (1794–1799), "Ideen Magazin für Liebhaber von Gärten, englischen Anlagen und für Besitzer von Landgütern" (1796–1808), "Archiv für die Baukunst und ihre Hülfswissenscheften" (1818), "Monatsblatt für Bauwesen und Landesverschönerung" (1821–1830). For an overview of the German nineteenth-century architecture periodicals, see Rolf Fuhlrott, *Deutschsprachige Architektur-Zeitschriften. Entstehung und Entwicklung der Fachzeitschriften für Architektur in der Zeit 1789–1918*, München, Verlag Dokumentation, 1975.

[6] In this regard, a special contribution is given by the policies that the Prussian *Ober-Bau-Department* undertakes since the last decades of the 18th century in order to preserve forests by supporting the use of masonry instead of timber. Prussian efforts to improve building conditions show positive consequences already at the end of the 18th century, as the words of an English traveller entering in Prussia via Hannover and Brunswick in 1799 demonstrate: "Roads, plantations, net cottages, pleasant country estates, well built towns and good inns take the place of the appearance of poverty and depopulation" (David Watkin, *German Architecture and the Classical Ideal*, Boston, MIT Press, 1987, p. 2).

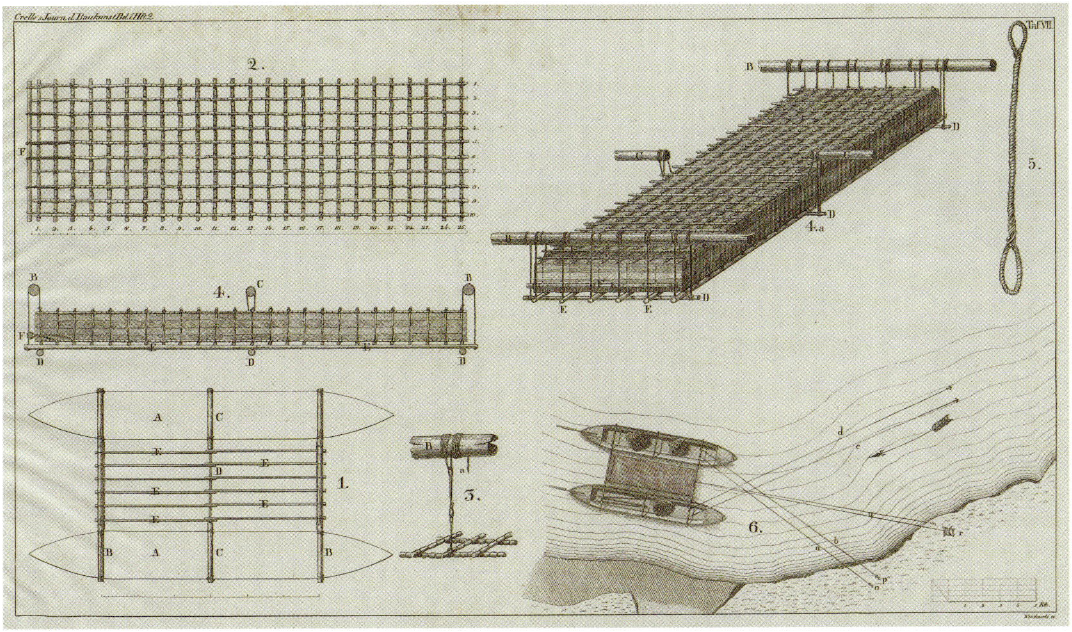

1. Spielhagen, traditional, hydraulic foundation made of bundle and ropes. Construction drawings (Spielhagen, *Beschreibung der bei den Bau des Ueberfalles in der Elbe bei Magdeburg angewendeten Sinkstücke, nebst einleitenden Nachrichten vom Zustande der Elbe bei Magdeburg vor dem Bau des Ueberfalles, und von dem Durchbruche desselbe*, in, "Journal für die Baukunst," vol. I, 1829).

2. J. A. Eytelwein, retaining walls and mill to grind bricks. Construction drawings (J. A. Eytelwein, *Praktische Anweisung zur Wasserbaukunst, welche eine Anleitung zum Entwerfen, Veranschlagen und Ausführen der am gewöhnlichsten vorkommenden Wasserbaue enthält*, Berlin, Realschulbuchhandlung, vol. III, 1805).

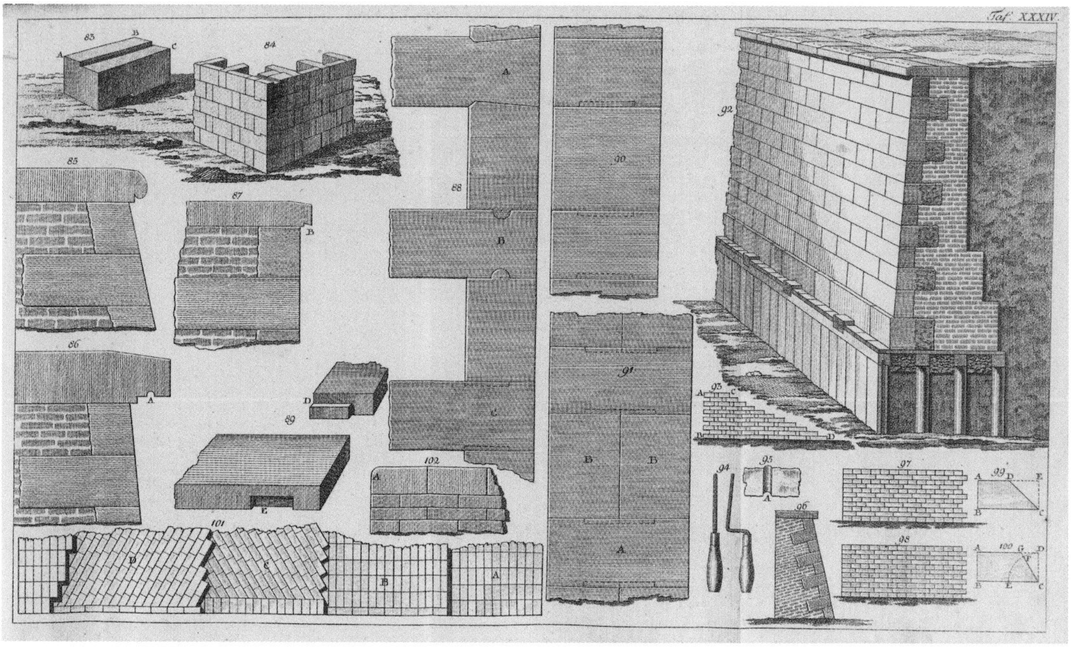

3. J. A. Eytelwein, retaining walls. Construction drawings, brick and stone bonds (J. A. Eytelwein, *Praktische Anweisung zur Wasserbaukunst, welche eine Anleitung zum Entwerfen, Veranschlagen und Ausführen der am gewöhnlichsten vorkommenden Wasserbaue enthält,* Berlin, Realschulbuchhandlung, vol. III, 1805).

"Taking the best care to choose good stones is not enough to build a good retaining wall; the same care should be taken in selecting materials to produce mortar."[7] Contemporaneously, several German master builders survey building materials and techniques abroad in more developed neighbouring countries, while the scarcity of natural stones in several German regions, together with the cost of transporting them in light of the lack of modern roads, waterways and railways, results in efforts to develop mortar suitable to mould artificial stones, decorations and ordinary objects such as basins and troughs.

The development of German, modern concrete is part of this general renewal process in construction practices, which involves the improvement of building materials already in use, along with the introduction of new ones. This is made possible through an extraordinary combination of studies about lime and mortar, surveys into ancient and recent building materials and techniques, early experimental uses of concrete and attempts at producing moulded artificial stones, decorations and ordinary objects. The earliest phase of this process, approximately from the late 18th century to the end of the 1810s, is still evolving in the context of the 18th-century building culture. Studies and surveys into lime and mortar principally analyse the state-of-the-art knowledge, while some early rudimentary kinds of concrete and mortar, used to mould artificial stones and objects, are produced with typical materials of the 18th century.

[7] "Bei aller Sorgfalt welche auf die gute Auswahl der Steine verwandt seyn kann wird man sich doch nicht schmeicheln können eine tüchtige Futtermauer zu erhalten, wenn nicht das Material zum Mörtel mit der großten Aufmerksamkeit ausgewählt und zubereitet wird" (Johann Albert Eytelwein, *Praktische Anweisung zur Wasserbaukunst, welche eine Anleitung zum Entwerfen, Veranschlagen und Ausführen der am gewöhnlichsten vorkommenden Wasserbaue enthält*, Berlin, Realschulbuchhandlung, vol. III, 1805, p. 39).

[8] The notion of substance of heat originates from the concept of phlogiston that Johann Joachim Becher (1635–1682) and Georg Ernst Stahl (1659–1734) develop in the second half of the 17th century.

[9] About the notion of chemical affinity, see Ferdinando Abbri, *La chimica del Settecento*, Torino, Einaudi, 1978, chap. IV.

[10] Paul Ludwig Simon, *Ueber die Natur des Kalksteins*, in, "Sammlung nützlicher Aufsätze und Nachrichten die Baukunst betreffend," III, 1799, vol. I, pp. 96–112; Id., *Ueber die Natur des Kalksteins. Chemisches Verhalten des Kalkstein, Theorie des Kalks*, in, "Sammlung nützlicher Aufsätze und Nachrichten die Baukunst betreffend," IV, 1800, vol. I, pp. 53–78; Id., *Ueber die Natur des Kalksteins. Analyse der*

The state-of-the-art knowledge about lime, between alchemical heritage and scientific achievements

Early nineteenth-century knowledge about lime originates from ancient Roman building practices, which the eighteenth-century passion for archaeology pushed to study, and from scientific findings about the nature and properties of lime made possible by the young science of chemistry – despite the survival of some alchemical beliefs. Lime is thought to be a substance produced from raw materials containing a combination of carbonic acid, crystallized water and lime earth, which is also called carbonated lime. Limestone is the most frequently used raw material in producing lime for constructions. Fired in kilns, it is thought to lose water and carbonic acid, while lime earth combines with a supposed existing substance of heat and forms quick lime.[8] It is believed that this tends to build its original state up again, combining with the same kinds of substances that are lost during burning. For this reason, quick-lime is supposed to combine with water, release the substance of heat and produce the so-called slaked lime, which is then thought to combine with carbonic acid from the air to provoke the phenomenon of hardening that changes it into a kind of reconstituted limestone.[9]

Most German master builders, architects, engineers and even chemists from the early 19th century consider such chemical explications reliable. In an extended survey about limestone, lime and mortar, which is published in several issues in the "Sammlung nützlicher Aufsätze und Nachrichten, die Baukunst betreffend," between 1799 and 1800,[10] Paul Ludwig Simon (1771–1815), a notable master builder and teacher at the Berlin *Bauakademie*, maintains that "just the inclination [of lime] to combine with the water that has been lost during burning, and the capability to fix it as crystallized water, make the use of lime for the production of mortar possible."[11] The celebrated master builder David Gilly (1748–1808), a prominent member of the *Ober-Bau-Department* and the Berlin *Bauakademie*, agrees with Simon. In 1805, he describes chemical combinations concerning lime as follows: "Through burning, lime earth undergoes an essential transformation. It is set free from carbonic acid and water, with which it was initially combined by nature, while the substance of heat pervades it. Later on, during the slaking process, the substance of heat parts with lime, which, instead, combines with water, due to great affinity between lime and water. Slaked lime has the property of absorbing carbonic acid from the air, adhering to surfaces of smooth bodies, and hardening to the point of becoming like a stone."[12]

Eytelwein essentially shares the same belief, although with a slight difference. He believes in the affinity between water contained in slaked lime and carbonic acid, rather than in the affinity between lime and carbonic acid. "Pure lime earth," he writes, "strongly tends to combine with water, but it has a quite low tendency to combine with carbonic acid. Instead, water has a

Kalkstein - Arten, Verwendung des Kalks zum Mörtel, in, "Sammlung nützlicher Aufsätze und Nachrichten die Baukunst betreffend," IV, 1800, vol. II, pp. 84–119.

[11] "Eben diese Verwandtschaft sich mit dem beym Brennen verloren Antheil Wassers wieder zu vereinigen, diese Fähigkeit, das Wasser zum Chrystallisations - Wasser zu binden liegt vorzüglich bey der Verwendung des Kalks zum Mörtel zum Grunde" (Simon, *Ueber die Natur des Kalksteins. Analyse der Kalkstein*...cit., p. 103).

[12] "Durch dieses Brennen erleidet die Kalkerde eine wesentliche Veränderung. Sie wird nämlich von der im rohen Zustande darin gebundenen Luftsäure und dem Wasser befreit, dagegen aber von der Wärmematerie durchdrungen. Letztere trennt sich wieder von dem gebrannten Kalke beim Löschen, vermöge der nähere Verwandtschaft des Wassers zur Kalkerde, und dieser gelöschte Kalk erhält sodann die Eigenschaft, sich aus der atmosphärischen Luft von neuen nach und nach mit Luftsäure zu sättigen, an der Oberfläche glatter Körper zu haften, und damit steinartig zu verhärten" (David Gilly, *Handbuch der Land-Bau-Kunst, vorzüglich in Rücksicht auf die Construction der Wohn- und Wirthschafts-Gebäude für angehende Cameral-Baumeister und Oeconomen*, vol. I, Berlin, Friedrich Vieweg der alter, 1797, p. 109).

strong inclination to join with carbonic acid and, since the atmosphere is full of it, water contained in slaked lime first combines with carbonic acid in the air and, only after it is entirely vaporized, carbonic acid and lime combine and produce limestone."[13]

Despite such minor divergences, convictions about the chemical processes concerning lime are quite coherent with each other and have two important implications in building practices. The former is that lime is mainly considered an aerial binder that hardens only in the presence of air, while the latter concerns the fact that only limestones that contain much carbonated lime are considered appropriate raw materials for producing lime. They are defined "pure limestones" (*reine Kalksteine*) and are described as particularly white and hard. "Pure raw limestone […] contains no other components than lime earth, carbonic acid and water," Eytelwein asserts in this regard, and Gilly furthermore maintains: "the more limestone is hard, the more it gives good lime […]; therefore marble gives the best, hardest, and finest lime."[14] There are quite pure limestones in quarries near the village of Rüdersdorf, in East Berlin. In his studies about lime, Simon analyses Rüdersdorf limestones and establishes that the carbonated lime can reach 97% of the entire composition.[15]

Carbonated lime is also the main component of shells, which are therefore used to produce a kind of lime called *Muschelkalk*. The practice of producing lime from shells is very common in the Netherlands. In Germany, it is known through accounts given in some German technical literature of the second half of the 18th century. Engineer Friedrich August Alexander Eversmann (1759–1837) describes the Dutch production of *Muschelkalk* in a book published in 1792.[16] Economy professor Johann Beckmann (1739–1811) also writes about Dutch *Muschelkalk* in his treatise *Anlteitung zur Technologie*, which is published in six editions between 1777 and 1809.[17] Following such observations, *Muschelkalk* is produced in German Friesland, which is located near the Netherlands. It is documented that at least two lime kilns are producing lime from shells in the 1820s, the *Roßmann und Kipp* in the village of Stade, and the *Brunkhorst und Westphalen* in Buxtehude near Hamburg.[18]

Despite all theoretical beliefs about chemical processes pertaining to lime, practice shows that limestone also contain other substances besides carbonated lime. The most common ones are silica, clay and metal oxides. Since they are not supposed to have any chemical role in lime transformations,

[13] "So groß auch das Bestreben der reinen Kalkerde ist, sich mit dem Wasser zu verbinden, so ist doch ihre Verwandtschaft zur Kohlensäure nur äußerst gering. Dagegen hat das Wasser eine große Verwandtschaft zur Kohlensäure, und weil sie sich überall in der Atmosphäre verbreitet findet, so vereinigt sie sich leicht mit dem Wasser des gelöschten Kalks und verbindet sich nach der Abdünstung des Wassers mit der Kalkerde zum Kalkstein" (Eytelwein, *Praktische Anweisung zur Wasserbaukunst*...cit., vol. III, p. 40).

[14] "Der rohe Kalkstein […] enthält wenn er rein ist keine Bestandteile als Kalkerde, Kohlensäure und Wasser" (ibid, p. 39). "Je härter die Kalkstein sind, desto mehr und dest bessern Kalk geben sie […]; daher giebt der Marmor den besten, härtesten und feinsten Kalk" (Gilly, *Handbuch der Land-Bau-Kunst*...cit., p. 107). In this regard, see also, Simon, *Ueber die Natur des Kalksteins. Chemisches Verhalten*...cit., pp. 76–78; Johann Beckmann, *Anleitung zur Technologie, oder zur Kenntniß der Handwerke, Fabriken und Manufakturen, vornehmlich derer, welche mit der Landwirtschaft, Polizei und Cameralwissenschaft in nächster Verbindung stehen*, Göttingen, Vandehoeck und Ruprecht, (1777), 1802^5, pp. 309–311; Heinrich Gustav Flörcke, *Dr. Johann Georg Krünitz's Oekonomische Encyklopädie oder allgemeines System der Staats- Stadt- Haus- und Landwirthschaft*, vol. XCIV, Berlin, Joachim Pauli, 1804, p. 171.

[15] Simon, *Ueber die Natur des Kalksteins. Analyse der Kalkstein*...cit., pp. 101–103.

[16] See Friedrich August Alexander Eversmann, *Technologische Bemerkungen auf einer Reise durch Holland*, Freyberg, Annaberg, Crazische Buchhandlung, 1792, pp. 173–179.

[17] Beckmann, *Anleitung zur Technologie*...cit., p. 311. See also, Otto Christian Friedrich Reinhold, *Sammlung practischer Erfahrungen und Vorschriften, Cemente, Mörtel und Bétons betreffend*, in, "Journal für die Baukunst," vol. XV, 1841, (pp. 88–100, pp. 131–150), pp. 133–134.

[18] Friedrich Wilhelm Otto Ludwig Freiherr von Reden, *Bericht über die von dem Gewerbe-Vereine für das Königreich Hannover in den Monaten Mai und Juni 1835 veranstaltete erste Ausstellung inländischer gewerblicher Erzeugnisse*, in, "Mittheilungen des Gewerbe-Vereins für das Königreich Hannover," 1835, coll. 527–529.

these substances are considered as impurities. Too much clay is even considered harmful, since clay is thought to glaze around lime particles, hindering the combination of lime with water. In such cases lime is called *todgebrannter Kalk* (dead-fired lime).[19] These convictions notwithstanding, some kinds of limestone containing clay – called *Märgel* (marls) – are used to burn lime in regions where pure limestone is not available, as is the case in northern Brandenburg and Pomerania.[20] This kind of lime is called *Märgelkalk* and is considered a low-quality material.

The fact that hydraulic lime could be produced from limestone containing clay is generally not recognized at that time, apart from a few exceptions. One dates back to the time between the end of the 1770s and the beginning of the 1780s, as the chemist of the village of Bahrnberg in the principality of Branschweig-Lüneburg, a man named Jordan, claims to have succeeded in producing hydraulic lime from local marls. He envisions applying it to produce mortar suitable for hydraulic and ordinary works, to plaster walls and floors, and to produce moulded "bricks and tiles without fire."[21] However, Jordan's finding has no practical consequence apart from a couple of mentions in technical literature. The applications he imagines remain theoretical forerunners of what will eventually be done with hydraulic lime fifty to sixty years later.[22] A second exception concerns the mention of some kinds of natural hydraulic lime that the eminent hydraulic engineer Carl Friedrich Wiebeking (1762–1842) provides in his treatise *Allgemeine auf Geschichte und Erfahrung gegründete theoretisch-practische Wasserbaukunst*.[23] Wiebeking develops considerable knowledge about lime, mortar and even certain uses of concrete in different European countries. In fact, he often travels to examine the most advanced hydraulic works of that time and serves as chief hydraulic engineer for the Austro–Hungarian Empire from 1802 to 1805, before moving in 1806 to Bavaria, where he works as *Geheimer Rat, und Chef des technischen Geheimen Zentralbüros im Straßen- und Wasserbauwesen* (Councillor and Direktor of the Technical, Central Office for Roads and Waterways). He furthermore studies ancient and recent technical literature, especially Vitruvius' *De Architectura* and the *Architecture hydraulique* by Bernard Forest de Bélidor.[24] Wiebeking's vast knowledge and experience leads him to consider hydraulic lime as an exception to Vitruvius' rule establishing that the best lime is produced from pure limestone. "Vitruvius," he writes in 1805, "prefers lime

[19] See Eytelwein, *Praktische Anweisung zur Wasserbaukunst*...cit., vol. III, p. 39; Simon, *Ueber die Natur des Kalksteins - Chemisches Verhalten*...cit., pp. 77–78.

[20] See Gilly, *Handbuch der Land-Bau-Kunst*...cit., p. 105; Johann Georg Krünitz, *Oekonomische Encyklopädie oder allgemeines System der Staats- Stadt- Haus- und Landwirthschaft*, vol. XXXII, 1784, p. 658. Krünitz notes the preference for pure limestone asserting that, if transport of stones from Rüdersdorf were not expensive, it would be advisable to stop the production of lime from marls (*ibid.*, p. 658).

[21] Jordan, *Nachricht*, in, "Hannoversches Magazin," XX, 1782, coll. 301–304.

[22] Krünitz mentions the possibility of producing hydraulic lime from marls (Krünitz, *Oekonomische Encyklopädie*...cit., vol. XXXII, 1784, p. 730) and Heinrich Gustav Flörcke, who edits the ninety-fourth volume of Krünitz's encyclopaedia in 1804, even mentions Jordan's hydraulic lime (Flörcke, *Dr. Johann Georg Krünitz's*... cit., p. 341). However, in a 1796 book about the manufacturing and trade in the principality of Branschweig-Lüneburg, Jordan is listed among the producers of chemical products but not as a producer of hydraulic lime. In this regard, see Christian Ludwig Albrecht Patje, *Kurzer Abriß des Fabriken-, Gewerbe-, und Handlungszustandes in den Chur Braunschweig-Lüneburgischen Landen*, Göttingen, Vandenhoek und Ruprecht, 1796, p. 373.

[23] Carl Friedrich Wiebeking, *Allgemeine auf Geschichte und Erfahrung gegründete theoretisch-practische Wasserbaukunst*, Darmstadt, Johann Franz Peter Stahl, vol. I, 1798, vol. II, 1799, vol. III, 1801, vol. IV, 1805, vol. V, 1807.

[24] The eminent translation of Vitruvius' *De Architectura* by August von Rode (1751–1837) had just been issued in 1796 (see Marcus Vitruvius Pollio, *Des Marcus Vitruvius Pollio Baukunst*, ed. August von Rode, Leipzig, Georg Joachim Göschen, voll. I , II, 1796). See also, Bernard Forest de Bélidor, *Architecture hydraulique*, 2nd part, vol. II, Paris, Charles-Antoine Jombert, 1753.

burnt from hard limestone to build masonries, and suggests using lime burnt from soft limestone for plastering. This statement is still true in most cases, but there are some exceptions […]. Some kinds of lime need to be used right away after slaking, like lime from Padua, which contains a large amount of silica, whereas others need to be slaked in pits and stay there for one year. Lime from Chorowitz in Moravia also needs to be used quickly, just like lime from Liège, which is well known in the Netherlands, and all other kinds of lime that contain a large amount of silica clay and iron, and are therefore suitable to compose hydraulic mortar."[25]

Nevertheless, Wiebeking's knowledge about hydraulic lime remains pioneering at the turn of the 19th century, as most master builders persist in praising the quality of aerial lime produced from pure limestone.

Cement, the practices of producing hydraulic mortar and the importance of trass

Coherent with the convictions about the effectiveness of aerial lime, German master builders of the early 19th century believe that the best kind of mortar is composed of only aerial lime and sand. They even think that it is waterproof, once hardened. "Only through the combination with carbonic acid," Simon writes, "slaked lime is anew transformed into raw limestone, and it can neither be dissolved in water, nor combine with other substances."[26] Gilly furthermore asserts: "Lime mixed with sand gives the best mortar, which would even be suitable for hydraulic works, if such works could dry for at least one year long before they come into contact with water."[27] When it is not possible to let mortar harden out of water, lime is usually mixed with some burnt clayish substances, thought to absorb water and accelerate the hardening process. "Since it is not always possible to let mortar set and harden [in a dry place] for one year," recommends Gilly, "it is necessary to use other materials instead of sand, which are partly made of burnt clayish substances, absorb water and cause a faster lime hardening."[28] While in the context of the eighteenth-century construction culture, these materials are generally called cements, in the German technical literature of the turn of the 19th century, the term cement is often used to mean mixtures of lime and burnt clayish substances. "We call cement," Gilly states, "mortar

[25] "Schon Vitruv Cap. V. Lib. II. giebt dem aus harten Kalksteinen gebrannten Kalk vor den aus weichen gebrannten zu den Mauern den Vorzug und will den letzten zur Bekleidung verwendet wissen. In der Regel ist diese Angabe auch noch jetzt richtig, aber es giebt Ausnahmen […]. Einige Kalkarten müssen gleich nach geschehener Löschung verarbeitet werden, wie z. B. der Paduanische Kalk, welcher viele Kieselerde enthält; andere müssen gelöscht in der Kalkgrübe ein Jahr und länger bleiben. Der Mährische-Chorowitzer muß z. B. bald verbraucht werden; eben so wie der in Holland bekannte Lütticher Kalk und alle Kalkarten, die zu Wassermörtel gut sind, d. i. viel Kiesel-, Thon- und Eisenerde enthalte" (Wiebeking, *Allgemeine auf Geschichte*...cit., vol. IV, 1805, p. 307).

[26] "Nun durch den Beytritt der Kohlensäure, der gelöschte Kalk wieder zum rohen Kalk umgeändert wird, und in diesem Zustande weder Auflösbarkeit, noch Hang zur Verbindung mit andern Körper wahrnehmen lässt" (Simon, *Ueber die Natur des Kalksteins. Analyse der Kalkstein*...cit., p. 108).

[27] "Kalk mit Sand vermischt, giebt den besten Mörtel, der auch selbst zum Wasserbau brauchbar seyn Würde, wenn ein damit gemachter Bau nicht wenigstens ein ganzes Jahr Zeit haben müßte, um gehörig auszutrocknen, ehe man denselben dem Einwirken des Wassers bloßsetzt" (Gilly, *Handbuch der Land-Bau-Kunst*... cit., p. 118).

[28] "Da aber dieses nicht allemal möglich ist, dem Mörtel ein Jahr zum Trocknen und Binden zu lassen, so werden zu dessen Zubereitung anstatt des Sandes andere Materien gewählt, welche zum Theil aus thonartigen gebrannten Theilen bestehen, das Wasser sehr anziehen, und daher verursachen, daß der damit vermengte Kalk eher erhärtet" (Gilly, *Handbuch der Land-Bau-Kunst*...cit., p. 118). In this regard, see also, Eytelwein, *Praktische Anweisung zur Wasserbaukunst*...cit., vol. III, pp. 48–50; Simon, *Ueber die Natur des Kalksteins. Analyse der Kalkstein*...cit., p. 114; Friedrich Ludwig Aster, *Gesammelte Nachrichten von dem Cemente aus Trasse und wasserdichte Mauerwerke der Holländer*, Dresden, Leipzig, Samuel Gerlach, 1791³.

[29] "[Mörtel], welche aus Kalk, der nicht mit Sand, sondern mir andern Materien verfertigt worden, und zum Wasserbau gebraucht wird, und diesen nennt man gemeiniglich Cement" (Gilly, *Handbuch der Land-Bau-Kunst*...cit., p. 119).

[30] "Alles Mauerwerk, welches entweder stets im Wasser bleibt oder oftmals davon benetzt wird, muss mit Cement-Mörtel gemauert werden" (Wiebeking, *Allgemeine auf Geschichte*...cit., vol. IV, 1805, p. 458).

[31] The modern use of

that is made of lime and other substances instead of sand, in order to make it suitable for hydraulic works."[29] Wiebeking concurs, "All masonry works, which either always stand in water, or often get wet, need to be built with cement-based mortar."[30]

Several German master builders believe that the Italian pozzolana, a kind of volcanic sand used in Ancient Rome, is the best material for producing cement.[31] Wiebeking observes the use of pozzolana in Venice, as he works for the Austrian Hungarian Empire and analyses a local kind of seawall called *murazzi*.[32] He even conceives the construction of a masonry pier bound with pozzolana-based mortar for the port of Trieste.[33] "Pozzolana," he relates in 1805, "is a kind of volcanic earth that can be found in the regions of Civitavecchia, Rome, Pozzuoli and Naples; from the former places, we import a dark red sort to Venice, Trieste and Fiume, for our hydraulic engineering works."[34] In Germany, however, pozzolana is not used because the importation of it from Italy would be too expensive. "We may not make use of pozzolana earth, which can only be found in Italy," Gilly affirms, "since it is too expensive for us."[35] In the same regard, Eytelwein states: "We do not prepare cement with pozzolana because it is too expensive."[36] Despite these assertions, the decisive reason that pozzolana is not used in Germany is the local availability of a similar kind of sand, produced by grinding volcanic tuff stones from the region of the Vulkan Eifel in Rhineland. The ancient Romans probably discovered this kind of tuff when building a military road from Trier to Andernach. They used it between the years 50 and 100 AD to prepare hydraulic mortar and even concrete.[37] Following the decline of the Roman Empire, expertise in preparing hydraulic mortar and concrete with volcanic tuff from the Eifel declines through the centuries. During the Middle Ages, tuff stones from the Eifel are still used along the Rhine, but mainly to produce ashlars and grindstone rather than using milled powder to prepare mortar.[38] It is not until the early 17th century that they regain a renewed understanding of the production of hydraulic mortar, thanks to the increasing interest of Dutch builders.[39] From that time onwards, most tuff stones quarried in Rhineland are exported to Holland, where, by means of typical mills, they are ground into a kind of sand called *trass* or *terras*,

Pozzolana spreads as of the 18th century in Mediterranean regions, as well as along the Atlantic coast of Portugal and France. On this matter, see Roberto Gargiani, *Concrete from archaeology to invention 1700-1769*, Lausanne, EPFL Press, 2013, pp. 41–57.

[32] About the *murazzi*, see *ibid.*, pp. 186–199, 205–213.

[33] See Wiebeking, *Allgemeine auf Geschichte*...cit., vol. IV, (pp. 21–25), p. 24; Id., *Beyträge zur Hafen-, Seeufer- und Flußbaukunde, oder: Abhandlung über die Verbesserung der Häfen von Venedig, und über die Erhaltung der Inseln vor Venedig; über die Verbesserung des Laufs der Brenta, des Bacchiglione, und der Schiffahrts-, so wie der Entwässerungs-Kanäle zwischen Venedig, Padua, Verona und der Etsch: nebst einem Vorschlage zur Anlage eines Seehafens vor Triest, auf genaue vom Verfasser angestellte Lokaluntersuchungen gegründet; und eine Beschreibung des Hafens Cronstadt*, München, Franz Hübschmann, 1810, (pp. 37–42), p. 42.

[34] "Puzzolana ist eine vulkanische Erde, die in den Gegenden von Civita Vecchia, Rom, Pozzuoli e Neapel gefunden wird. Von den ersten Gegenden beziehen wir sie nach Venedig, Trieste und Fiume, die braunrötlich ist, zu unsern Wasserbauten" (Wiebeking, *Allgemeine auf Geschichte*...cit., vol. IV, p. 309, n. 2).

[35] "Der Pozzolanerde, die nur in Italien gefunden wird, können wir uns, als einer zu kostbare Sache, nicht bedienen" (Gilly, *Handbuch der Land-Bau-Kunst*...cit., p. 126).

[36] "Die Bereitung eines Cements mit einem Zusatz von Puzzolane [...] ist bei uns ihrer Kostbarkeit wegen nicht im Gebrauch" (Eytelwein, *Praktische Anweisung zur Wasserbaukunst*...cit., vol. III, p. 50).

[37] See Gustav Haegermann, *Vom Cæment zum Zement*, Wiesbaden, Berlin, Bauverlag GmbH, 1964, p. 31.

[38] *Ibid.*, pp. 31–32; Friedrich Quietmeyer, *Die Mörtelkunde von ihren ersten Anfängen bis zur zielbewußten Herstellung des Portlandzementes*, in, Hans Riepert, ed., *Die deutsche Zementindustrie*, Charlottenburg, Zementverlag GmbH, 1927, (pp. 1–87), pp. 18–26.

[39] See Anton Hambloch, *Der Trass, seine Entstehung, Gewinnung und Bedeutung im Dienste der Technik*, Berlin, Julius Springer, 1909, p. 24.

which is considered the northern pozzolana.[40] In the early 1800s, most tuff quarried in Rhineland is still ground in Holland and, in part, reimported to Germany, transported along the Rhine or, via Bremen, along the River Weser.[41] Wiebeking condemns such practice and, in 1805, he complains: "Tuff stones are transported to Holland where they are ground; then, strangely enough, they are sometimes sold in Germany along the Rhine, as if grinding tuff were a witchcraft!!"[42]

From the Dutch ways of preparing trass-based cement, German master builders draw instructions for the preparation of two kinds of different quality hydraulic mortar, *Trassmörtel* and *Bastard-Mörtel*. "The *Trassmörtel* that is used to build the foundations of most new locks in Holland," Wiebeking explains, "is made of two parts of sieved lime from Liège and one part of trass." Instead, *Bastard-Mörtel*, which is used for walls that are not in water, "is made either of three parts lime, one part trass and two parts sand, or of two parts lime and one parte sand."[43] A cheaper but less effective type of trass is the so-called *wilder Trass* or *Tau*. It is a kind of volcanic ash found along the Rhine, especially near the town of Brohl, that is used locally for affordable constructions.

The increasing reputation and demand of trass leads to the search for similar and alternative materials. This primarily happens in regions far away from the Rhine, like Silesia, where Adolph Traugott von Gersdorf, in the 1780s, tries to use local basalt instead of trass, after having translated into German the book by Barthélemy Faujas de Saint-Fond, *Mémoire sur la manière de reconnaître les différentes espèces de pouzzolane*. Some substitutes of Dutch trass are also known in the early 1800s; they are the so-called *weisser Trass* produced near Braunschweig, a milled tuff produced in the village of Zabenstadt near Madgeburg, and the black volcanic sand from Weinfelder Kapelle in the Eifel region.[44] Since it is easily available and affordable, the most common substitute for trass is brick dust (*Ziegelmehl*), which is produced by grinding old bricks, tiles and every kind of terra cotta pieces by means of hammers and other similar hand tools. To demonstrate the efficacy of brick dust-based cement, Eytelwein, Gilly and Wiebeking note that Jean-Rodolphe Perronet had used such a cement to build the *Pont de*

[40] The terms *trass* and *terras* probably derive from the Italian word *terrazzo*, meaning a typical flat roof from the region of Naples, which is usually built with pozzolana-based cement. On this matter, see Hambloch, *Der Trass, seine Entstehung...* cit., p. 12; Gargiani, *Concrete from archaeology to...*cit., p. 68. Concerning the fact that trass is considered the northern pozzolana, see Christian Ludwig Ziegler, *Von dem Terras, der Puzzolana, und deren Sobstitutionsarten*, in, "Hannoversches Magazin," XI, 1773, (coll. 81–94), col. 85; Flörcke, *Dr. Johann Georg Krünitz's...*cit., p. 285. To know more about the export of tuff stones to Holland, see Haegermnn, *Vom Cœment...*cit., p. 32. To know more about the mills to grind trass, see Heinrich Wolffram, *Mittheilungen über Tuffstein, Trass und einige andere Baumaterialien der Vulkanische Eifel*, Diez an der Lahn, Franz Schickel, 1885, p. 6; Anton Hambloch, *Der rheinische Trass als hydraulischer Zuschlag in seiner Bedeutung für das Baugewerbe*, Andernach am Rhein, Selbstverlag des Verfassers, 1903, p. 8; Id., *Der Trass, seine Entstehung*...cit., p. 24.

[41] Zimmermann, *Einige Bemerkungen über die Festigkeit, Mischungsverhältnisse und Zubereitung des Bétons, oder des Mauerwerkes aus klein geschlagenen, mit Mörtel untermengten Steinen, dessen man sich zuweilen, um Fangdämme und Wasserschöpfen zu sparen, zur Fundamentierung von Bauwerken unter Wasser bedient*, in, "Journal für die Baukunst," vol. III, 1830, (pp. 1–31), p. 14.

[42] "So wird er in Steinen nach Holland verfahren und dort zu Pulver zerstossen und vermahlen; dann sonderbar genug, nicht selten wieder in Deutschland und am Rhein verkauft, als wenn eine solche Trassmühle eine Hexerey wäre!!" (Wiebeking, *Allgemeine auf Geschichte*...cit., vol. IV, pp. 309, n. 1).

[43] "Der Trassmörtel besteht bey dem Grundbau der neuesten Schleusen in Holland aus zwey Theilen gesiebten Lütticher Kalk und einem Theil Trass" (Wiebeking, *Allgemeine auf Geschichte*...cit., vol. IV, p. 313). "Zu dem innern Gemäuer, welches nicht im Wasser steht, wird der sogenannte Bastard–Mörtel gebraucht. Erselbe besteht entweder aus drey Theil Kalk, ein Theil Trass und zwey Theile Sand, oder aus zwey Theil Kalk und ein Theil Sand" (*ibid.*, p. 314).

[44] See Gilly, *Handbuch der Land-Bau-Kunst*...cit., pp. 127–128; Harmen Jan Freiherr van der Wyck, *Uebersicht der Rheinischen Eifeler erloschenen Vulkane und der Erhebung-Gebilde, welche damit geognostischer Verbindung stehen, nebst Bemerkungen über den technischen Gebrauch ihrer Produkte*, Bonn, Eduard Weber, 1826, p. 95.

[45] See Gilly, *Handbuch der Land-Bau-Kunst*...cit., p. 129;

Neuilly in Paris. They also recall the masonry locks of the Bromberg canal in East Pomerania, which are built between 1792 and 1800 using trass-based mortar for the lowest masonry joints, and mortar containing brick dust for the rest of the masonry.[45]

In addition to trass and brick dust, several other substances are considered suitable for preparing hydraulic mortar, although they are used less often. The most widely recognized are iron powder, ash from peat burning and the so-called Dornick ash, which is actually a residue from coal-burning in lime kilns.[46] Most proficient master builders are also aware of the special mortar developed in the second half of the 18th century by the French inventor Antoine–Joseph Loriot, known as *Loriotsche Mörtel* (Loriot's mortar).[47] This cement-based compound is prepared with brick powder and some unslaked lime, added shortly before using the mortar. Unslaked lime used in Loriot's mortar is thought to absorb water and accelerate mortar hardening, just like trass and all other similar products.

The use of cement out of water is not admitted and is even considered disadvantageous.[48] Furthermore, German master builders generally mistrust mixtures of different substances assembled to improve the quality of mortar. In this regard, Simon writes: "Many people suggest using this or that substance to improve the quality of mortar, sometimes even with emphasis, but none of such substances really offers the promised advantages. The wish of unknown people to become famous or, at least, to be praised by inexperienced people for having contributed to fulfil this or that [building] need, lies behind these suggestions. Is the solidity of good mortar really scanty to the point of making necessary additives to improve it? Ancient expertise shows that mortar made of sand and lime lasts for a long time, exactly because of its simple composition [...]. Thus, why try to improve the solidity of mortar by adding further products, since its simplest kind is absolutely enough for us?"[49]

Rudimental traces of concrete in foundations, between local experiments and references to antiquity

On the basis of the current knowledge about lime and mortar, rudimentary kinds of concrete are already being produced by the end of the 18th century. They consist of rubble, pebbles and mortar, and they are principally used as

Eytelwein, *Praktische Anweisung zur Wasserbaukunst*...cit., vol. III, p. 47; Wiebeking, *Allgemeine auf Geschichte*...cit., vol. IV, p. 317. Concerning the construction of the masonry locks along the Bromberg canal, see [Johann Christian] Wutzke, *Bemerkungen über den Bromberger Kanal*, in, "Preussische Provinzial-Blätter," vol. VII, 1832, pp. 337–348.

[46] The Dornick ash is the same as the ash from Tournai. Dornick is an old German name for the town of Doornik, which is called Tournai in French language. Regarding the use of the Dornick ash, see Gilly, *Handbuch der Land-Bau-Kunst*...cit., p. 225; Eytelwein, *Praktische Anweisung zur Wasserbaukunst*...cit., vol. III, p. 50.

[47] See Simon, *Ueber die Natur des Kalksteins. Analyse der Kalkstein*...cit., p. 112; Gilly, *Handbuch der Land - Bau - Kunst*...cit., p. 121; Eytelwein, *Praktische Anweisung zur Wasserbaukunst*...cit., vol. III, p. 50; Wiebeking, *Allgemeine auf Geschichte*...cit., vol. IV, p. 319. Regarding Loriot's studies about mortar in the second half of the 18th century, see Gargiani, *Concrete from archaeology to*...cit., pp. 340–351.

[48] "Mortar prepared with trass or pozzolana does not harden so well and completely as good ordinary mortar, if it is used out of water, just exposed to the air" ("Allein braucht man die Pozzolanerde oder Traß auch im Mörtel an Gemäuern, die außer dem Wasser der freyen Luft ausgesetzt sind, so ist die Erhärtung nie so groß und so vollkommen, als die von bloßem guten Mörtel") (Flörcke, *Dr. Johann Georg Krünitz's Oekonomische Encyklopädie*... cit., p. 198).

[49] "Fehlt es nicht an Vorschlägen, die bald diesem, bald jenem Stoff als ein vorzügliches Mittel, die Bindbarkeit des Mörtels zu erhöhen, empfehlen, jedoch kann im Allgemeinen davon angeführt werden, dass keiner darunter den Vortheil leistet, den ihm der Erfinder öfters mit so vielem Nachdruck zuschrieb. Dergleichen Vorschläge entspringen mehrentheils von dem Hang, den ihre Urheber haben, sich einen Namen zu machen, und wenigstens von unerfahrenen das Lob einzuernten, zur Verbesserung dieser oder anderer Bedürfnisse vieles beytragen zu haben. Allein ist denn die Festigkeit, die ein gut behandelter Mörtel erlangt, so gering, dass es uns besonders um Zusätze zur Vermehrung desselben zu thun wäre? Die ältesten Erfahrungen belehren uns, dass der Mörtel aus blossem Kalk und Sand dem nagenden Zahn der Zeit trotzet, und der vorzügliche Grund dieser Unveränderlichkeit ist wohl in der Einfachheit seiner Bestandtheile zu suchen [...]. Warum also zum Mörtel einen Zusatz zur Vermehrung der Festigkeit, wo uns das einfache vollkommen genüget?" (Simon, *Ueber die Natur des Kalksteins. Analyse der Kalkstein*...cit., pp. 116–117).

filling materials in different kind of foundations. A unique case pertains to the special well foundations developed in Berlin on the initiative of a spirit maker and trader named Benjamin Georg (1739–1823). Georg is the owner of a wide estate situated between today's Friedrichstrasse, Georgenstrasse and Reichstagufer, which he intends to develop with new buildings.[50] Because of the proximity to the River Spree, the soil of the area is quite marshy and foundations for buildings need to rest at a depth of 22 feet in the ground.[51] After having used a traditional timber piling foundation to build a storehouse, in 1798 Georg starts building linear foundations composed by a number of underground masonry pillars arranged in rows and connected at the top by means of arches, which are intended to support the bearing walls of the planned building. Pillars consist of circular walls made of bricks and filled with rubble; brick dust-based, hydraulic mortar is used to bind both the bricks and the rubble pieces. The circular walls are built following the construction technique of the sinking wells, which is in use in several countries to build ordinary wells for water supply.[52] First, a pit between 6 and 7 feet deep is dug into the soil, then a wooden crown laid on the bottom of it and a portion of circular wall between 5 and 6 feet high is built on the wooden crown. After that, a rope around the external surface of the wall fixes several vertical wooden bars, evenly spaced from one another, to form a sort of reinforcement chain. Workers dig out soil under the wooden crown inside the circular wall portion so that it sinks down and a further similar one is built on top of it. The sequence is repeated until the circular wall reaches the necessary depth, that is to say, the bearing soil. The interior space is then filled with rubble and mortar. Once this hardened, the ensemble forms a solid underground pillar. Following Georg's example, other landlords similarly implement this kind of foundation, which must have been quite known in Berlin around the turn of the 19th century. Gilly takes an interest in it and writes two reports appearing in 1798 and 1800 in the "Sammlung nützlicher Aufsätze und Nachrichten, die Baukunst betreffend," followed by a pamphlet in 1804.[53] Beyond a mere description of the building technique, Gilly also searches for its origin. In French technical literature, he finds three quite different and unrelated references, dating from different ages. The earliest concerns a supposed use of masonry well foundations by ancient Egyptians during the 12th century; the second describes the foundations of the *Château de Saint Maur-des-Fossés* near Paris, which are conceived by Philibert de l'Orme as cut stones pillars built deep into the ground and connected on the top through arches; and the third concerns an ancient Indian building technique that appears very similar to the one adopted by Georg.[54] It, in fact, involves the construction of circular walls that gradually sink into the ground to be then filled with rubble, earth and sand.[55] British colonialists working on improving the fortifications of the city of Arcat are thought to have discovered this construction technique, as the French official Alexandre le Goux de Flaix reports in his memoir written

[50] A portrait of Benjamin Georg, along with further information about the development of his estate in Berlin, are given in, Berndt Weber, *Georgstrasse. Nach wem wurde die Georgstrasse in Berlin-Mitte bennant?*, in, "Mitteilungen des Vereins für die Geschichte Berlins," CI, 2005, n° 1, also published in, www.diegeschichteberlins.de.

[51] See David Gilly, *Über die Gründung der Gebäude auf ausgemauerte Brunnen*, Berlin, Realschulbuchhandlung, 1804, p. 4.

[52] See Wiebeking, *Allgemeine auf Geschichte*...cit., vol. IV, 1805, p. 358.

[53] David Gilly, *Über die Gründung der Gebäude auf ausgemauerte Brunnen*, in, "Sammlung nützlicher Aufsätze und Nachrichten, die Baukunst betreffend," 1797, vol. I, pp. 137–138, 1798, vol. II, pp. 124–125; Id., *Über die Gründung der Gebäude*...cit., 1804. Gilly discusses well foundations built in the boulevard Unter den Linden and in the Jägerstrasse (*ibid.*, p. 2).

[54] The Château de Saint Maur-des-Fosses is built since 1541. "I let dig several holes," De

4. D. Gilly, well foundations. Construction drawings (D. Gilly, *Ueber die Gründung der Gebäude auf ausgemauerte Brunnen*, Berlin, Realschulbuchhandlung, 1804).

after having served for a long time in India as an engineer. Le Goux de Flaix's memoir is published in Paris in 1802, four years after Georg's first well foundation is completed. Solely on the basis of this observation, Gilly attributes a kind of German authorship of this building technique to Georg, and Wiebeking later shares this hypothesis.[56]

At the same time that the Berlin well foundations are developed, technical surveys into foreign constructions with the aim of improving inland building standards, along with the contemporary literary, philological and archaeological passion for antiquity, facilitate study of the use of concrete in ancient Roman constructions and in some recent hydraulic works built abroad. This gives rise to a heterogeneous ensemble of literary mentions about the topic. Wiebeking's *Wasserbaukunst*, referencing Bélidor's *Architecture hydraulique*, describes two concrete hydraulic foundations built in the harbour of Toulon in southern France around mid-18th century. The former consists of layers of pebbles and mortar poured underwater inside enclosures made of timber piles, while the latter is built by pouring concrete in floating caissons, which, consequently, sink.[57] In the wake of that interest in antiquity, Wiebeking compares these two building

l'Orme writes, "as if I had planned to build four or five feet wide wells [...] and I let fill the mentioned hole, or wells, with good masonry" ("je feis faire plusieurs trous ou pertuis, come si I' eusse voulu faire des puis de quatre ou cinqu pieds de large, [...] & faisois remplir lesdicts trous ou puis de bonne maçonnerie") (Philibert de l'Orme, *Le premier tome de l'architecture*, Paris, Federic Morel, 1567, pp. 45–47). See also Gilly, *Über die Gründung der Gebäude*...cit., 1804, pp. 4–7, 23–26.

[55] See Gilly, *Über die Gründung der Gebäude*...cit., 1804, p. 7; Alexander le Goux de Flaix, *Mémoires sur les travaux de constructions hydrauliques*, Paris, Fauvre, 1802; Anon., *D'une nouvelle méthode de fonder, en substituant des puits de maçonnerie aux pilotis; extraite d'un mémoire de M. A. Le Goux de Flaix*, in, "Mémorial de l'Officier du Génie," 1803, pp. 231–236.

[56] See Wiebeking, *Allgemeine auf Geschichte*...cit., vol. IV, p. 360.

[57] See Wiebeking, *Allgemeine auf Geschichte*...cit., vol. III, 1801, pp. 218, 254–255.

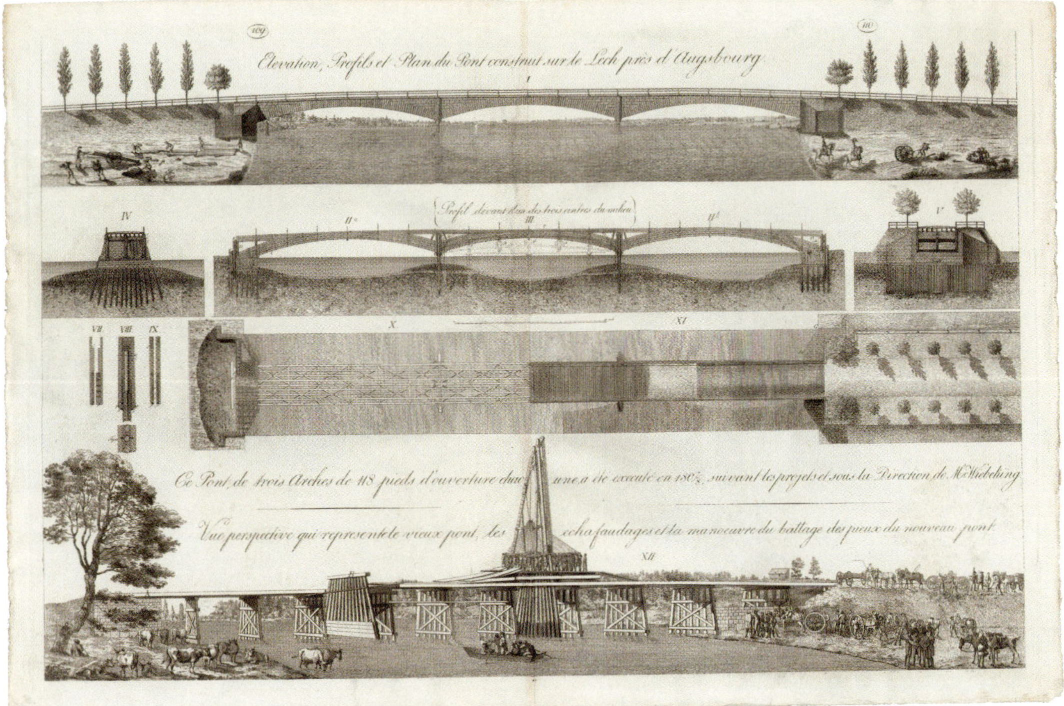
Elevation, Profils et Plan du Pont construit sur le Lech près d'Augsbourg
Ce Pont, de trois Arches de 118 pieds d'ouverture chacune a été executé en 1807, suivant les projets et sous la Direction de Mr Wiebeking
Vue perspective qui represente le vieux pont, les echafaudages et la manoeuvre du battage des pieux du nouveau pont

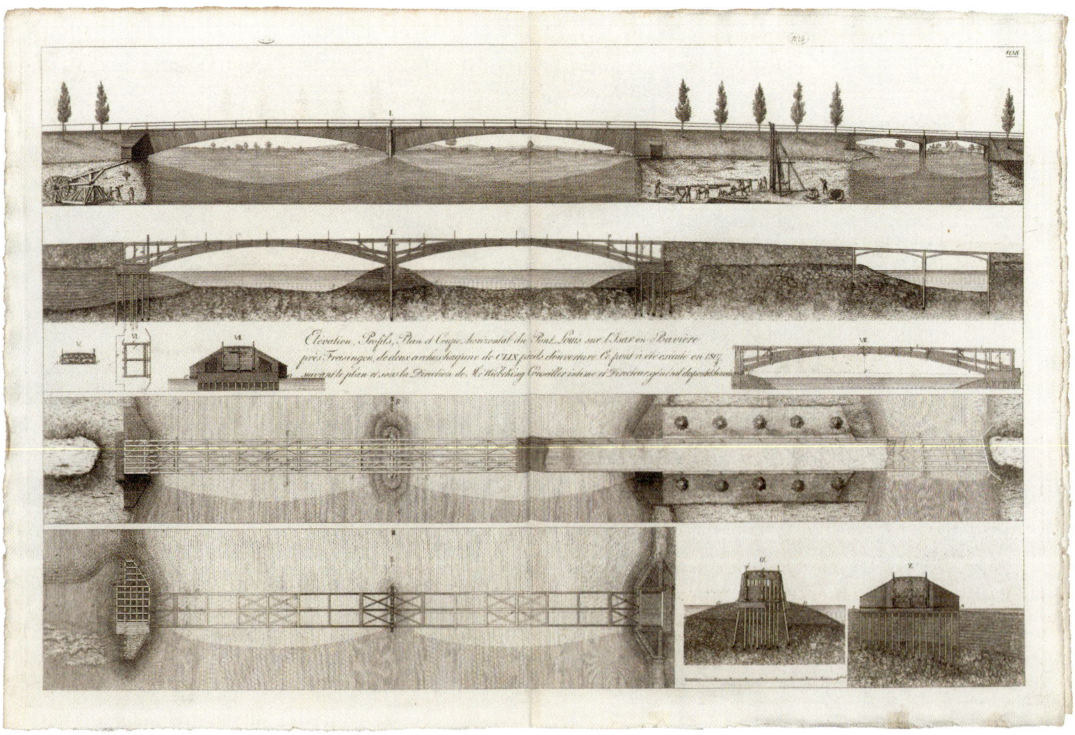
Élévation, Profils, Plan et Coupe horizontal du Pont Louis sur l'Isar en Bavière
près Freisingen, de deux arches chacune de CLIX pieds d'ouverture. Ce pont a été executé en 1807
suivant le plan et sous la Direction de Mr Wiebeking Conseiller intime et Directeur général

5-7. C. F. Wiebeking, bridge over the River Loch, near Augsburg, 1807–1808; bridge over the River Isar, near Freysingen, 1807–1809; bridges over the Rivers Wertach, near Ettringen, Biesenhofen and Irsingen, 1808–1809. Elevations, sections, plans and views (C. F. Wiebeking, *Theoretisch- Practische Wasserbaukunst*, München, Joseph Zängl, vol. III, 1814).

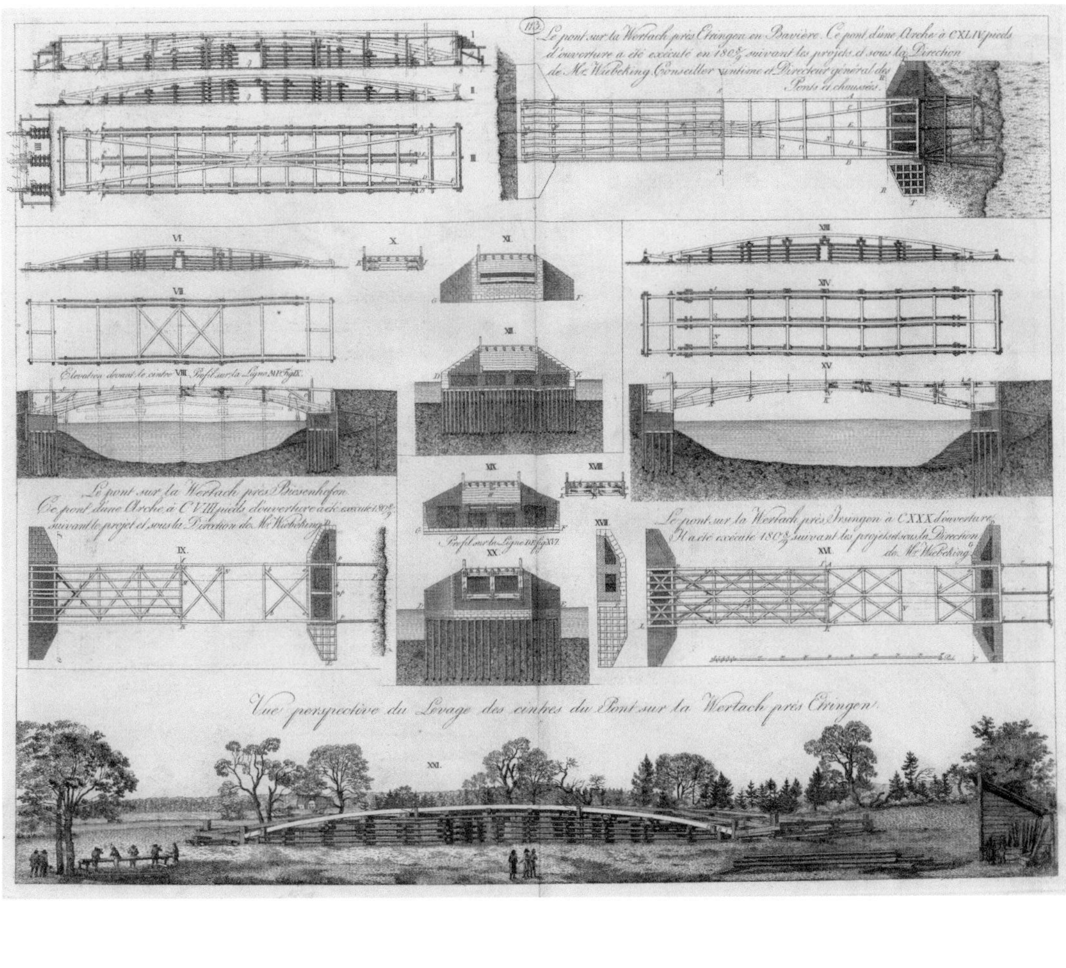

Tab: 104 b

Ansicht des Hafens und der Stadt Lindau vom Vorhafen aus genomen.

104 a

Vüe perspective du port près de Lindau prise du côté du lac.

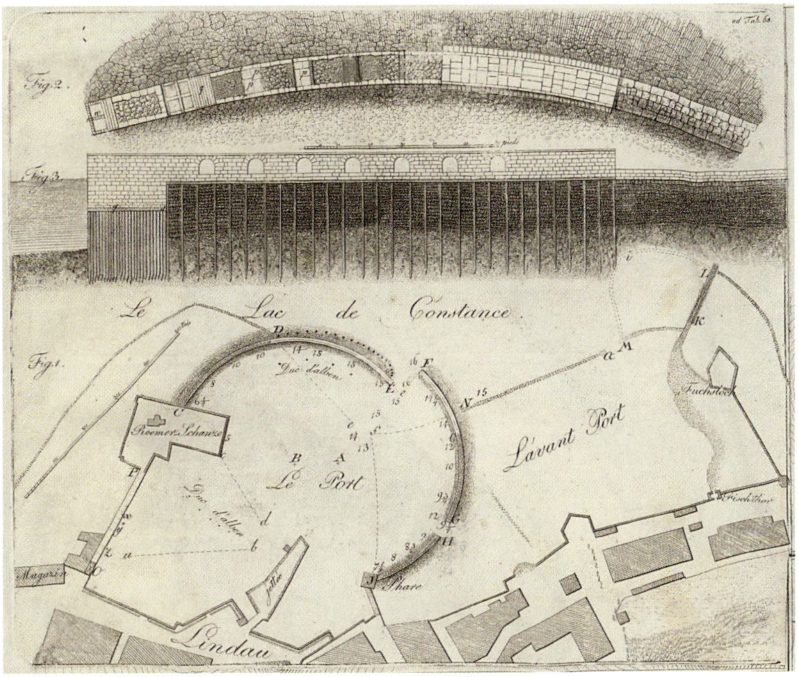

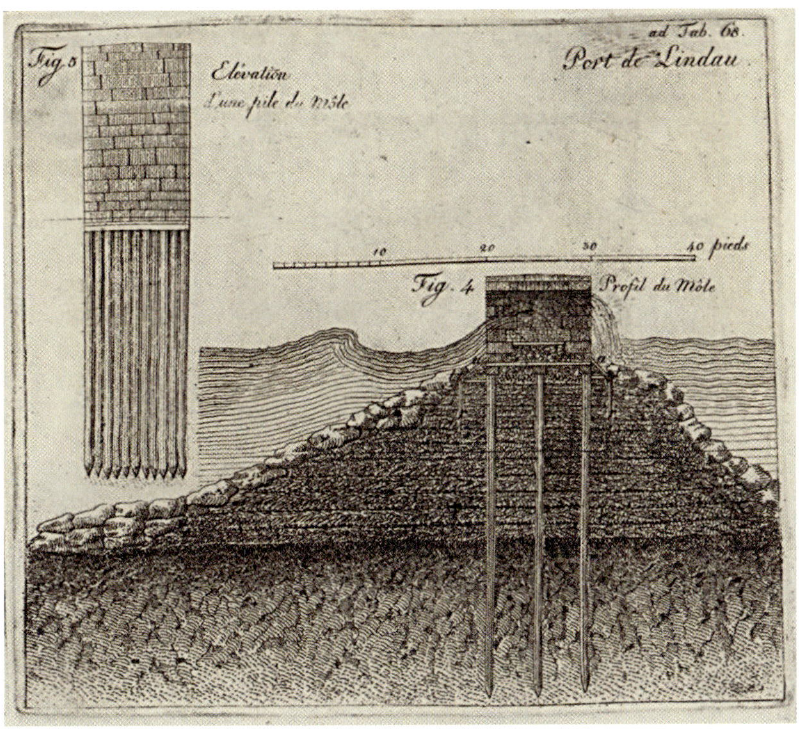

8-11. C. F. Wiebeking, the Port of Lindau on the Lake of Costance, 1812. Views, plan, sections, plan of the site (C. F. Wiebeking, *Theoretisch- Practische Wasserbaukunst*, München, Joseph Zängl, vol. II, 1812).

12-13. A. Hirt, ancient masonry walls. Construction drawings showing concrete filling, stone and brick bonds (A. Hirt, *Die Baukunst nach den Grundsätzen der Alten*, Berlin, Realschulbuchhandlung, 1809).

techniques to ancient Roman constructions. He mentions that the use of concrete poured underwater in enclosures of timber piles is described in Vitruvius' *De Architectura* and, concerning the technique of caissons filled with concrete, he reports that Romans "used to sink old ships filled with pebbles and mortar" to build foundations underwater.[58] A few years after the publication of Wiebeking's *Wasserbaukunst*, the interest in antiquity leads Alois Hirt (1759–1836), an eminent historian and archaeologist, to write of underwater foundations made of concrete poured in timber pile enclosures, in *Die Baukunst nach den Grundsätzen der Alten*, a work that retraces Vitruvius' cultural heritage.[59] Another kind of foundation that uses concrete and has relevance in German literature at the turn of the 19th century is the one with timber piles and concrete poured between the pile heads to form a solid basement supporting the walls above.

[58] See *ibid.*, pp. 212, 218. Wiebeking relates that the ship used to transport, from Egypt to Rome, the obelisk placed in the Circus of Caligula was later sunk in the port of Ostia at the time of the Emperor Claudius, in order to build the foundation of a quay. He actually writes: "the obelisk that stands in the Saint Peter square," since the Egyptian obelisk, which had been originally placed in the Circus of Caligula, was relocated in Saint Peter Square in 1586 (see Domenico Fontana, *Della trasportatione dell'obelisco vaticano et delle fabriche di nostro signore papa Sisto V*, Roma, Domenico Basa, 1590;

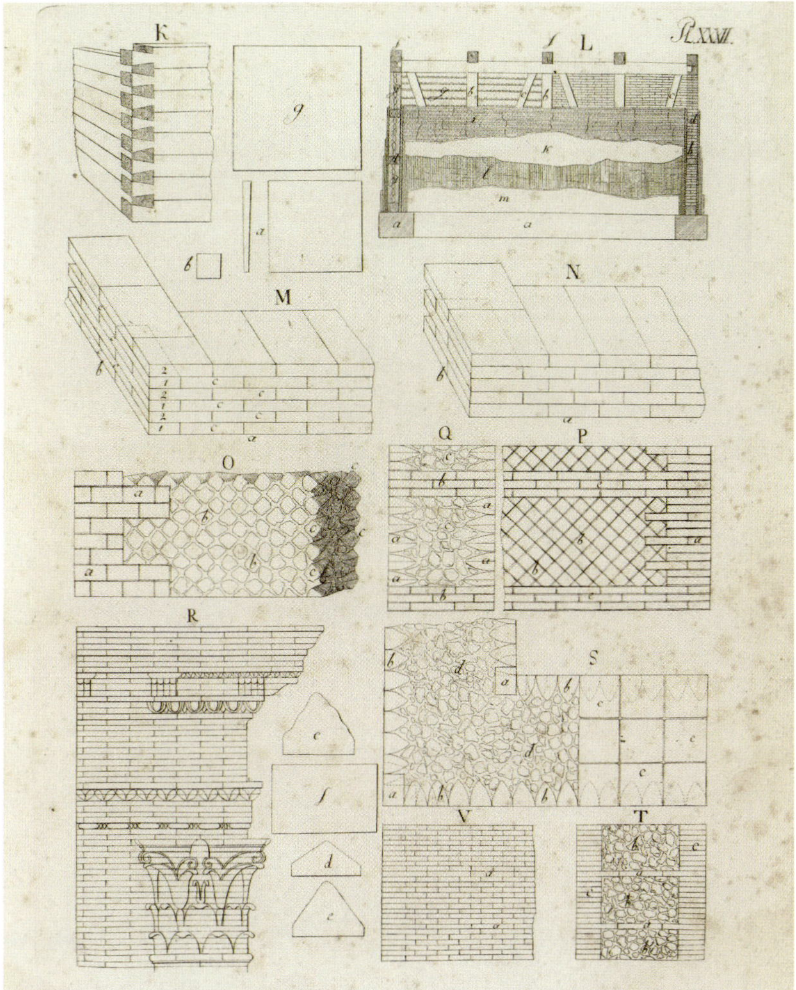

Again Wiebeking depicts this kind of foundation after learning about it from Venetian public engineers while working for the Austro-Hungarian Empire.[60]

Knowledge about ancient and more recent uses of concrete in foundations inspires Wiebeking to use mixtures of pebbles and mortar as filling materials in the timber piling foundations of the masonry abutments of several timber bridges that he builds in Bavaria between 1806 and 1813, as well as in the foundations of the abutments of the weir of Landshut and of the jetties at the entrance of the port of Lindau in Bavaria, which he builds between 1810 and 1812. These foundations consist of timber piles, framings of beams nailed on the piles, and mixtures of different kinds of aggregate and mortar poured inside the compartments of the framings, between the pile heads. Wiebeking adjusts the composition of the filling-mixtures on the basis of the local availability of

Carlo Fontana, *Il tempio Vaticano e sua origine con gli edifici più cospicui antichi, e moderni fatti dentro, e fuori di esso*, Roma, G. F. Buagni, 1694, Lib. III, Cap. IV–XIII, pp. 119–174).

[59] See Alois Hirt, *Die Baukunst nach den Grundsätzen der Alten*, Berlin, Realschulbuchhandlung, 1809. p. 121.

[60] See Wiebeking, *Allgemeine auf Geschichte*...cit., vol. IV, pp. 381–382. A similar kind of foundation is depicted in Bélidor's *Architecture hydraulique*, which Wiebeking knows well. See Bélidor, *Architecture hydraulique*... cit., p. 184.

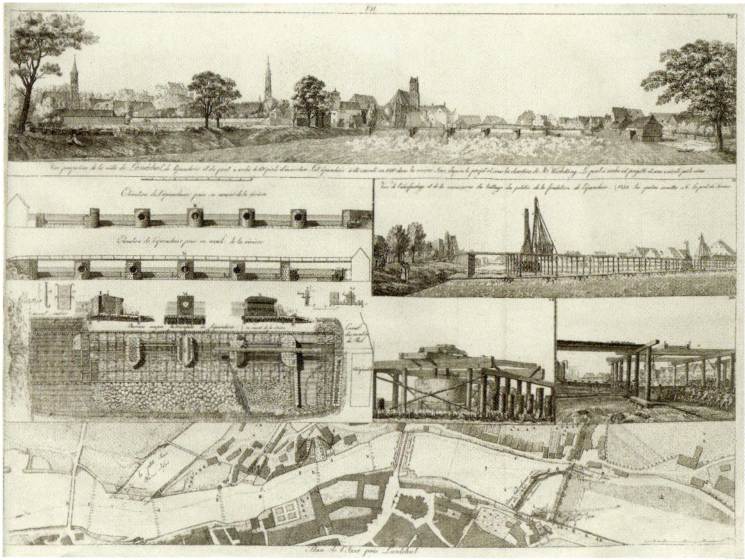

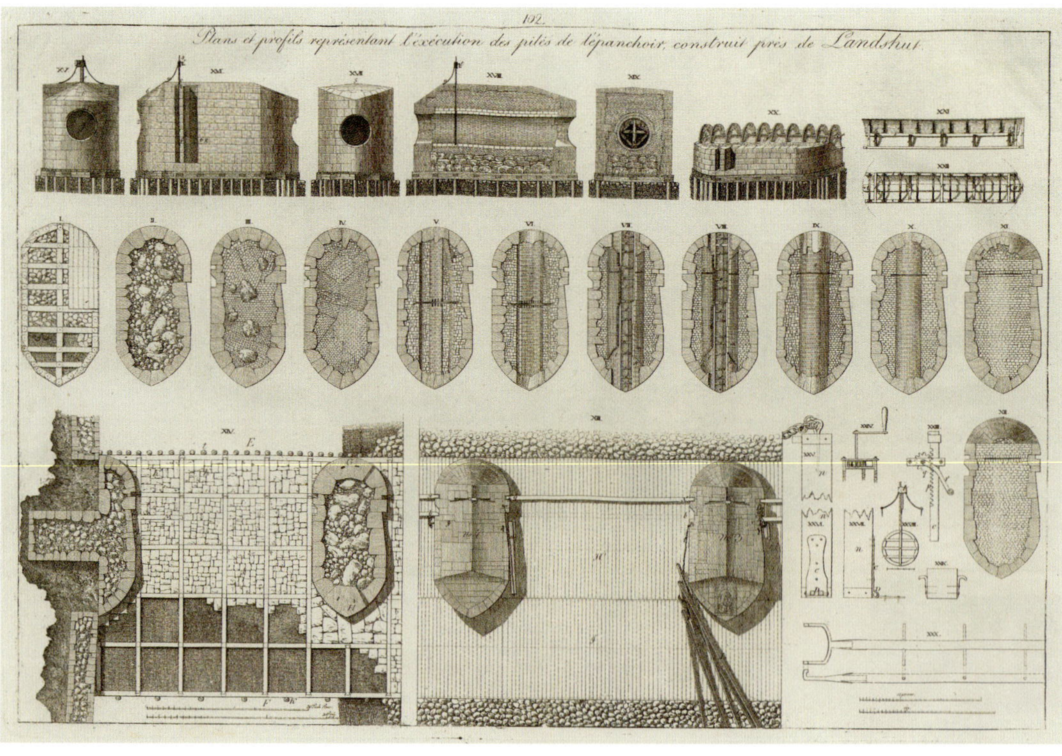

14-15. C. F. Wiebeking, the Weir of Landshut, 1810. View, elevations, plan and plan of the site, construction drawings (C. F. Wiebeking, *Theoretisch- Practische Wasserbaukunst*, Joseph Zängl, München, vol. II, 1812).

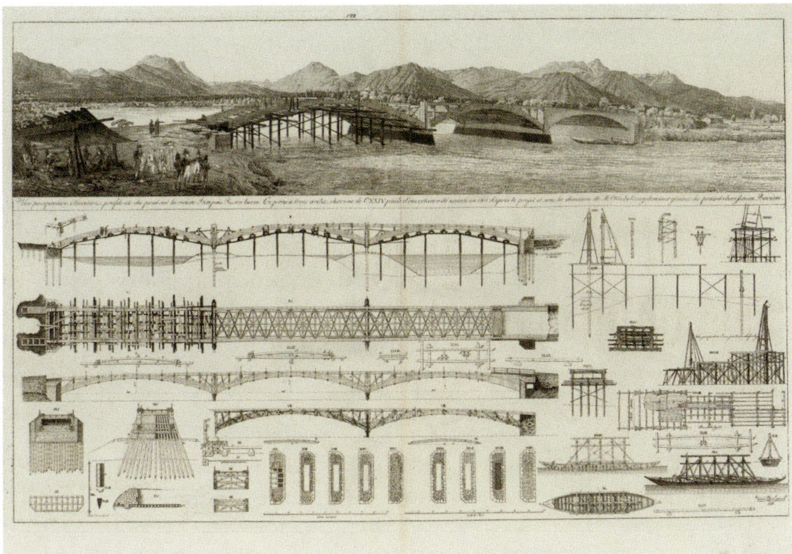

16. C. F. Wiebeking, the bridge over the River Inn, near Rosenheim, 1810-1811. Perspective view of the bridge in process of building, construction drawings (C. F. Wiebeking, *Theoretisch- Practische Wasserbaukunst*, München, Joseph Zängl, vol. III, 1814).

materials, using different proportions of pebbles, rubble, crushed bricks, sand and either cement-based mortar or unslaked lime. He does not consistently use cement and, in several cases, he uses mixtures that are only made of pebbles, sand and unslaked lime, asserting that they would become hard like stone once poured into water.[61] The use of unslaked lime is reminiscent of Loriot's mortar, although Loriot used to add cement – namely, brick dust – to his mortar.

Hollow masonry-faced walls filled with concrete

In the context of the German construction culture around the turn of the 19th century, walls made of two masonry faces with rubble fill provide further occasion, besides foundations, to develop observations about concrete. Amidst efforts to expand the use of masonry over timber, Gilly provides an account of different kinds of masonry bonds in his *Handbuch der Landbaukunst*. Among several cases, he also depicts stone-faced walls filled with crushed bricks, fieldstones, slags, coarse gravel, sand and similar loose materials, which are embedded in a casting of lime that is intended to bind the whole into a mass.[62] Gilly links this kind of masonry to medieval building practices in Poland, offering as a pertinent example the ruins of the Ritterschloß of Marienburg (Malbork Castle) in north-eastern Pomerania, which his son Friedrich studies and draws. He does not consider this "Polish bond" to be appropriate for the German building needs of his time. "This kind of masonry," he warns, "needs to be built with such slowness and patience that

[61] Wiebeking, *Beyträge zur Brückenbaukunde*...cit., p. 32. Just to give some examples: the timber piling foundation of the bridges near Neuötting (1807) is filled with pebbles, sand and unslaked lime, whereas the compartments of the framework are filled with bricks and trass-based mortar, and spaces among the piles of the in-between piers are filled with bricks rubble, pebbles and cement; for the bridge of Augsburg (1807–1808), Wiebeking fills the timber piling foundations of the abutments and the in-between piers with mixtures of brick rubble, pebbles and unslaked lime; the filling material for the foundation of the bridges over the River Rott near the town Schärding (1808–1809) is composed of coarse unrefined stones and mortar, the one for the bridge in Ettringen (1808–1809) is made of pebbles and unslaked lime, the one for the bridge in Irsingen (1808–1809) is made of pebbles, cement and unslaked lime, the one for the bridge in Freysingen (1808–1809) is made of gravel and unslaked lime, and the one for the bridge in Altenmarkt (1809) is made of gravel and cement-based mortar - i.e. two parts of river sand, one part of brick dust and one part of unslaked lime.

[62] See David Gilly, *Handbuch der Land-Bau-Kunst*...cit., pp. 238–239, fig. 121.

it ends up being inappropriate for the current state of our buildings and for future expectations."[63]

Masonry-faced walls filled with concrete are also depicted in Wiebeking's *Wasserbaukunst* and in Hirt's *Baukunst nach den Grundsätzen der Alten*. Unlike Gilly, however, the two authors, who use Vitruvius as an essential reference, relate this kind of walls to the Greek *emplecton* and to Roman building practices. Wiebeking furthermore experiments with the construction of such walls when building both the abutment of the aforementioned Bavarian bridges and the piers of the weir of Landshut. In most cases, he builds hollow masonry walls of brick, which he then fills with different mixtures of rubble, pebbles and mortar.[64] However, for the piers of the weir of Landshut, as well as for the abutments of the bridge of Rosenheim, he builds stone-faced walls as the ancients used to do, namely by erecting exterior closed walls made of cut stones with unrefined internal faces, and by filling the in-between spaces with pebbles and mortar. These remain quite rare cases of masonry-faced walls filled with concrete in Germany. This kind of construction does not find significant application, as Gilly's pupil Karl Friedrich Schinkel (1781–1841) confirms about twenty years later. "Most ancient walls," he writes, following in the footsteps of his mentor, "especially Gothic walls, are built this way, but this kind of construction is less, or even not at all, used nowadays, since it needs to be built very slowly, due to the irregular setting; moreover, there is a risk that, over time, the faces of the walls detach from the coarse masonry work inside."[65]

French concrete foundations in Rhineland

During the occupation of Rhineland, approximately 1794 to 1814, French engineers build some hydraulic concrete foundations that German master builders later use as models for similar constructions around the late 1820s. The most important case concerns the masonry works of the so-called *Grand Canal du Nord*, which the French engineers of the *Corp des Ponts et Chaussées* attached to the *Department de la Roer*, conceive between the end of the 18th and the beginning of the 19th centuries to connect the town of Venlo in the Netherlands, with Neuss in Rhineland. Works begin in 1808 and stop in 1811, when France achieved supremacy over Holland and the canal is no longer considered necessary.[66] At that time, only a few works have been completed but the technical information about the canal project is later spread thanks to a detailed book by Aimable Hageau (1756–1836), an engineer involved in the planning and in the partial construction of the canal.[67] According to Hageau,

[63] "Dergleichen Bauart kann aber nur mit einer Langsamkeit und Geduld ausgeführt werden, welche den jetzigen Verfassungen und Absichten unserer Bauten nicht angemessen seyn würde" (Gilly, *Handbuch der Land-Bau-Kunst*...cit., p. 239).

[64] Hollow masonry filled with pebbles and mortar mixtures are in the abutments of the bridges near Augsburg (1807–1808), Ettringen and Biesehof (1808–1809), Freysingen (1807–1809), Mühldorf (1812–1813).

[65] "Die meisten alten, namentlich gothischen Gemäuer, sind auf diese Art verbunden; dagegen macht man heut zu Tage wenig oder gar keine Anwendung von dieser Construction, weil sie, des ungleichförmigen Setzens wegen, nur mit großer Langsamkeit ausgeführt werden kann, und dabei doch noch befürchten läßt, daß sich die gerade Steinverkleidung, mit der Zeit, von dem innern rauhen Mauerwerk ablöset" (Karl Friedrich Schinkel, *Grundlage der praktischen Baukunst*, vol. I, *Mauerwerkkunst*, Berlin, Schenk & Gerstäcker, 1834, p. 2, pl. IV, fig. 4).

[66] The *Grand Canal du Nord* was intended to create a direct waterway between France (Venlo is under French control from 1794) and Prussia, avoiding the port of Rotterdam, which is not yet under the direct control of France at the time when the canal is conceived. The construction is stopped in 1811, after the king of Holland, Louis Napoléon Bonaparte, has abdicated and Holland becomes part of the French Empire, making the construction of the canal unnecessary (see Hans Scheller, *Der Nordkanal zwischen Neuss und Venlo*, Neuss, Stadtarchiv, 1980, pp. 4–7).

[67] Aimable Hageau, *Description du canal de jonction de la Meuse au Rhin*, Paris, Chez l'Auteur, 1819.

[68] "On well-rammed soil, a first group of masons spread, ram and level a 35 centimetres thick layer of concrete. A second group of masons covers this layer with a second one of the same thickness, which need to be spread, rammed and levelled as well. Finally, a third group covers the second layer with a third one, which is 25 centimetres thick. Concrete is

the foundations of all masonry along the Canal – walls, locks, lift bridges and aqueducts – are conceived as beds of concrete. The foundations of the locks, in particular, should consist of 85 cm thick beds of concrete, poured into 1.5 meter deep pits, with concrete composed of lean lime from Namur, trass from Brohl, sand, gravel and crushed bricks.[68]

Besides the project and the partial completion of the *Grand Canal du Nord*, some embankments along the Rhine are built on foundations consisting of trass-based concrete poured underwater into pits that are dug along the riversides and isolated from the stream by means of single rows of sheet piles. The embankments at the entrance of the Cologne port, for instance, are built in 1812 on 3-feet thick beds of concrete composed of two parts lime, three parts trass, one part sand, two parts pebbles and two parts sharp-cornered pebbles.[69] For the local availability of trass and the French technical heritage, Rhineland later assumes a leading role for the development of German hydraulic concrete constructions.

The ambition to produce mortar-based artificial stone

Starting the 1810s and following in the French tradition of producing artificial stones known as *pierres factices*, some research and efforts to produce mortar mixtures suitable to be moulded as ashlars, architectural decorations and ordinary objects for different kinds of usage are underway.[70] They represent a significant corollary to the search for new building materials and techniques, and they offer opportunities to experiment in domains contiguous to concrete.

A first German attempt at manufacturing mortar-based artificial stone originates in a coincidence that occurred during the period of the French occupation of Prussia. In May 1807, French soldiers take some residents of the city of Halle hostage and deport them to the town of Pont-à-Mousson in France, where they stay for about two months. Among the hostages are the land councillor Von Wedell, the police chief councillor Gabriel Wilhelm Gottlieb Keferstein (1755–1816) and the savant August Hermann Niemeyer (1754–1828), who later writes a book about the deportation, *Beobachtungen auf einer Deportationsreise nach Frankreich im Jahr 1807*, which is published in two volumes in 1824 and 1826.[71] Considered state prisoners, they are put up in a local inn called Hôtel Imperial. The host, a man named Mr. Montagne, with a "passion for construction," has built a pavilion in his garden, "with several bells hanging from the roof," which brings "the bells of the Magic Flute to mind," as Niemeyer recounts in his book.[72] A local mathematician

made using lime from Namur, the leanest available, or lime from Trier. It is composed of two parts of lime, one part and half of best quality trass from Brohl, one part and half of sand taken from the river or from quarries, one part gravel coming from the excavation of the foundation pit, two parts rubble and two parts egg-sized crushed bricks" ("Sur ce sol bien battu, une première bande de maçons étendra, battra et arasera, dans la superficie des fondemens, une première couche de maçonnerie de béton de trente centimètres d'épaisseur. Une deuxième bande de maçons recouvrira cette couche de béton, d'une seconde couche de même épaisseur que la précédente, qu'on étendra, battra et arasera comme dessus. Enfin, une troisième bande de maçons recouvrira cette seconde couche de béton d'une troisième, à laquelle on donnera 25 centimètres d'épaisseur uniforme. Cette maçonnerie de béton sera faite avec de la chaux vive de Namur, la plus maigre qu'on pourra se procurer, ou de la chaux de Trêves. Elle sera composée de deux parties de cette chaux; d'une partie et demie de trass de Brohl de première qualité; d'une partie et demie de sable de rivière, ou de mine équivalent; d'une partie de gravier sorti de la fouille; de deux parties d'éclats de pierre, et de trois parties d'éclats de briques, réduits à la grosseur d'un œuf") (*ibid.*, pp. 52–53).

[69] See Elsner, *Über die Anwendung des Béton-Mörtels zum Fundamentierung unter Wasser*, in, "Journal für die Baukunst," I, 1829, (pp. 236–245), pp. 242–243; Julius von der Orbach, *Tufstein, Trass und hydraulischer Mörte. Gesammelte Bemerkungen*, Coblenz, Karl Bädecker, 1849, pp. 27–28.

[70] About the *pierre factice*, see Gargiani, *Concrete from archaeology to*...cit., pp. 25.

[71] August Hermann Niemeyer, *Beobachtungen auf einer Deportationsreise nach Frankreich im Jahr 1807*, Halle, Buchhandlung des Waisenhauses, vol. I, 1824, vol. II, 1826; see in particular, vol. I, pp. 28–31.

[72] "So stand im Hofe ein kleines Lusthaus, an dessen Dach sich ringsum Glocken bewegten. Ich dachte an die Glöckchen in der Zauberflöte" (*ibid.*, p. 196).

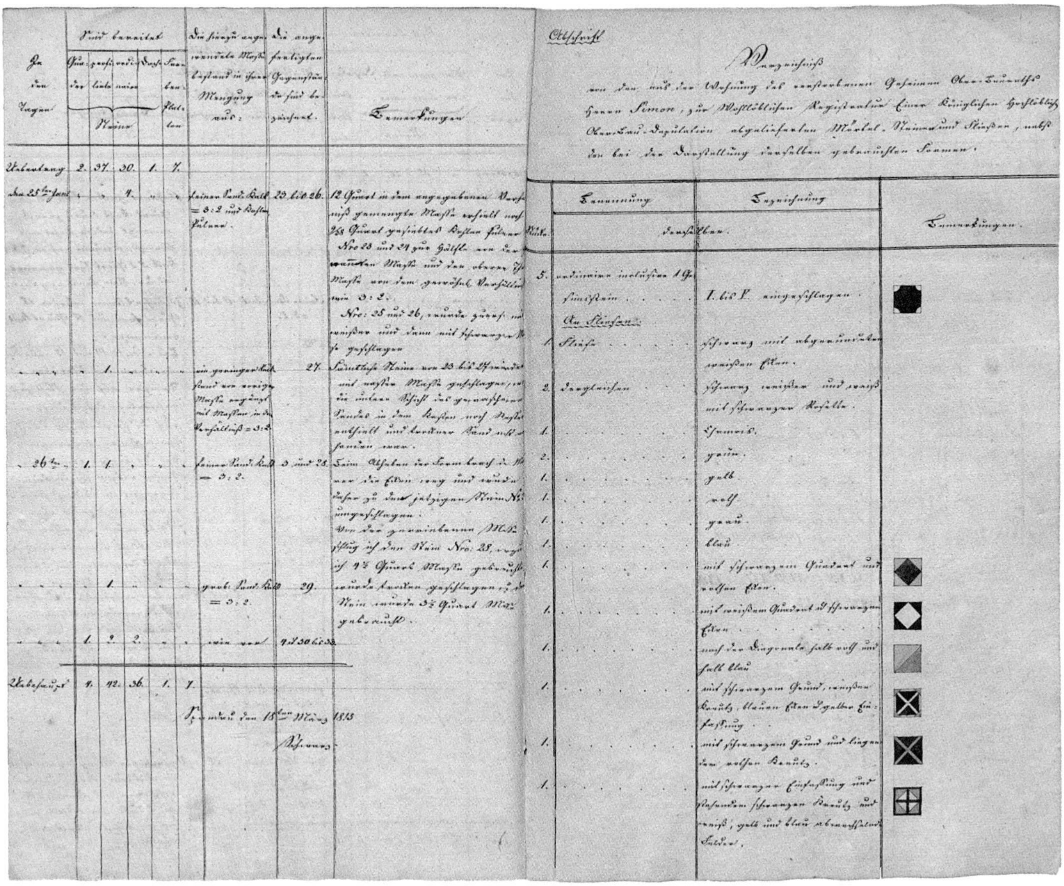

17. Schwarz, tables reporting the results of the studies about Simon's attempts at producing artificial stones, 1815 (Schwarz, "Uebersicht von den im Monat Juni 1812 in dem Hause des verstorbenen Geheimen Ober-Bauraths Herrn Simonangefertigten Mörtelsteinen," March 18, 1815, GstA PK I.HA Rep. 93B, Ministerium für öffentliche Arbeiten, n° 1394, n. fol.).

named Professor Fleuret helped Montaigne build the pavilion. Niemeyer describes Fleuret as a former teacher of the military school who is living a sad existence in Pont-à-Mousson, after having survived persecution during the French Revolution as a result of his fervent religious convictions.[73] Still, according to Niemayer, he planned a summer residence for the "Emperor Paul," but the project faded into oblivion after the emperor was murdered.[74] Apart from Niemayer's memories, Fleuret is actually almost known in the region of Pont-à-Mousson – and even in Paris – for the manufacturing of pipes moulded from a special kind of mortar. He produces such pipes, among other objects, in an atelier where the three German deportees observe the manufacturing process and express interest in it. Von Wedell learns Fleuret's method, aim-

[73] *Ibid.*, pp. 196–197.

[74] Paul I, Tsar of Russia, was murdered in 1801.

[75] Fleuret, *L'art de composer des pierres factices aussi dures que le caillou: et recherches sur la manière de bâtir des anciens, sur la préparation, l'emploi et les causes du durcissement de leurs mortiers*, Pont-à-Mousson, Chez l'Auteur, 1807; Niemeyer, *Beobachtungen...* cit., pp. 135, 198.

[76] "Wenn eine bestimmte Masse von Sand, gemischt mit zerschlagenen Feldsteinen, kleinen Kieseln und zerstoßnen Scherben zubereitet war, so ward eine verhältnißmäßige Menge gebrannter Kalksteine in einen von Weiden geflochtenen Korb geschüttet, und dieser in einen mit Wasser gefülltes Gefäß gestellt, bis sich Blasen zeigten. Man ließ sodann das Wasser ablaufen, schüttete die getränkten Steine auf die angefeuchtete Sandmasse, und bedeckte sie damit einige Zoll hoch, so daß nicht die kleinste Öffnung blieb und jeder Zutritt der äußeren Luft abgeschnitten wurde. Nach 12–16 Stunden ward sodann mit eisernen Schaufeln der zerfallene kalk mit der Sandmasse durch einander gerührt, und so viel Wasser darauf gegossen, daß der nun gewonnene Mörtel, so lange er weich blieb, gleich dem Thon, in jede Form verarbeitet werden konnte" (*ibid.*, p. 199).

[77] See *ibid.*, p. 200; Georg Christoph Hamberger, Johann Georg Meusel, Joahnn Wilhelm Sigismund Lundner, *Das*

ing to reproduce it once back in Prussia, while Keferstein begins to translate into German a book about the manufacturing and use of artificial stones that Fleuret has just completed and published in 1807.[75]

The mixture used by Fleuret to produce artificial stones is essentially made of lime, sand and rejects from iron manufacturing or crushed terracotta used as cement. Niemeyer carefully describes the manufacturing process that he observed in Pont-à-Mousson. "Once a certain amount of wet sand, mixed with crushed stones, flints and crushed terracotta had been prepared," he writes, "an appropriate amount of burnt lime was put into a wicker and plunged into water, until the lime produced bubbles. At this point, water flew away and the resulting soaked lime was thrown into the previously prepared heap of sand and stone, to be then covered with part of it, taking care to close all holes in order to prevent any contact between lime and air. After 12–16 hours, the decomposed lime, sand and stones were mixed by means of iron shovels, adding as much water as necessary, in order to get a kind of soft mortar, suitable to be poured in all kinds of moulds, as if it were clay."[76]

Toward the end of July 1807, following the peace treaties of Tilsit, the German deportees are released and, after a short stay in Paris, they come back to Prussia. Keferstein publishes the translation of Fleuret's book in 1808, Niemeyer writes his memoir and Von Wedell performs trials to produce artificial stone pipes using tools that he has brought from France.[77] Faced with the difficulty of exactly reproducing Fleuret's methods in Halle and lacking all the necessary raw materials, Von Wedell changes Fleuret's recipe and produces a kind of lean aerial mortar made of 1/5 lime and 4/5 sand, rather than the mixture made of three equal parts of lime, sand and cement that he has learned in Fleuret's atelier.[78] Despite the altered recipe, Niemeyer reports that Von Wedell's trials are successful. Von Wedell's recipe for lean mortar actually seems to forerun the mixture of very lean mortar used in various rural villages to build walls and artificial stones starting in the early 1840s.

Von Wedell probably sends a copy of Fleuret's book to the *Ober-Bau-Deputation* in Berlin and informs this office of his experiments. The idea to carry out similar trials in Berlin emerges at this time, as evidenced in a passage of a letter that the *Ober-Bau-Deputation* addresses to King Friedrich Wilhelm III in December 1810. "Seeing the lack of sandstone in most parts of the Kingdom, which compels us to spend money abroad," the passage explains, "and considering the availability of limestone from the quarries of Rüdersdorf near the capital, this kind of manufacturing could offer great advantages for all building activities, if the trials' outcomes were good."[79] By order of Friedrich Wilhelm III, the *Ober-Bau-Deputation* contacts Von Wedell, who claims to be ready to participate in the experiment, as long as it takes place between March and May 1811, during a planned stay in Berlin. The *Ober-Bau-Depu-*

gelehrte Teutschland, oder Lexikon der jetzt lebenden teutschen Schriftsteller, vol. XXIII, Lemgo, Meyetsche Hof-Buchhandlung, 1834, pp. 106–107.

[78] "Since the quality of materials more or less changes from one country to another, it was not possible to fulfil all instructions, and a different composition was unavoidable" ("Die Qualität des Materials welches in jedem Lande mehr oder weniger eine andere ist, ist Grund von Nothwendigen Abweichungen jener Vorschriften und machte andere Compositionen nothwendig") (Königliche Ober-Bau-Deputation, letter to the Departement für die Gewerbe und den Handel im Ministerio des Innern, December 25, 1810, GstA PK I.HA Rep. 93B, Ministerium für öffentliche Arbeiten, n° 1394, n. fol.). See also Niemeyer, *Beobachtungen*...cit., p. 200.

[79] "Jetzt da für den größten Theil der Monarchie der Sandstein fehlt, und das Geld dafür ins Ausland gehen müßte, und bey der Ergiebigkeit unserer wenigstens der Hauptstadt so nahe gelegenen Rüdersdorfer Kalksteinbrüche, möchte sich, wenn die Versuche genügende Resultate geben sollten, eine große und vortheilhafte Anwendung dieser Fabrikation für das ganze Bauwesen ergeben" (Königliche Ober-Bau-Deputation, letter to the Departement für die Gewerbe und den Handel im Ministerio des Innern, December 25, 1810, cit.).

tation charges the master builder Simon with the trials, most likely due to his previous studies concerning lime and mortar. Simon carries out the trials in his own house between 1811 and 1812, producing specimens and tiles in appropriate moulds. Von Wedel does not directly participate in the trials; he just corresponds with the *Ober-Bau-Deputation*.[80] Simon uses mixtures made only of sand and lime, taking more inspiration from Von Wedell's procedure than from Fleuret's instructions. The trials fail, and the *Ober-Bau-Deputation* writes to Fleuret in November 1812 asking his opinion on the matter. Fleuret replies one month later, supposing the quality of German lime to be the origin of the failure.[81]

Simon's death in 1815 gives further occasion to study some of the specimens that had been kept in his house. The master builder (*Bauconducteur*) Schwarz is charged with these studies, which then mark the conclusion of this early attempt at producing artificial stones in Prussia. The upshot is that Fleuret's artificial stone is considered unsuitable for Prussian needs and incomparable to the quality of local bricks. "We do not consider," claims Schwarz, "this invention particularly advantageous for Prussian lands, especially because it is not comparable with our burnt bricks, in what concerns durability and affordability."[82]

Two years later, on August 1817, when Friederich Wilhelm III visits the occupying army in Paris, the French architect Antoine Raynal shares a report with him about an Italian manufacturing process for artificial stones suitable for hydraulic works. Raynal has worked for fifteen years in Piedmont, a northwest Italian region, where he learned how to manufacture tetrahedral artificial stones, called *prismi*, that are suitable for building dams and embankments on foundations made of loose stones.[83] The mixture used to produce the *prismi* consists of hydraulic lime, sand and small stones. It is poured in special moulds dug into the soil and divided into several compartments by means of timber boards.[84] The king hands over Raynal's report to the Interior Minister, Friedrich Von Schuckmann, who then gives it to Heinrich von Bülow, a German diplomat in London, who later helms the department for industry and trade of the Prussian foreign office, before being appointed foreign affairs minister. Nevertheless, Raynal's report falls into oblivion, probably because it is linked to the previously failed attempts at reproducing Fleuret's artificial stones. Evidence of this conceptual relation is given in a short note about Simon's failure, found on a cover letter accompanying a copy of Raynal's report; the note is signed, inter alias, by Eytelwein.[85]

Despite the results, these events show that some interest in mortar-based artificial stone arises in the 1810s, as is also evidenced by an interesting description of artificial stone that Wiebeking introduces in the second edition of his *Wasserbaukunst*.[86] "Artificial stone," he writes, "can consist of a mass of lime and granite-sand or quartz-sand; crushed tuff stones, sandstones, pumice or volcanic ash and lime; marls, slaked lime and sand; melt iron slags poured into mould; clay and crushed pumice or volcanic ash; or, finally, clay and sand that actually produce bricks."[87] Wiebeking also depicts some artificial ashlars that he describes being produced in 1804 in Moravia by mixing "two parts of quartz-sand, one part of slaked lime and one part of milled lime."[88] This mass is then poured and pressed into moulds whose size could be adjusted through screws, and the resulting ashlars are reportedly used to build a wall in the Castle of Felsberg's park. This kind of mould are a pioneering example of the adjustable moulds for artificial stones later developed and even patented in the 1830s in the United Kingdom.[89]

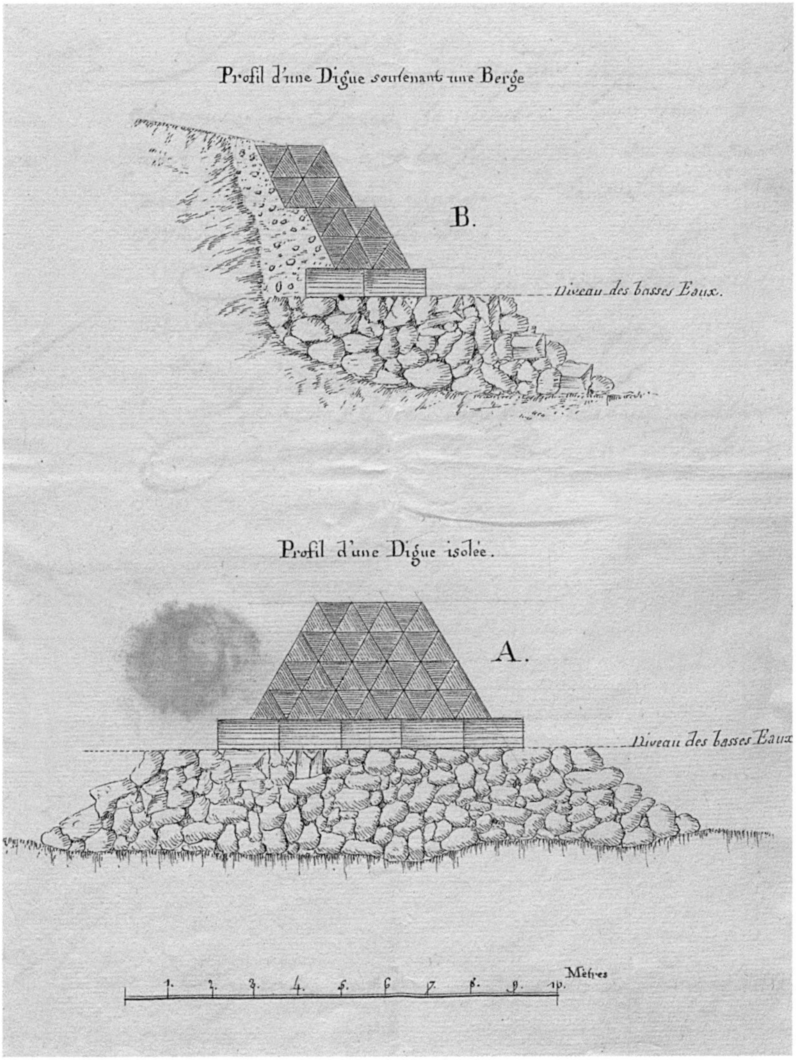

18. A. Raynal, drawings of an embankment and a dam built with artificial stones called *prismi*, 1817 (A. Raynal, "Mémoire sur la Fabrication des Pierres Artificielles Connues en Piémont sous le nom de Prismes. Pour servir a la Construction des Digues et autres ouvrages de maçonnerie exposés à la rapidité des Courantes," GstA PK I.HA Rep. 93B, n° 1394, August 19, 1817, n. fol.).

[80] See Schwarz, "Uebersicht von den im Monat Juni 1812 in dem Hause des verstorbenen Geheimen Ober-Bauraths Herrn Simonangefertigten Mörtelsteinen," 1815, (GstA PK I.HA Rep. 93B, n° 1394, n. fol.).
[81] See *ibid.*
[82] "Nach allem diesem scheint es uns nicht so sehr vortheilhaft diese Erfindung für die Preußische Länder insbesondere in Anwendung zu bringen, vorzüglich da sie in keinem Vergleich mit der Dauer und Wohlfeilheit unserer gebrannter Ziegel steht" (*ibid.*).
[83] See Antoine Raynal, "Mémoire sur la Fabrication des Pierres Artificielles Connues en Piémont sous le nom de Prismes. Pour servir a la Construction des Digues et autres ouvrages de maçonnerie exposés à la rapidité des Courantes" (GstA PK I.HA Rep. 93B, n° 1394, August 19, 1817, n. fol.).
[84] Prismi are actually a variation of another kind of artificial stone called "cantoni di smalto," whose use spread in Tuscany since the 17th century. In this regard, see Gargiani, *Concrete from archaeology to...*cit., pp. 34–39.
[85] Heinrich von Schuckmann, letter to Heinrich von Bülow, January 16, 1818 (GstA PK I. HA Rep. 93B, n° 1394, n. fol.).
[86] Carl Friedrich Wiebeking, *Theoretisch-practische Wasserbaukunst*, München, Joseph Zängl auf Kosten des Verfasser, vol. I, 1811; vol. II, 1812; vol. III, 1814, vol. IV, 1817.
[87] "Die Künstlichen steine bestehen entweder aus einer Masse von Granitsand oder Quarzsand und Kalk; aus gestossenen Tuffsteinen, Sandsteinen, Bimmstein oder vulkanische Asche und Kalk, oder aus Kalkmergel, frisch gelöschten Kalk und Bausand; aus geschmolzenen in Formen gegossenen Eisenschlacken; aus Thon und zerstossenen Bimmstein oder vulkanische Asche; oder endlich aus Lehm, Thon und Sand, welche letztere man *Ziegel* nennt" (Wiebeking, *Theoretisch-practische Wasserbaukunst*...cit., vol. II, p. 523). See also Wiebeking, *Allgemeine auf Geschichte*...cit., vol. III, pp. 322–323.
[88] "Zur Verfertigung der künstlichen Sandsteine wird zum sichersten eine aus zwey Theil Quarzsand, ein Theil frisch gelöschten Kalk, und ein Theil zerriebenen Kalk genommen" (Wiebeking, *Theoretisch-practische Wasserbaukunst*...cit., vol. II, p. 524).
[89] This refers to the concrete artificial stones by William Ranger from Brighton.

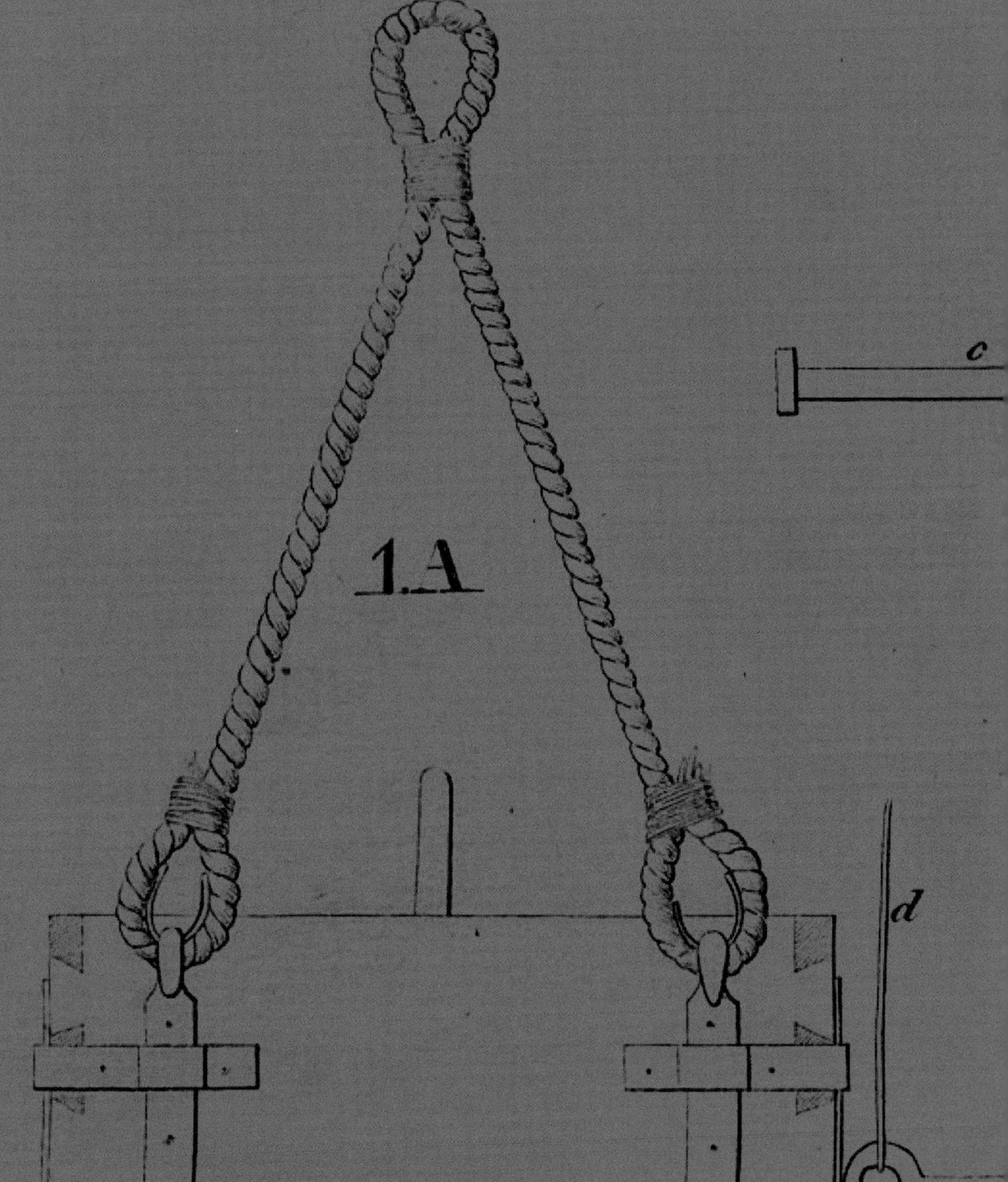
1.A
c
d
c
a
a

chapter two
new knowledge about hydraulic lime, cement and mortar

The *Theorie des besten Mörtels* by John

The belief that aerial lime is more effective than other binders begins to waver towards the end of the 1810s, following the achievements of some chemical studies about the nature and properties of hydraulic lime, whose effectiveness in constructions is finally scientifically proven. A new phase in the German construction history begins. It is characterized both by a significant interest in the scientific analysis of hydraulic binders and by their gradually increasing use in construction. This brings far-reaching consequences for the production and use of concrete, first applied to foundations and later to above-ground construction, as of about the end of the 1820s, following French and British models.

The symbolic starting point of this new phase can be identified in the research about lime and mortar that the chemist Johann Friedrich John (1782–1847) completes in 1817. John actually takes part in a competition promoted by the Royal Holland Society of Sciences and Humanities in 1810, with the aim of discovering the "chemical reason that lime burnt from limestone makes masonries more solid and durable than masonries built using lime produced from shells, and how to improve this last one."[1] At the time he enters this competition, John has been studying the composi-

[1] "Quelle est la cause chimique, que la chaux de pierre fait sur le total une maçonnerie plus solide et plus durable, que la chaux de coquilles et quels sont les moyens de corriger à cet égard la chaux de coquilles" (see Johann Friedrich John, *Ueber Kalk und Mörtel im Allgemeinen und den Unterschied zwischen Muschelschalen- und Kalksteinmörtel insbesondere: nebst Theorie des Mörtel; eine von der Holländischen Gesellschaft der Wissenschaften gekrönte Preisschrift*, Berlin, Duncker und Humboldt, 1819, p. V).

tion of several natural substances for roughly four years.[2] He has analysed, inter alias, a kind of limestone called *Lucullan* and several organic calcareous substances such as eggshells and human bones.[3] These studies surely help his investigation of lime. At first John imagines that shells and limestone contain two different kinds of lime earth and therefore produce lime of different qualities. Rather, the opposite is true: chemical analyses show that shells and limestone contain the same kind of lime earth. "Marble and all kinds of limestone," John then writes, "react in this respect like shells; there is therefore no difference between the nature of organic and inorganic limes."[4] As a second step, John investigates the process of mortar hardening. Like most of "the new chemists and architects," he believes that "the phenomenon of mortar hardening originates from the regeneration of limestone, namely that slaked lime loses water and absorbs carbonic acid from the air, forming a hardening mass that adheres to sand contained in mortar, and to the surface of bricks and stones."[5] However, the observations by chemists Louis-Bernard Guyton de Morveau and Johann Wolfgang Döbereiner (1780–1849) about supposed, natural affinities among earths cast doubts on John's convictions about lime and mortar.[6] De Morveau and Döbereiner believe that lime tends to combine with silica and clay by nature, and that all these substances tend to combine in a kind of insoluble substance, as observed by the French chemist Claude-Louis Berthollet.[7]

Searching for possible combinations of earths in mortar, John first investigates whether a combination between lime and silica contained in sand is possible. He tries to dissolve a quartz crystal, which is a siliceous substance, in limewater, but a negative outcome pushes him to review the problem and search further. He thus examines several mortar samples from different ages, which he takes from notable German buildings, namely the Berlin Royal Palace, the Berlin Saint Peter Church, the Brandenburg Cathedral and the ruins of the Roman walls in Cologne. John attaches special importance to the study of Roman mortar, which he considers the "ideal" sample to analyse. The Cologne walls consist of rough masonry with sizeable layers of coarse-grained

[2] Coherently with the general cultural inclination that urges the reconciliation of theory and practice, John believes that theoretical chemistry has practical implications, and he therefore shows a major interest for useful substances; in this regard he affirms: "the conviction of the importance of chemical analysis for medicine, mineralogy, applied chemistry, agronomy, technology, mining activities, botany, and for art and science urged me to undertake studies about these matters" ("die Ueberzeugung von dem Einflusse chemischer Analysen auf Medicin, Mineralogie, Cammeralchemie, Agronomie, Technologie, Bergwerkskunde, Botanik u.s.w. überhaupt auf Künste und Wissenschaften, munterten mich auf, sie fortzusetzen") (Johann Friedrich John, *Chemische Untersuchungen mineralischer, vegetabilischer und animalischer Substanzen. Fortsetzung des chemischen Laboratoriums*, Berlin, Friedrich Mauerer, 1810, p. IX). See also Id., *Chemisches Laboratorium. Oder Anweisung zur chemischen Analysen der Naturalien. Nebst Darstellung der nötigsten Reagenzien*, Berlin, Friedrich Maurer, 1808, pp. IX–XII. John uses the expression *Cammeralchemie* [sic] for applied chemistry; a definition of *Cammeralchemie* is to find in, Franz Karl Alexander Döbereiner, *Cameralchemie für Land- und Forstwirthe, Techniker, Sanitäts-, Cameral- und Justizbeamte, sowie überhaupt für alle Freunde der Naturwissenschaft*, vol. I, Dessau, Gebrüder Rats, 1855, p. 2.

[3] John, *Chemische Untersuchungen*...cit., 1810, pp. 219–251; Id., *Chemische Untersuchungen mineralischer, vegetabilischer und animalischer Substanzen. Fortsetzung des chemischen Laboratoriums*, Berlin, J. E. Hitzig, 1811, pp. 22–24, 88–93.

[4] "Marmorkalk und jeder Kalkstein verhält sich in dieser Hisicht, wie Muschelschalenkalk, so dass darin kein Unterschied des Kalks der organischen und unorganischen Natur zu suchen ist" (John, *Ueber Kalk und Mörtel*...cit., p. 23).

[5] "Die neueren Chemiker und Architekten sind der Meinung, Dass die Mörtelhärtung eine Regeneration des Kalksteins sey; dass das Kalkhydrat die Feuchtigkeit verliere und Kohlensäure aus der Luft anziehe, und dass die dadurch erhärtete Masse sich an den Sand des Mörtelgemenges und die Oberflächen der Mauersteine fest anlege" (*ibid.*, p. 27).

[6] Except for water, substances composing mortar are considered different kinds of earth at that time.

[7] In 1801, De Morveau writes: "when we had observed that two infusible earths, such as silica and lime, if fired in kilns, combine into an homogenized material, whose components cannot be disjoined except through chemical ways, we had to admit that at a certain temperature a mutual reaction occurs, due to a kind of real affinity" ("Mais lorsqu'on eut observé que deux terres infusible seules au feu de nos fourneaux, telles que la silice et la chaux, y couloient assez facilement en une matière

mortar, and they therefore appear as giant conglomerate masses, similar to concrete. In this regard, the description given by John himself is enlightening. "These walls," he writes, "are built with bricks, clayish porphyry, several kinds of limestone, a kind of dark smoke grey sandstone, and so on. Mortar is produced with 3–4 inch-sized pebbles and sand with various grain sizes, from the finest to the coarsest ones. Since the quantity of lime is scarce in comparison to that of sand and stones, [...] mortar is wastefully employed and sometimes forms ¼ feet thick layers [...]. Roman mortar in the city walls [of Cologne] looks like molasses and it is so hard that it can be sometimes compared to puddingstone or to some special silica conglomerates."[8]

John expects to find much carbonic acid in Roman mortar since, according to the current belief of that time, the more lime is pure, the more carbonic acid it absorbs, and the more it hardens. Quite the contrary, chemical analyses show an extremely small amount of carbonic acid and an unexpected, considerable amount of silica.[9]

Searching for the origin of silica discovered in Roman mortar, John analyses further limestone samples and finds out that they always contain a certain amount of silica clay and metal oxides, besides lime earth. He realizes that these three substances combine with part of lime earth during limestone burning, and that lime contained in mortar hardens partially through the combination with carbonic acid and partially through the contribution of silica, clay and metal oxides, which are supposed to act as glue, making mortar extremely hard and water-resistant. John calls this explication *Theorie des besten Mörtels*. He defines silica, clay and metal oxides the real cement in mortar and argues: "The larger the amount of cements is, the harder and better mortar becomes, whereas the amount of carbonic acid is small."[10] At the same time, however, John warns against an excessive amount of cement, which is thought to prevent lime slaking.

On the basis of the principles of the *Theorie des besten Mörtels*, it is assumed that the lack of cement causes the lower quality lime from the shells. As such, limestone containing clay is considered a better raw material than

homogène, dont les principes ne pouvoient plus être séparés que par des moyens chimiques, il fallut bien reconnoître une action réciproque qui devenoit efficace à une certaine temperatue, c'est-à-dire, une veritable affinité") (Guyton de Morveau, *Sur les affinités que les terres exercent les unes sur les autres, soit par la voie humide, soit par la voie sèche*, in, "Annales de chimie ou Recueil de mémoires concernant la chimie et les arts qui en dépendent," vol. XXXI, 1799, (pp. 246–268), pp. 246–247). With reference to De Morveau, in 1814, Döbereiner writes: "This eminent chemist found out that limewater [...] set silica earth and clay from soda apart – this is due, according to Berthollet, to the inclination of lime, silica and clay to combine and form an insoluble substance (artificial wollastonite etc.)" ("Dieser achtungswerthe Chemiker fand nämlich, dass Kalkwasser [...] Kiesel- und Thonerde vom Kali trennt - ein Erfolg welcher nach *Berthollet*'s Ansicht seinen Grund hat in der Tendenz des Kalks und Kiesel, oder Thonerde, sich mit einander zu einem unauflöslichen Körper (zu künstlichem Tafelspath u.s.w.) zu verbinden") (Johann Wolfgang Döbereiner, *Ueber das Verhalten des Kalks zu dem Kiesel- und Thonkali auf nassem Wege und über andere verwandte Gegenstände*, in, "Journal für Chemie und Physik," 1814, vol. X, p. 114). See also Claude Louis Berthollet, *Recherches sur les lois de l'affinité*, Paris, Baudouin imprimeur de l'Institut national des sciences & des arts, année IX, [1801].

[8] "Diese Gesteine sind theils mit gebrannten Ziegelsteines, theils mit Ziegelsteinen, Thonporphyr, verschiedenen Arten Kalksteins, einem dunkel rauchgrauen Sandstein u. s. w. construirt. Der Mörtel selbst ist mit Geschieben von 3 bis 4 Zoll Durchmesser, dem feinsten Staube und Geschieben mit Sand von jeder zwischen beiden Extremen gelegenen Größe und Feinheit verfertigt. Ob nun gleich das Verhältniss des Kalkes zu dem Sande und den Geschieben, [...] gering ist, so ist der Mörtel doch sehr verschwenderig angewandt, denn er bildet oft ¼ Fuß dicke Lagen [...]. Der Römische Mörtel in den Stadtmauern hat das Ansehen einiger Nagelfluhgebirge und ist von so einer ungemein großer Härte, daß das ganze Gemäuer oft mit dem Puddingstein, oder einigen besonderen Kieselconglomeraten verglichen werden könnte" (John, *Ueber Kalk und Mörtel*...cit., pp. 34–35).

[9] See also Friedrich Quietmeyer, *Zur Geschichte der Erfindung des Portlandzementes*, Berlin, Tonindustriezeitung, 1911, (pp. 73–78), p. 74.

[10] "Je größer daher die Menge jenes Cäments ist, desto härter, und vorzüglicher ist auch Mörtel und desto geringer der Gehalt an Kohlensäure Kalk" (John, *Ueber Kalk und Mörtel*... cit., p. 58).

pure limestone, suitable to producing a kind of lime which is appropriate for building underwater and above ground (hydraulic lime). Thus, John completely subverts the traditional convictions about lime and mortar, providing a chemical explanation of the role of silica and clay in hydraulic lime for building. The concept itself is not new. British engineer John Smeaton has already proven that hydraulic lime contains clay, while Guyton de Morveau has formulated early hypothesis about supposed chemical reactions involving the different components of hydraulic lime. Moving from such premises, however, and being firmly convinced of the importance of applied chemistry, John develops a general theory of lime and mortar relating to the theoretical domain of chemistry and the practical domain of constructions. He even develops a method for improving the quality of shells as raw material for lime by adding some silica, which is actually a method to produce a kind of artificial, hydraulic lime. At the end of 1817, John submits the final report of his research to the Royal Holland Society of Sciences and Humanities, and, in 1818, he is awarded the gold medal.

A new awareness about hydraulic lime and mortar

John's report is published as a book in 1819.[11] Shortly thereafter, Chrètien Constant Vauvilliers, a French engineer of the *Corps des Ponts et Chaussées*, based in Strasburg, translates it into French and sends a copy to Louis-Joseph Vicat, an engineer of the same Corps who has published, in 1818, an essential book about the experimental researches on lime, mortar and concrete that he conducted over six years.[12] Vicat writes a report about John's book which appears in the "Annales de Chimie et de Physique" in 1821 and is then followed by several other articles, reviews or simply mentions of John's work in different periodicals and books.[13] While essential concepts of John's theory begin to circulate in France, similarly the outcomes of Vicat's researches begin to be known in Germany through the translations of several French articles, primarily published in the "Polytechnisches Journal."[14]

Actually, John and Vicat undertake the study of lime and mortar from two different points of view. As a scientist, John largely explores the chemical combinations that he considers to be the origin of lime and mortar properties, whereas Vicat, being an engineer, gives more emphasis to practical experi-

[11] See *ibid.*

[12] Louis-Joseph Vicat, *Recherches expérimentales sur les chaux de construction, les bétons et les mortiers ordinaires*, Paris, Goujon, 1818.

[13] Louis-Joseph Vicat, *Sur la Chaux et le Mortier en général et en particulier; sur la différence entre les mortiers de chaux de coquilles de moules et de pierres calcaires, avec théorie des mortiers*, in, "Annales de Chimie et de Physique," vol. XIX, 1821, pp. 15–25. Other writings or mentions about John's work are to find in: Johann Friedrich John, *Sur la chaux et le mortier en général, et en particulier sur la différence entre les mortiers de chaux dé coquilles de moules et de pierres calcaires, avec la théorie des mortiers*, ed. Louis-Joseph Vicat, in, "Annales des mines, ou Recueil de mémoires sur l'exploitation des mines, et sur les sciences qui s'y rapportent," vol. VII, 1822, pp. 467–482; André Étienne Justin Pascal Joseph François d'Audebert de Férussac, ed., *Bullettin des sciences technologiques*, vol. II, Paris, Au bureau du Bullettin (etc.), 1824, p. 106; Plusieurs Professeurs du Jardin du Roi, et des principales Écoles de Paris, ed., *Dictionnaire des Sciences Naturelles es d'après l'état actuel de nos connoissances, dans lequel on traite méthodiquement des différens êtres de la nature, considérés soit en eux-mêmes, soit relativement a l'utilité qu'en peuvent retirer la médecine, l'agriculture, le commerce et les arts*, vol. XXXIII, Strasbourg, Paris, F. G. Levrault, Le Normant, 1824, p. 32; Jean-Henry Hassenfratz, *Traité théorique et pratique de l'art de calciner la pierre calcaire et de fabriquer toutes sortes de mortiers, cimens, bétons, etc., soit à bras d'hommes, soit à l'aide de machines*, Paris, Carilian-Goeury, Mme Huzard, 1825, p. IX; Louis Jacques Thénard, *Traité de chimie élémentaire, théorique et pratique*, vol. I, Bruxelles, Canongette et compagnie, 1829[7], p. 347; Clément-Louis Treussart, *Mémoire sur les mortiers hydrauliques et sur les mortiers ordinaires*, Paris, Chez Carillan-Goeury, Anselin, Mahler et Compagnie, 1829, pp. 6–7, 49, 79–80, 142–148, 182, 214.

[14] Louis-Joseph Vicat, *Ueber Kalk und Mörtel. Ein Auszug aus Hrn. Vicat's, Brücken- und Strassenbau-Inspektors, Recherches expérimentales sur les chaux de construction, les bétons et les mortiers ordinaires; et du Supplément inédit relatif à la fabrication des pouzzoulanes artificielles*, in, "Polytechnische Journal," vol. IV, 1821, pp. 280–293 (Id., *Extrait de*

ments and to a series of strength tests on mortar and concrete specimens that he conceives in order to trace practical rules for the production of different kinds of mortar and concrete. Despite these two different approaches, John and Vicat share essential principles about the nature and properties of hydraulic lime. They both believe hydraulic lime to be more suitable than aerial lime for both hydraulic works and for above-ground constructions. Vicat also develops a method to produce artificial, hydraulic lime by firing mixtures of fat lime and clay, which is essentially similar to the one that John conceives to improve the quality of lime produced from shells.

Vicat's book impresses Moritz Karl Ernst von Prittwitz und Gaffron (1795–1885), a military engineer and captain of the *Schlesische Festungspionier-Compagnie* (Silesian Fortress-Engineer-Unit) who serves in France from 1815 to 1818. Once back in Germany, Von Prittwitz is posted at the

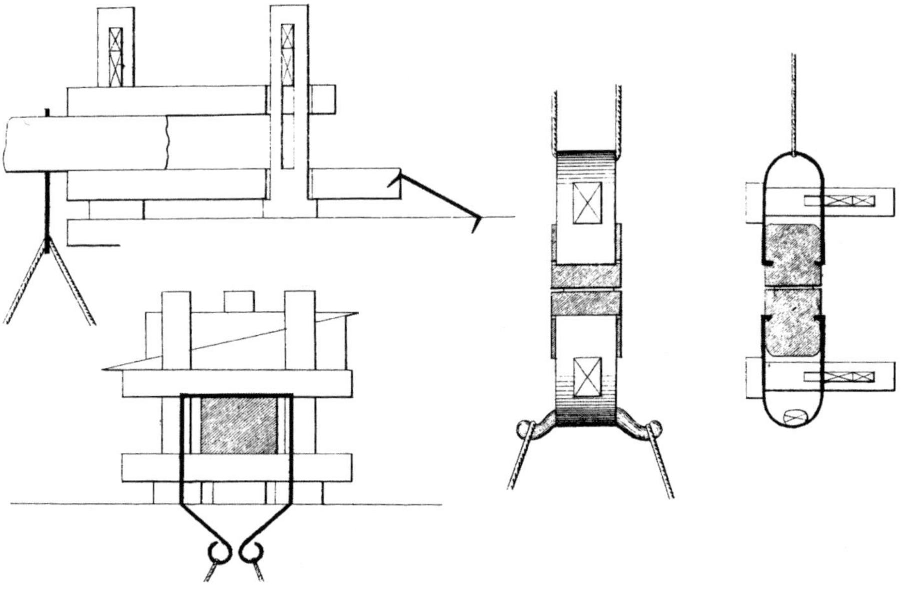

19. M. K. E. von Prittwitz, contraptions for strength tests, 1820–1822. Longitudinal section and front elevation of the contraption for relative strength tests; side elavation and cross section of the contraption for absolute stength tests (M. K. E. von Prittwitz, ed., *Neue Versuche über den Kalk und Mörtel von L. J. Vicat und Andern*, Berlin und Posen, Ernst Sigfried Mittler, 1825).

l'ouvrage intitulé: Recherches expérimentales sur les chaux de construction, les bétons et les mortiers ordinaires; et du Supplément inédit relatif à la fabrication des pouzzolanes artificielles, in, "Annales de Chimie et Physique," vol. XV, 1820, pp. 365–379); Id., *Bemerkung über einige Erscheinungen bei dem Brennen des gemeinen und künstlichen Kalkes*, in, "Polytechnisches Journal," vol. XII, 1823, pp. 429–433 (Id., *Note sur quelques phénomenes que présente la caisson de la chaux ordinaire et de la chaux artificialle*, in, "Annales de Chimie et Physique," vol. XXIII, 1823, pp. 424–429); Id., *Untersuchungen über die Anwendung des hydraulischen Mörtels statt des Gipses zum Modelliren*, in, "Polytechnisches Journal," vol. XV, 1824, pp. 186–207 (Id., *Recherches sur la substitution du mortier hydraulique au plâtre, dans l'art du mouleur*, in, "Bulletin de la Société d'Encouragement pour l'Industrie Nationale," XXIII, 1824, pp.132–148); Id., *Neuer hydraulischer Mörtel*, in, "Polytechnisches Journal," vol. XVIII, 1825, p. 480 (Id., *Théorie des ciments*, in, "Nouveau Bulletin des Sciences par la Société Philomathique de Paris," 1825, pp. 184–185); Anon., *Ueber Theorie der Mörtel von Hrn. Vicat*, in, "Polytechnisches Journal," vol. XX, 1826, pp. 106–107 (Id., *Théorie des ciments*, in, "Nouveau Bulletin des Sciences, par la Société Philomatique de Paris," 1825, pp. 184–185); Louis-Joseph Vicat, *Neue Thatsachen zur Theorie der Kalkmörtel*, in, "Polytechnisches Journal," vol. XXI, 1826, pp. 432–437 (Id., *Nouveaux faits pour server à la théorie des Cimens calcaires*, in, "Annales de Chimie et Physique," XXXII, 1826, pp. 197–204); Id., *Nachträgliche Bemerkung über die Mörtel*, in, "Polytechnisches Journal," vol. XXII, 1826, pp. 166–167 (Id., *Note de M. Vicat, Ingénieur en chef des Ponts et Chaussées, relativement à son dernier Mömoire sur les Mortiers*, in, "Annales de Chimie et Physique," XXXII, 1826, p. 443). The "Bavarian Kunst- und Gewerb-Blatt des polytechnischen Vereins für Bayern" also publishes two articles about the different ways to adopt for obtaining hydraulic mortar, including the use of natural and artificial hydraulic lime; see Anon., *Ueber den Mörtel und dessen Verbesserung*, in, "Neues Kunst- und Gewerbblatt," IX, 1823, pp. 109–114; Anon., *Ueber Wasser-Mörtel*, *ibid.*, pp. 285–287.

building of the Koblenz fortress until 1824.[15] During this period, he translates Vicat's book and writes an abstract from the *Traité sur l'art de faire de bons mortiers et notions pratiques pour en bien diriger l'emploi*, by Antoine Raucourt de Charleville, another engineer of the *Corps des Ponts et Chaussées*.[16] Between 1820 and 1822, taking Vicat's tests and John's theory as a reference, Von Prittitz carries out a series of strength tests on a number of mortar specimens made of mixtures containing different proportions of lean lime from Trier, fat lime from Mainz, trass from Andernach, crushed bricks, and different kinds of sand.[17] At first, he plans to test tensile strength, which is called, at that time, *absolute Festigkeit*.[18] For this purpose, he devises a special contraption to pull two stones connected by means of mortar: one stone hanging from a metal bracket, the other holding a bucket gradually filled with bricks.[19] The real aim of this kind of test is to investigate the adhesion between mortar and stones, rather than the tensile strength of mortar specimens. This is plausible for two reasons: the Koblenz fortress is mainly built in stone and brick masonry, and Von Prittwitz could have known of similar mortar adherence tests during his stay in France, since they are described in eminent French technical books of the early 19th century.[20] However, the tests do not work as hoped; they give incoherent outcomes and Von Prittwitz replaces them with compression and cantilever strength tests (*rückwirkende* and *relative Festigkeit*). He derives the procedure and the design of an appropriate machine to do cantilever tests from Vicat's book. This test consists of a workbench where prismatic mortar specimens are fixed so that they partially protrude from the bench and can be incrementally charged with suspended loads until the specimens cracked.[21] Von Prittwitz's tests confirm the superiority of hydraulic lime over the aerial one, as maintained by John and Vicat,[22] but they nevertheless remain unknown until 1825, when Johann Gustav Georg von Rauch (1774–1841), head of the Royal Corps of Prussian Engineers, promotes the publication of a collection of writings that include Von Prittwitz's translation of Vicat's book,

[15] See Historische Commission der Königliche Akademie der Wissenschaften, ed., *Allgemeine Deutsche Biographie*, vol. XXVI, Leipzig, Duncker & Humboldt, 1888, entry "Prittwitz, Moritz Karl Ernst v. P. und Gaffron," pp. 609–611. See also Otmar Schäuffelen, *Die Bundesfestung Ulm und ihre Geschichte: Europas grösste Festungsanlage*, Ulm, Vaas, 1980, p. 25. Koblenz fortress was built between 1815 and 1834 using, inter alias, trass-based mortar; see Petra Egloffstein, *Entwicklung von Kalk-Trass-Mörtlen im preußischen Festungsbau am Beispiel der Festung Ehrenbreitstein, Koblenz*, in, Institut für Steinkonservierung e. V., ed., *Hydraulische Bindmittel im 19. Jahrhudert auf dem Gebient der heutigen Bundesländer Hessen, Rheinland-Pfalz, Saarland und Thüringen*, Mainz, Institut für Steinkonservierung e. V., 2012, pp. 15–20.

[16] Antoine Raucourt de Charleville, *Traité sur l'art de faire de bons mortiers et notions pratiques pour en bien diriger l'emploi: précédé d'expériences récentes faites sur les Chaux de France et de Russie*, Saint-Pétersbourg, Voie de la Communication, 1822. A second edition of Raucourt's treatise is published in 1828 (Id., *Traité sur l'art de faire de bons mortiers et notions pratiques pour en bien diriger l'emploi ou méthode générale pratique pour fabriquer en tous pays la chaux, les cimens et les mortiers les meilleurs et les plus économiques*, Paris, De Mahler et Cie, 1828).

[17] See Moritz Karl Ernst von Prittwitz, *Versuche, angestellt in Coblenz, in den Jahren 1820 bis 1822*, in, Id., ed., *Neue Versuche über den Kalk und Mörtel von L. J. Vicat und Andern*, Berlin, Posen, Ernst Sigfried Mittler, 1825, pp. 55–75.

[18] For the 19th century nomenclature of different kind of strength, see Franz Joseph Ritter von Gerstner, *Handbuch der Mechanik*, vol. I, Prag, Johann Spurny, 1833², p. 241.

[19] A detailed description of the machinery is not to find, but an illustration is published by Von Prittwitz in 1825; see Von Prittwitz, *Versuche*...cit., p. 56, pl. III, fig. 4.

[20] Louis-Charles Boistard (1763–1823), engineer of the *Corps de Ponts et Chaussées*, carries out mortar adherence tests in 1800. These tests are described by Boistard himself in a book dating from 1804, and later mentioned in the first volume of the *Traité de la Constructions des Ponts* by Émiland–Marie Gauthey (1732–1806), published posthumous in 1808 by his nephew Claude-Louis-Marie-Henry Navier (1785–1836); see Louis–Charles Boistard, *Expériences sur la main d'oeuvre de différens travaux dépendans du service des ingénieurs des ponts et chaussées, sur l'adhérence des mortiers de sable et de ciment employés à l'air et sous l'eau et sur l'usage des machines à épuiser*, Paris, J.–S. Merlin, 1804, p. 51, cit. in, Émiland–Marie Gauthey, *Traité de la Constructions des Ponts*, ed. Claude Louis Marie Henry Navier, Paris, Firmin Didot, 1809, p. 327. See also Louis-Charles Boistard, *Recueil d'expériences et d'observations faites Sur différens Travaux exécutés pour la construction du pont de Nemours, pour celle de l'arsenal et du port militiare d'Anvers, et pour la reconstruction du port de Flessingue; dans lequel on a traité la théorie de l'équilibre*

the abstract from Raucourt de Charleville's treatise and a detailed report about the tests of Koblenz.[23] The publication is announced in the "Verhandlungen des Vereins zur Beförderung des Gewerbfleißes in Preußen" and in the "Allgemeine Literatur-Zeitung."[24] A passage taken from the review in this latter paper sounds prophetic regarding the scientific approach that characterizes the study of lime and mortar in Germany in the following decades: "These essays shed new light on the preparation of different kinds of mortar, the ones that harden in the air and the ones that harden underwater, showing how future studies should be developed in order to achieve precise and reliable results about this important subject."[25]

The contemporary spread of John's theory and Vicat's experimental research topples the old belief in the superiority of aerial lime and mortar. Von Prittwitz seizes the importance of such a change. "It is extraordinary," he writes, "that almost at the same time when Mr. Vicat was carrying on his tests in France, Professor John in Berlin was studying the same topic and, although he adopted a radically different method, based just on chemical analysis, he came to similar results, through which the ancient theory of mortar has been thrown away and knowledge has been considerably improved."[26]

As soon as one question is answered, another one emerges about the eventual chemical role of clayish-siliceous substances like pozzolana, trass and brick dust in mortar. Having failed in trying to dissolve a quartz crystal in limewater, John maintains that siliceous substances must be burnt together with lime to act as a cement and that, therefore, the role of pozzolana, trass and all similar substances in mortar is merely to absorb water. Instead, Vicat imagines that lime and pozzolana react together to produce good hydraulic mortar. A real controversy about John's hypothesis arises in the first half of the 1820s in France, involving Vicat and Pierre Berthier, a mineralogist of the *École des Mines*. The two scholars write each other by means of articles published in the "Annales de Chimie et Physique," which are then promptly translated into German and published in the "Polytechnische Journal." The

des voutes, Paris, J.–S. Merlin, 1822, p. 125).

[21] Vicat, *Recherches expérimentales*...cit., pl. I. Von Prittwitz could have also read of cantilever beam tests in French technical literature of the early 19th century, which widely deals with this topic as part of the widest theory of beam. The question is analysed, in 1773, by Charles-Augustin Coulomb and Coulomb's solution spreads later through Navier's teaching at the *École des Ponts et Chaussées* in the early 19th (see Jacques Heyman, *The strength of beams: Galilei to Navier*, in, Roberto Gargiani, ed., *L'architrave, le plancher, la plate-forme. Nouvelle histoire de la construction*, Lausanne, Presses Polytechniques et Universitaires Romandes, 2012, (pp. 310–315), p. 312). Navier deals with this matter in a note to the 1813 edition of *La science des ingénieurs dans la conduite des travaux de fortification et d'architecture civile* by Bélidor, which is published at the time as Vicat undertakes his experiments and tests (see Bernhard Forest de Bélidor, *La science des ingénieurs, dans la conduite des travaux de fortification et d'architecture civile. Par Bélidor*, ed. Claude Louis Marie Henry Navier, Paris, Firmin Didot, 1813, (pp. 336–342), p. 337, n. 1, pl. XII).

[22] Von Prittwitz observes that lean lime from Trier is moderately hydraulic and could be improved through the addition of trass or bricks powder, and that fat lime from Mainz has no hydraulic property but produces a good hydraulic mortar if mixed with trass from Andernach.

[23] Von Prittwitz, *Neue Versuche*...cit.

[24] See Anon., *Auszug aus den Protokollen der monatlichen Versammlungen in den Monaten Juli und August des laufenden Jahres*, in, "Verhandlungen des Vereins zur Beförderung des Gewerbfleißes in Preußen," IV, 1825, (pp. 129–133), p. 133; Anon., *Technologie*, in, "Allgemeine Literatur-Zeitung," vol. II, 1827, col. 480.

[25] "Durch diese Aufsätze wird ein ganz neues Licht, über die Bereitung der verschiedenen Arten von Mörtel, sowohl solcher, die in der Luft, als solcher, die auch unter dem Wasser erhärten, verbreitet, und der Weg gezeigt, welcher bey künftigen Forschungen, über diesen so wichtigen Gegenstand eingeschlagen werden muss, wenn man endlich zu bestimmten und sichern Ergebnissen gelangen will," (*ibid.*).

[26] "Es ist daher eine auffallende Erscheinung, daß ungefähr zu derselben Zeit, wo Herr Vicat diese Versuche in Frankreich anstellte, Herr Professor John in Berlin sich mit demselben Gegenstände beschäftigte, und auf einem dem vorigen entgegensetzten Wege, nämlich durch bloße chemische Analyse, mit sehr geringen Abweichung ganz auf die nämlichen Resultate geführt wurde, als Vicat, wodurch die alte Theorie des Mörtels über den Haufen geworfen, und die Kenntniß desselben auf einen durchaus neuen Standpunkt gestellt worden ist" (Von Prittwitz, *Versuche, angestellt in Coblenz*...cit., pp. 63–64).

confrontation begins in 1823, when Berthier announces his agreement with John regarding the inefficacy of siliceous substances in mortar.[27] "Mr. Vicat," he writes, "assumes that the chief cause of mortar solidity is the chemical action of lime on siliceous substances," while, "Mr. John believes that the substances that we mix together with slaked lime to produce mortar, which he names admixtures, are essentially inert."[28] Given this divergence, Berthier resolutely states: "Mr. Vicat's assertion about a chemical action of lime on siliceous substances has no factual base." He adds, furthermore: "I agree with Mr. John about the fact that admixtures do not play any chemical role in mortar."[29] Vicat answers Berthier twice, listing a long series of facts taken from his own experience to corroborate his belief in the existence of chemical combinations between lime and siliceous materials.[30] Far from being of secondary importance, the issue raised by John, Vicat and Berthier connects to questions about the chemical nature and properties of siliceous substances later studied in Germany.

Cement from England and policies for the *Gewerbeförderung*

Together with the achievements concerning the nature and properties of hydraulic lime, the introduction of British cement on the German market of building materials contributes to increase the use of hydraulic binders. Early importation traces are evident in Hamburg, where a dealer named Benjamin Dodson trades in the so-called Roman cement as early as 1817.[31] It is a kind of extremely effective hydraulic lime that does not need to be slaked. The English lime producer James Parker discovered it in the early 1790s while burning marls from the Sheppey Island.[32] Dodson sells this kind of cement on behalf of a London manufacturer, along with "the tools that allow the use of cement" – probably moulds.

In addition to trade, Roman cement becomes known in Germany through the so-called *technologische Reise*. This expression refers to research journeys undertaken by technicians, businessmen or civil servants to investigate new industrial products and manufacturing processes abroad. England is the main destination for these journeys; most travellers are principally interested in mechanical topics but some attention is also paid to new building techniques and materials, and it would be nearly impossible for German travellers not to notice the Roman cement plasters and decorations that spread all over London after the lapse of the Parker's patent in 1810.

[27] Pierre Berthier, *Analyses de différentes Pierres à Chaux*, in, "Annales de Chimie et Physique," vol. XXII, 1823, pp. 62–91 (Id., *Analyse verschiedener Kalksteine in Bezug auf den besten Mörtel*, in, "Polytechnisches Journal," vol. XI, 1823, pp. 350–363).

[28] "M. Vicat admet, comme cause principale de la solidification des mortiers, l'action chimique que la chaux exerce sur les matières siliceuses" (Berthier, *Analyses...* cit., pp. 85). "M. John pense, au contraire, que les substances que l'on mêle avec la bouillie de chaux pour faire les mortiers, et qu'il nomme alliages, sont tout-à-fait passives" (*ibid.*, 86).

[29] "L'assertion de M. Vicat, que la chaux en général et la chaux hydraulique en particulier agissent chimiquement sur les matières siliceuses, n'est appuyée sur aucun fait (*ibid.*, p. 88). "Je pense avec M. John que les alliages ne jouent aucun rôle chimique dans les mortiers" (*ibid.*).

[30] Louis-Joseph Vicat, *Notes en réponse à un Article de M. Bérthier, Ingénieur en chef des Mines, sur la théorie des mortier*, in, "Annales de Chimie et Physique," vol. XXIII, 1823, pp. 69–80 (Id., *Bemerkungen, als Antwort auf den (vorstehenden) Aufsaz des Hrn. Berthier, Ingénieur en chef des Mines, über die Theorie des Mörtels*, in, "Polytechnisches Journal," vol. XI, 1823, pp. 363–373); Louis-Joseph Vicat, *Nouveaux faits pour éclairer la théorie des cimens calcaires*, in, "Annales de Chimie et Physique," vol. XXVIII, 1825, (pp. 142–147), p. 144 (Id., *Thatsachen zur Aufhellung der Theorie über Kalkmörtel*, in, "Polytechnisches Journal," vol. XVII, 1825, pp. 481–84).

[31] See Anon., *Hamburgisches Address–Buch für das Jahr 1817*, Hamburg, Hermann am Fischmarkt, 1817, p. 70; Id., *Hamburgisches Address–Buch für das Jahr 1818*, Hamburg, Hermann am Fischmarkt, 1818, p. 74; Id., *Hamburgisches Address–Buch für das Jahr 1819*, Hamburg, Hermann am Fischmarkt, 1819, p. 76. The port of Hamburg can be considered the German gate for British wares. The city trades with England since the Middle Age and a British community lives there permanently at least since 1611, as the Company of Merchant Adventures obtained

Likely, one of these travellers provides Prince Leopold of Bavaria with a report about Roman cement taken from the London Building Department. The prince hands the report over to Christoph Arzberger (1772–1822), a mathematician and high-ranking civil servant in Coburg. Arzberger analyses the report and writes a short article about it, which is then published in the "Kunst- und Gewerbe- Blatt des polytechnischen Vereins für das Königreich Bayern" in 1819.[33] Arzberger describes Roman cement as a fine milled powder that is produced by firing and grinding clayish limestone, and is appropriate for composing mortar suitable for hydraulic works, plaster and mouldings. Arzberger also presumes that Roman cement could be reproducible in Germany by burning local marls. This idea will be further developed in the following years and several efforts will be undertaken to reproduce Roman cement in Germany.

Another significant investigation of Roman cement is the one by [Johann Heinrich] Krahmer, a master builder working for the Prussian *Ober-Bau-Deputation*, who embarks on a research journey to England in 1821.[34] During a stay in London, Krahmer visits the house of Charles Francis, owner of the greatest Roman cement factorie at the time, together with his associate John Bazley White (Francis & White). Upon his return to Berlin, Krahmer reports his observations to the *Ober-Bau-Deputation*. He systematically describes the properties of Roman cement and lists all the applications he observed in Francis's garden: several plaster works on walls, staircases, pavements, basins of fountains and many mouldings. Krahmer also reports having seen some mouldings in Gothic style, in the newly restored New College Chapel in Oxford. While, on one hand, Krahmer is apparently impressed with the quality of the Roman cement-based mortar and plaster he has observed, on the other hand, he complains about the difficulty in obtaining detailed information about the raw materials and the manufacturing process of cement. As with Arzberger before, he also believes in the feasibility of a German production of Roman cement, once the composition of the English product has been discovered through chemical analysis. For this reason he brings two casks of Roman cement to Berlin and hands them over to Eytelwein, head of the *Ober-Bau-Deputation*, and to Peter Christian Wilhelm Beuth (1781–1853), head of the *Technische Deputation* (Technical Committee) and of the *Verein zur Beförderung des Gewerbfleißes in Preußen* (Society for Supporting the Development of Manufacturing and Trade in Prussia, or briefly *Gewerbverein*).

the permission to settle; see Ortwin Pelc, *Hamburg und die Engländer im 19. Jahrhundert*, in, Ortwin Pelc, Susanne Grötz, ed., *Konstrukteur der modernen Stadt. William Lindley in Hamburg und Europa 1808 – 1900*, München, Hamburg, Dölling und Galitz, 2008, (pp. 48–57), p. 49. The importation of Roman cement probably begins soon after the removal of the Continental Blockade – an embargo in force from 1806 to 1814 in order to prevent trade and trips between the United Kingdom and European countries during the Napoleonic wars.

[32] Regarding the invention and the early exploitation of Roman cement in England, see Arthur James Francis, *The Cement Industry 1796-1914: A History*, London, Newton Abbot, North Pomfret, Vancouver, David & Charles, 1977, pp. 26–42.

[33] See Christoph Arzberger, *Römisches Cement*, in, "Kunst- und Gewerbe- Blatt des polytechnischen Vereins für das Königreich Bayern," V, 1819, (coll. 598–600), col. 600.

[34] See [Johann Heinrich] Krahmer, letter from London to the Ministerium für den Handel das Gewerbe und das Bau–Wesen, August 28, 1821 (GStA PK I.HA Rep.120 D XII 1, n° 3, vol. 1, n. fol.); Id., "Roman cement" (GStA PK I.HA Rep.120 D XII 1 n° 3 vol. 1, 1r–9v, partly transcribed in, Thomas Brunsch, *Die historische Verwendung zementgebundener Kunststeine im Außenraum im 19. und frühen 20. Jahrhundert unter besonderer Berücksichtigung Berlins und Brandenburgs*, doctoral dissertation, Technische Universität, Berlin, 2006, pp. 56–58). See also [Johann Heinrich] Krahmer, *Bemerkungen über die Dauer und Festigkeit des in England erfundenen Roman-Cement. Nebst Nachrichten von einigen andern Cementen*, in, "Journal für die Baukunst," III, 1830, pp. 476–483. About the full name of Krahmer, see Reinhart Strecke, ed., *Inventar zur Geschichte der preußischen Bauverwaltung 1723-1848*, Berlin, Geheimes Staatsarchivs Preußischer Kulturbesitz, 2005, p. 549.

20. Krahmer's receipt for the importation of two casks of Roman cement and one cask of Hamelin mastic from the United Kingdom, Hamburg, 1821 (GStA PK I.HA Rep.120 D XII 1 n° 3 vol. 1, n. fol.).

The two institutions led by Beuth are committed to supporting Prussian manufacturing and trade, this task is known as *Gewerbeförderung*. The *Technische Deputation* is a public institution while the *Gewerbverein* is a society of private entrepreneurs, which, inter alias, encourages the art of building.[35] Roman cement soon becomes the object of their interest. In 1822, the *Gewerbverein* launches a competition divided into several sections, including one pertaining to the development of a compound to be produced with local raw materials and suitable for moulding solid and weather-proof architectural decorations and objects.[36] The idea of such a competition probably originates from Krahmer's experience in London and from the observation of a similar competition that the *Société d'Encouragement pour l'Industrie nationale* has promoted in France one year earlier, with the aim of encouraging the search for a new material that should be mouldable as gyps and solid like limestone.[37] The Prussian competition, moreover, is an attempt to resume the search for material suitable to produce mortar-based artificial stone after the failure of Simon's trials. This is evidenced by the invitation for applicants to take into account both Simon's experiments and a short list of books about hydraulic lime and pozzolana, including the one by John, the one by Vicat, and the *Recueil des diverses Mémoires sur le Pouzzolanes naturelles et artificielles*, from 1805, written by Gratien Le Père.[38]

The competition does not yield the expected results – at least in the short term.[39] Nevertheless, together with the text by Arzbeger, it probably contributes to raising interest in Roman cement outside Prussia. In 1823, the king of Württemberg, Wilhelm I, asks Friedrich Schmidt, a sort of industrial spy from Wurttemberg based in England, to search for information about Roman cement, focusing on its quality, its price and its suitability to plaster buildings in Stuttgart.[40] In June 1823, Schmidt sends the king two positive reports. As with Arzberger and Krahmer, Schmidt believes that Roman cement could be reproduced in Germany, providing an occasion for the development of new manufacturing activities in Wurttemberg. He therefore suggests searching for

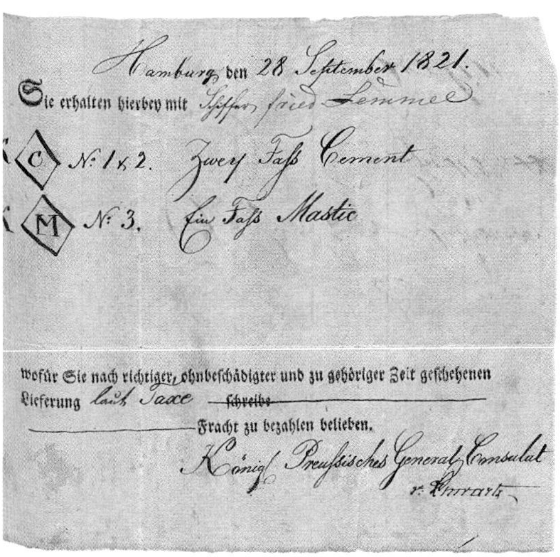
Hamburg den 28 September 1821.
Sie erhalten hierbey mit Schiffer
C N: 1 & 2. Zwey Faß Cement
M N: 3. Ein Faß Mastic
wofür Sie nach richtiger, ohnbeschädigter und zu gehöriger Zeit geschehenen Lieferung laut Taxe ~~schreibe~~
Fracht zu bezahlen belieben.
Königl. Preußisches General Consulat

[35] See "Statut für den Verein zur Beförderung des Gewerbfleißes in Preußen" II Abtheilung, Bildung des Vereins, Art. 7, in, Anon., *Ueber die Stiftung eines Vereines zur Beförderung des Gewerbfleißes in Preußen, nebst dessen Statut*, in, "Polytechnisches Journal," vol. IV, 1821, (pp. 486–494), p. 490.
[36] See *Preisaufgaben des Vereins. Vierte Preisaufgabe*, in, "Verhandlungen des Vereins zur Beförderung des Gewerbfleißes in Preußen," I, 1822, p. 22.
[37] The competition promoted by the *Société d'Encouragement pour l'Industrie nationale* suggests trying either modifications of common gyps mortar composition or inventing a new material similar to Roman cement or to the cement from Boulogne-sur-Mer. The activities undertaken by the *Société d'Encouragement pour l'Industrie nationale* must be known in the circle of the *Gewerbverein*, since the French society is taken as a model for the foundation of the *Gewerbverein* itself (see Anon., *Ueber die Stiftung eines Vereines*...cit., p. 486). Moreover, the announcements of the competitions promoted by the *Société d'Encouragement pour l'Industrie nationale* are regularly translated and published in the "Polytechnisches Journal," as is also the case for the competition in question; see *X. Preisaufgaben der Société d'Encouragement pour l'Industrie nationale aus der General-Sitzung vom 3. Oktober 1821*, in, "Polytechnisches Journal," vol. VII, 1822, (pp. 89–127), pp. 125–126.
[38] See *Preisaufgaben des Vereins. Vierte Preisaufgabe*, cit. The *Recueil des diverses Mémoires sur le Pouzzolanes*

clayish limestone along rivers; however, the search, which takes place around 1823, does not give the expected results. The available amount of clayish limestone is judged too small, and so no cement manufacturing is undertaken then in Wurttemberg.[41]

While early awareness about Roman cement is spreading in Germany, English cement makers, such as James Frost, Edgar Dobbs, William Lockwood, James Pulham and Joseph Aspdin develop new kinds of cement by burning artificial mixtures of chalk and clay, similar to the method of John and Vicat for producing artificial, hydraulic lime. Aspdin, in particular, patents a new kind of cement in 1824 and names it Portland cement, since it is thought to endow mortar with the characteristics of Portland stone.[42] An early short report about Aspdin's patent is taken from the "London Journal of Arts and Sciences," translated into German and published on the "Polytechnisches Journal" in 1825. It is followed, in 1826, by a second short report giving an account of the controversy in Britain surrounding the originality of Aspdin's patent, because of the closeness to Vicat's method for producing artificial hydraulic lime.[43] The announcement of this new finding, however, does not seem to bring any real consequence until the early 1840s, when a Hamburg dealer becomes licenced to sell Portland cement on the continent.

Artificial stones by Sachs and moulded statues by De la Rivallière-Preignac von Frauendorf

Two different kinds of mortar-based mouldable materials are developed in the first half of the 1820s in Prussia. Moving from the construction technique based on the use of rammed earth and with the aim of producing artificial stones, the Berlin master builder Salomo Sachs (1772–1855) conceives a mixture made of three measures of clay and one measure of mortar, this one being made of one measure of fat lime and two measures of sand.[44] He names the artificial stones he produces with this mixture *Mörtel-Steine* (mortar stones), and is able to test their solidity when, around 1821–1823, he replaces part of a 120-foot long enclosure wall made of rammed earth that a hurricane has

naturelles et artificielles is erroneously attributed also to Guyton de Morveau in the competition announcement; this is probably due to the fact that, just in 1805, Le Père et De Morveau publish together the *Recueil des rapport et observations sur les expériences faites a Cherbourg pour remplacer la pouzzolane dans les constructions hydrauliques.*

[39] However, the *Gewerbverein* receives proposals of methods for producing artificial mouldable masses (see *Auszug aus den Protokollen der monatlichen Versammlungen in den Jahren 1823*, in, "Verhandlungen des Vereins zur Beförderung des Gewerbfleißes in Preußen," III, 1824, (pp. 13–19), p. 16.

[40] See Helmut Albrecht, *Vom Caementum zu Zement*, in, Id., ed., *Kalk und Zement in Württemberg: Industriegeschichte am Südrand der Schwabischen Alb*, Ubstadt-Weiher, Landesmuseum für Technik und Arbeit in Mann, Verlag Regionalkultur, 1991, (pp. 117–130), pp. 125–129.

[41] Just as it happened at the beginning of the 19th century in Boulogne-sur-Mer in France; see Charles-François Mallet, *Rapport fait par M. Mallet, au nom d'une Commission spéciale, sur le ciment découvert par M. Lacordaire, ingénieur des ponts et chaussées, à Pouilly, en Auxois, département de Saône-et-Loire*, in, "Bulletin de la Société d'Encouragement pour l'Industrie Nationale," XXVIII, 1829, (pp. 317–337), pp. 323–326. See also Edmé Liégeard, Henry, Nicolas Jean-Baptiste Anselin, Blanchart, Pierre-Charles Lesage, *Rapport fait à la société d'Agriculture, de Commerce et des Arts, de Boulogne-sur-Mer, le premier floréal an 10; au nom d'une Commission chargée d'examiner les propriétés d'un plâtre-ciment*, in, "Journal des Mines, ou Recueil de Mémoires sur l'exploitation des Mines, et sur les Sciences et les Arts qui s'y rapportent," vol. XII, année X, [1802], pp. 459–489; Anon., *Note sur le plâtre-Ciment, découvert à Boulogne-sur-Mer*, in, "Bulletin de la Société d'Encouragement pout l'Industrie Nationale," année X, [1802], pp. 30–32.

[42] See Francis, *The Cement Industry*...cit., pp. 73–80.

[43] See Anon., *Neue Verbesserung in Verfertigung künstlicher Steine, worauf Joh. Aspdin, Maurermeister zu Leeds in Yorkshire, am 21. Oktober 1824. sich ein Patent geben ließ*, in, "Polytechnisches Journal," 1825, vol. XVI, pp. 304–305; Anon., *Hrn. Aspdin's künstliche Steine (Portland-Cement)*, in, "Polytechnisches Journal," 1826, vol. XIX, pp. 588–589.

[44] Rammed earth-based, artificial stones are developed in the context of the *pisé* building technique; on this, see François Cointeraux, *Ecole d'Architecture rurale*, Paris, Chez l'Auteur, 1791; Id., *Schule der Landbaukunst*, Hildburghausen, Johann Gottfried Hanisch, 1793. About Sachs' artificial stones, see Salomo Sachs, *Anleitung zur Erd-Bau-Kunst, (Pisé-Bau); mit Anwendung auf alle Arten von Stadt- und Land-Bauten, nebst einer vollständigen Lehre von der Konstruktion der Tonnen- Kappen- und Kreuzgewölbe in reinem Lehm und von der Anfertigung feuersicherer Dächer ohne alles Holzwerk, auch einer Anweisung, die Fundamente bis auf den Baugrund in bloßen Lehm anzufertigen*, Berlin, Carl Friedrich Amelang, 1825, pp. 39–51.

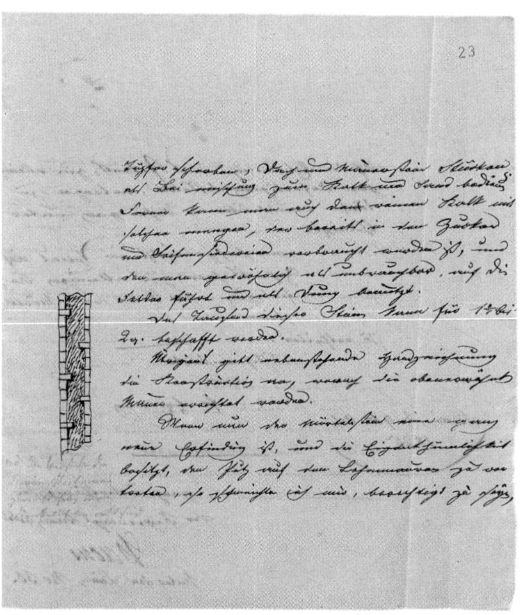

21. S. Sachs, wall made of artificial stone faces and rammed earth filling. Sketch of the horizontal section (letter to the Ober-Bau-Deputation, October 17, 1825, GstA PK I.HA Rep.120 D XII 1 n° 12, vol. 1, n. fol.).

damaged.[45] Sachs also envisages the construction of walls consisting of two faces made of mortar stones and rammed earth fill. He describes this kind of wall in a book dealing with building techniques using rammed earth, which is published in 1825.[46] A sketch of the same kind of wall is in the letter that Sachs addresses to the *Ober-Bau-Deputation* to present a copy of his book.[47]

The second mortar-based mouldable material is developed by Charles de la Rivallière-Preignac, an ex-member of the Napoleonic army, in the wake of the French tradition of the *pierre factice*. De la Rivallière-Preignac arrives in Prussia at the time of the French occupation and buys the estate of Frauendorf situated between Berlin and Dresden, where he settles after the war.[48] He has artistic ambitions and wants to open an art gallery with paintings of battles and busts of generals. It is likely for the purpose of producing moulded statues that he develops a method "to produce hydraulic lime and artificial pozzolana, along with new hinges, a rolling mill, and other machines and tools to mould the soft mass."[49] He then applies for a patent, which is granted in 1825.

De La Rivallière-Preignac mentions among his inventions artificial pozzolana, which deserves special attention. It is actually burnt clay and is known in Germany through Dutch and French technical literature, which is probably De La Rivallière-Preignac's source of knowledge about the topic. An article dealing with an artificial pozzolana produced by firing clayish and rotten soil from the Amsterdam canals, extracted from a Dutch report of 1809, is published in 1812 in the "Allgemeine Literatur-Zeitung."[50] Further explanations about artificial pozzolana are also in a number of articles by Vicat, Berthier and Treussart. First appearing in France in the early 1820s, they are translated soon thereafter into German and published in the "Polytechnisches Journal." "Since chemical analyses show that pozzolana is composed of silica, clay, iron oxide and a little lime," Vicat writes in 1821, "it is easy to think that our clay, which consists of a similar composition, could be fired into artificial pozzolana."[51] Raucourt de Charleville also addresses artificial pozzolana in his 1822 treatise about mortar, and Von Prittwitz carefully discusses it in his abstract published

[45] See Johann Daniel Friedrich Engel, *Der Kalk-Sand-Pisébau und die Kalksand-Ziegelfabrikation*, Leipzig, C. H. Seemann, Leipzig, 1865, pp. 4–5.
[46] Sachs, *Anleitung zur Erd-Bau-Kunst*...cit., pp. 29–33.
[47] See Salomo Sachs, letter to the Ober-Bau-Deputation, October 17, 1825 (GstA PK I.HA Rep.120 D XII 1, n° 12, "Verfertigung von künstlichen Bausteinen Patente," vol. I, n. fol.).
[48] For a biographical outline of Charles de la Rivallière-Preignac, see the autobiography of "Karl Franz Jakob Heinrich Schumann (1767–1827)," in, Reimar F. Lacher ed., *Kunstler(auto)biographien*, Berlin, Berlin-Brandenburgische Akademie der Wissenschaften, 2005, (pp. 65–72), pp. 71–72, n. 154.
[49] "Auf eine neue Methode, hydraulischen Kalk und künstlichen Puzzolane zu verfertigen, ingleichen auf die für neu erkannten Charniere, an den zu dieser Fabrikation, außer andern Maschinen, bestimmten Walzwerk und auf das Instrument zum Formen der weichen Masse" (*Verzeichniß der im Königreiche Preußen im Jahre 1825 ertheilten Patente*, in, "Verhandlungen des Vereins zur Beförderung des Gewerbfleißes in Preußen," V, 1826, p. 51; and in, "Polytechnisches Journal," 1826, vol. XX, p. 193).
[50] It is the so-called Amsterdam cement, patented by the cement handler Adrian de Booys in the 1780s, see Anon., *Veflag aan Zijne Excellentie den Minister van Eeredienst an Binnenlandsche Zaken, door de Eerste Klasse van het Koninklijk Instituut van Wetenschappen, Letterkunde en Schoone Kunsten, betrekkelijk den haar opgedragen last, nopens het fluk der buitenlandse, en de sedert 1789 uitgevondene*

in 1825.[52] In 1821 and 1823 Vicat also explores the question of how to mould decorations using hydraulic mortar instead of plaster in two reports on the matter sent to the *Société d'Encouragement pour l'Industrie nationale*. For the latter he even receives a gold medal on the proposal of Léonor Mérimée.[53]

Towards the mid-1820s, De La Rivallière-Preignac runs into financial trouble. He tries in vain to accomodate the works already completed for the art gallery in the building of the University of Frankfurt (Oder). In 1825, he declares bankruptcy and the collection is auctioned off to pay creditors. Apart from a mention in the "Verhandlungen des Vereins zur Beförderung des Gewerbfleißes in Preußen," his patent falls into oblivion. It is once again mentioned in 1828 in an epistolary exchange between the city of Frankfurt (Oder) and the Home Office in Berlin, but without any practical implication.[54]

Roman cement applications and hydraulic lime manufacturing

Since at least the mid 1820s, some German master builders begin to use Roman cement imported from England to plaster walls and floors and to seal masonry joints exposed to water or humidity. Around 1826, Krahmer experiments with the use of Roman cement in a suburban villa nearby Berlin.[55] He plasters some exterior walls, staircases and pavement, and he seals the joints of a dock along the River Spree. In the same years, Roman cement is also used to seal the brick masonry joints of the locks along the Friedrich-Wilhelm Canal and the joints of the new lighthouse in Kap Arkona on the island of Rügen. The definitive plans of the lighthouse are elaborated by August Adolph Günther (1779–1842), probably with the contribution of Schinkel, around 1825.[56] The building is carried out between 1826 and 1827 under the supervision of the master builder Michaelis, who, in a later account, reports that the "3 floors of the lighthouse are built with fired bricks arranged in running bonds, the surface and the embossed joints being sealed with the so-called Roman cement, mixed with one part of sand and lime, as ordered from Newcastle."[57] The Roman cement for the construction of the lighthouse

binnenlandse Tras of Cement (Bericht an Se. Excell. Den Minister der Gottesverehrung und der innern Angelegenheit, von der ersten Klasse des Königl. Instituts der Wissenschaften, Literatur und schönen Künste, betreffend den an sie gelangten Auftrag in Ansehung des ausländischen, und des seit 1789 erfundenen inländischen Trasses oder Cement), in, "Allgemeine Literatur-Zeitung," vol. III, n° 212, 1812, coll. 1–8.

[51] "Da ferner die chemische Analyse uns Kieselerde, Thonerde, Eisenoxyd und ein wenig Kalk als Bestandtheile der Puzzolana kennen lehrte, so war es sehr natürlich zu glauben, daß unsere Thonarten, die auf ähnliche Weise zusammengesezt sind, durch Brennen sich in künstliche Puzzolanen verwandeln lassen könnten" (Vicat, *Ueber Kalk und Mörtel. Ein Auszug aus Hrn. Vicat's*...cit., p. 288) ("De même, puisque l'analyse chimique donne pour principes des pouzzolanes naturelles, la silice, l'alumine, le fer oxidé et un peu de chaux, il était tout simple de penser que nos argile, dont la composition est toute semblable, pourraient se transformer, par la cuisson, en pouzzolanes artificielles") (Vicat, *Extrait de l'ouvrage intitulé*...cit., p. 374).

[52] Raucourt de Charleville believes that air plays a crucial role in the production of pozzolana during clay burning; see Antoine Raucourt de Charleville, *Versuche mit verschiedenen Kalkarten in Rußland*, in, von Prittwitz, ed., *Neue Versuche über den Kalk und Mörtel*...cit., p. 83.

[53] See Léonor Mérimée, *Rapport sur le prix pour la découvert d'une matière plastique se moulant comme le plâtre et pouvant résister à l'air autant que la pierre*, in, "Bulletin de la Société d'Encouragement pour l'Industrie Nationale," XXII, 1823, pp. 281–284. The "Polytechnische Journal" reports on that in 1824; see Vicat, *Untersuchungen über Anwendung des hydraulischen Mörtels statt des Gypses*...cit.

[54] See Brunsch, *Die historische Verwendung*...cit., p. 61.

[55] See Krahmer, *Bemerkungen*... cit.

[56] See Reinhart Strecke, *Schinkel und der Leuchtturm auf Kap Arkona*, in, *Jahrbuch Preußischer Kulturbesitz*, Berlin, Gebr. Mann Verlag, 1996, vol. XXXII, pp. 297–319.

[57] "Diese vorgenannten 3 Etagen des Turms sind von guten gebrannten Ziegelsteinen im Kreuzverbande ausgemauert und auf die Außenfläche mit erhabenen Fugen mit sogenanntem römischen Zement, welcher aus Newcastle verschrieben wurde, unter Zusatz voneinem Teil Sand und Kalk abgefugt" (Landbaumeister Michaelis, "Erläuterung zu den Zeichnungen über den Bau des Leuchtturms zu Arkona, Bau- und Erläuterungsbericht des Landbaumeisters Michaelis vom 8. Januar 1830," GSTA I.HA Rep. 93D Lit Ok Tit. VIII n° 4, vol. 1; trans. in, Strecke, *Schinkel und der Leuchtturm auf Kap Arkona*...cit., p. 318). It is quite usual to seal external masonry joints with hydraulic mortar in the first half of the 19th century. For major buildings, entire external façades are sometime sealed by daubing a thin layer of solid mortar and marking embossed joints. This is due to the humble quality of bricks. Eminent examples of this façade treatment are the castle of Babelsberg in Potsdam by Schinkel, Ludwig Persius and Johann Heinrich Strack, and the Heilandskirche in Sacrow by Persius.

22. K. F. Schinkel, diary of the journey to France, England and Scotland, 1826. Sketches of the Thames Tunnel, sections (SMB-ZA, IV/NL Schinkel 94, fol. XXIX).

is, in fact, imported from Newcastle, although the soil of Kap Arkona essentially consists of marls, clay and chalk – appropriate raw materials to produce hydraulic binders. The importation of Roman cement from England to Kap Arkona demonstrates of the lack of knowledge and expertise in producing hydraulic binders at that time.

In 1826, Schinkel goes with his friend Beuth on a research journey to France, England and Scotland. In London, they both visit the building site of the Thames tunnel conceived by the engineer Marc Isambard Brunel, who personally leads them around. In his travel diary, Schinkel drafts two sections of the tunnel structure and notices that the main building materials are bricks and Roman cement-based mortar. One year later, Beuth publishes an article about the construction of the Thames tunnel in the "Verhandlungen des Vereins zur Beförderung des Gewerbfleißes in Preußen," highlighting the huge quantity of Roman cement that Brunel has been using since the beginning of the construction.[58] Cement-based mortar is also used to build the twenty-four turrets placed on the roof of Schinkel's Friedrichswerder Church in Berlin; they are completed in 1828, consisting of a brick masonry bond plastered with cement-based mortar.[59]

The spread of English Roman cement in the second half of the 1820s pushes the *Gewerbverein* to plan a competition focused on the reproduction of Roman cement. Aware of such intentions, the interior minister Friedrich von Schuckmann buys two casks of Roman cement from the factory of Francis and White in London and gives them to the *Gewerbverein* on April 1828, urging its technicians to investigate the properties of Roman cement.[60] Some tests transpire between 1828 and 1829. They are conceived to verify whether Roman cement-based mortar sets under water, binds bricks, is suitable to produce artificial stones and works efficiently as plaster under different conditions, notably in water containers.[61] Although most results are encouraging, in a report following the accomplishment of the tests, Roman cement is described as an only partially useful material. It is considered too expensive for hydraulic engineering and not needed for plastering ordinary building in Prussia, since the air is supposedly less humid than in England. Instead, it is considered valuable for sealing founding walls against rising damp, for binding the masonry of vaults that are exposed to rain, and for plastering wet walls.[62] The competition is launched in 1829. It focuses on the search for a kind of hydraulic mortar suitable for building both underwater and aboveground, but that should cost no more than six times the price of ordinary lime mortar.[63]

The competition declares no winner. Nevertheless, starting in the late 1820s, hydraulic lime and artificial pozzolana begin to be produced in the regions of the Joachimsthal and the Werbellinsee, northeast of Berlin, where considerable deposits of clay and lime are already being extracted to produce lime and bricks.[64] Gottfried Menzel (1792–1870), an inspector of the

[58] Peter Christian Wilhelm Beuth, *Der Tunnel unter der Themse*, in, "Verhandlungen des Vereins zur Beförderung des Gewerbfleißes in Preußen," VI, 1827, pp. 286–294.
[59] Only in 1843 are the turrets covered with zinc plates; see Stein, *Ueber den Bau eiserner Thurmspitzen*, in, "Zeitschrift für Bauwesen," X, 1860, (coll. 481–490), col. 484.
[60] Friedrich von Schuckmann, letter to the *Gewerbverein*, April 6, 1828, published in, Anon., *Ueber die mit Roman Cement, aus der Fabrik von Francis, White und Francis, in London, angestellten Versuche*, in, "Verhandlungen des Vereins zur Beförderung des Gewerbfleißes in Preußen," VIII, 1829, (pp. 123–126), p. 123.
[61] *Ibid.*
[62] *Ibid.*, p. 126.
[63] See *Preisaufgaben für die Jahre 1829 und 1830. Dritte Preisaufgabe*, in, "Verhandlungen des Vereins zur Beförderung des Gewerbfleißes

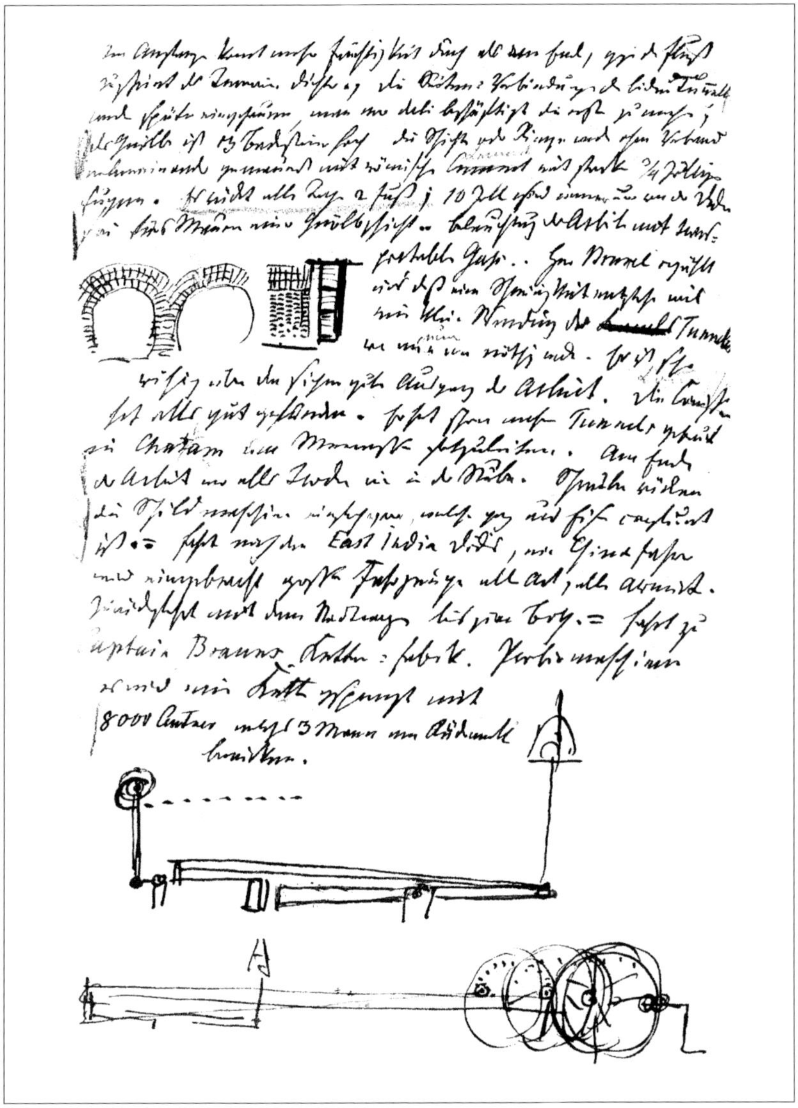

Royal brickyard situated in the Joachimsthal, Johann Bernoulli and Theodor Buschius, two lime producers from the southwest area of the Werbellinsee (called Wildau), undertake this initiative around 1827. Buschius produces artificial hydraulic lime and a kind of artificial pozzolana made of clay, silica and iron oxide, as described by Vicat.[65] The Royal brickyard produces ar-

in Preußen," VIII, 1829, p. 31.
[64] Gottfried Menzel, *Verfahren bei der Fabrication der Ziegel und des Mörtels auf der königl. preuß. Ziegelei bei Joachimsthal*, in, "Verhandlungen des Vereins zur Beförderung des Gewerbfleißes in Preußen," XXV, 1846, (pp. 52–73), pp. 72–73; and in, "Polytechnisches Journal," vol. CII, 1846, pp. 194–219.
[65] See Zimmermann, *Einige Bemerkungen über die Festigkeit, Mischungsverhältnisse und Zubereitung des Bétons, oder des Mauerwerkes aus klein geschlagenen, mit Mörtel untermengten Steinen, dessen man sich zuweilen, um Fangdämme und Wasserschöpfen zu sparen, zur Fundamentierung von Bauwerken unter Wasser bedient*, in, "Journal für die Baukunst," vol. III, 1830, (pp. 1–31), pp. 13, 15. The Buschius factory is also known, since 1846, as the Wildau Brick and Lime Kiln, from the name of the small local area of Wildau; see Königl. Regierung. Abtheilung des Innern, *Vermischet Nachrichten*, in, "Amtsblatt der Königlichen Regierung zu Potsdam und der Stadt Berlin," 1846, p. 166.

tificial hydraulic lime burning mixtures of 3 parts marl, 2 parts clay and a small amount of iron oxide. Marl is open-air quarried near the Werbellinsee during the fall, arranged in 2-foot high heaps and remains exposed to the elements during the winter. It is then dissolved in water, mixed with clay and iron oxide in a clay mixer, to be then moulded in the form of bricks, dried in huts and, finally, burnt in kilns. Like Roman cement, hydraulic lime from the Joachimsthal is not slaked. The fired blocks are crushed and ground into powder, first in a stamp mill, then in a grind mill – both powered by water. Lime powder is packaged in sacks and casks, and it is sold in the region of Berlin. Several masonry works along the Finow canal are built using a kind of mortar made of one part Joachimsthal lime and three parts brick powder, which is also produced at the royal brickyard of the Joachimsthal by grinding discarded bricks, while the joints are sealed with pure hydraulic lime mortar.[66]

Concrete hydraulic foundations: trass versus hydraulic lime

Since the mid 1820s, the effectiveness of hydraulic binders is more or less acknowledged and this gradually has favourable consequences for the use of concrete. Basically, two kinds of concrete mixtures are known at that time: the traditional one made of ordinary lime and trass, and the more recent one based on the use of hydraulic lime. At that time, British cement is still too expensive and is only used for fine works. Two different and reliable sources indicate that, in 1825, trass-based concrete is used to build the fountain on the Pfaueninsel in Potsdam, designed by Friedrich Martin Rabe (1765–1856). The Berlin master builder Braun insists that a 2-foot high bed of concrete supports the fountain basin made of fired bricks and trass-based mortar.[67] However, an 1826 drawing, notably a cross section over the fountain, shows a different – and far more complex – design. The fountain appears to be composed of two different and concentric basins, each one made of a horizontal layer of bricks laid on edge, with two circular walls also made of bricks. The two horizontal layers of bricks are spaced approximately 2 feet apart, whereas the two circular walls stand 2/3 foot apart. In the cavity between the two concentric basins are two different kinds of filling material. A finer conglomerate forms a layer of about 2/3 foot on the bottom and fills the space between the two circular walls, whereas a coarse conglomerate fills the rest of the space between the two horizontal layers of bricks. The finer filling material is probably trass-based mortar, whereas the coarse one looks like a mix of rubbles and mortar. The central sculptural column stands on a layer of bricks laid on edge, over an entire block of rubble and mortar.

A further and even more important example of concrete use in the second half of the 1820s is in the town of Hamm in Rhineland, where, in 1828, the master builder Zimmermann builds a concrete foundation for a lock along the River Lippe, after the previous lock, standing on a timber foundation, was dragged away by violent stream. Zimmermann's construction is

[66] Menzel, *Verfahren bei der Fabrication*...cit., pp. 72–73.
[67] See Braun, *Über Anwendung des Trass-Bétons zur Fundamentierung der Gebäude*, in, "Journal für die Baukunst," vol. III, 1830, (pp. 112–117), p. 113.
[68] He mentions Bélidor, Jean-Rudolph Perronet, Émiland–Marie Gauthey and Reinhard Woltmann.
[69] Zimmermann, however, does not explain which kind of filling he considers appropriate for caissons; he very probably means the traditional ones filled with masonry.
[70] "Die Nützlichkeit und Zweckmässigkeit des Bétons bei schwierigen Fundamentirungen, welche ohne Fangdämme und Ausschöpfung, oder nicht in Senkkasten ausgeführt werden

23. F. M. Rabe, Fountain on the Pfaueninsel, Potsdam, 1825. Sketch of the section, 1826 (copy from the *Stiftung Preußische Schlösser und Gärten*).

based on his knowledge of, and expertise with, concrete and hydraulic binders that have been developed in the previous years, as well as on the French examples of concrete foundations in Rhineland dating back to the turn of the century. Zimmermann is familiar with the writings of John, Vicat, Raucourt de Charleville, Treussart, Von Prittwitz, together with some essential eighteenth-century technical literature.[68] Moreover, he is aware of the hydraulic mortar that the Swedish chemist Gustaf Erik Pasch prepares using burnt and milled slate instead of pozzolana for the construction of the Göta Canal, and, above all, he knows Hageau's book on the construction of the *Grand Canal du Nord*, and likely uses Hageau's description of concrete foundations of locks as a main reference.

In the wake of eighteenth-century building culture, Zimmermann considers foundations made of concrete poured inside enclosures as an advantageous building technique when neither caissons, cofferdams nor drainage could be applied.[69] "In case of difficult foundations," he writes in 1829, "to build without resorting to caissons, cofferdams and drainage, concrete is appropriate and useful, as long as the foundation site is sheltered from stream."[70] Only in the case of marshy soil does Zimmermann advise against the use of concrete poured straight onto the soil, and recommends instead using timber piling or caissons. "I would not dare," writes Zimmerman on this topic, "build a concrete foundation on muddy and marshy soil [...] I would choose piles or caissons."[71]

Zimmermann attaches great importance to the use of egg-sized crushed bricks in concrete mixtures. Only in the absence of bricks does he suggest

sollen, ist mithin in die Augen fallend, unter der Voraussetzung, dass der Baugrund sicher und gegen Unterspühlung geschützt sei" (Zimmermann, *Einige Bemerkungen über die Festigkeit*...cit., p. 6).

[71] "Auf schlammigem, morastigen Grunde würde ich doch nicht wagen [...] einen Bau auf Béton zu gründe, sondern einen Pfahlrost oder senkkästen vorzuziehen" (*ibid.*, p. 8).

using crushed stones, and he recommends stones whose components have a chemical affinity to cement. He lists tuff, claystone and porphyry as pertinent examples, whereas he advises against the use of stones containing crystals, on the basis of John's failed attempt at dissolving a crystal in limewater. The surface roughness of crushed stones is also thought to play a crucial role in the quality of concrete, since it is supposed to improve the adherence to mortar. Zimmermann gives the adherence between mortar and crushed stones the attention typically given to the adherence between mortar and stones in case of traditional masonry, since he considers concrete blocks as a kind of masonry.[72]

Cement is the additional essential component of concrete. Zimmermann considers trass, as opposed to artificial hydraulic lime, to be the optimal material for producing high quality hydraulic mortar. As a matter of fact, a sort of controversy about the minor or major suitability of trass or hydraulic lime in concrete already arose in 1825; the eminent French engineer Clément-Louis Treussart asserted that traditional hydraulic mortar composed of lime and pozzolanic materials is more effective than mortar produced with Vicat's artificial hydraulic lime.[73] The long tradition in trass exploitation and the easy availability of this material in Rhineland very probably play a crucial role in determining Zimmermann's assessment in favour of it. The Rhine master builder, however, also bases his preference on the outcomes of strength tests that he himself executes. He considers the trass milled near Koblenz and Andernach, which is easily transportable on boats along the Rivers Rhine and Lippe, as the best, most affordable one available in his district.[74] Among materials that could replace it, he mentions burnt and milled slate, brick powder and the artificial pozzolana produced by Buschius in Berlin, besides *wilder Trass*, which is cheaper but "half so strong as ordinary trass."

Before he begins to build the lock in Hamm, Zimmermann carries out a series of strength tests on mortar specimens made of different mixtures of aerial and hydraulic lime from limestones quarried along the River Lippe, artificial hydraulic lime and artificial pozzolana produced by Buschius in Berlin, trass from Brohl, bricks powder, slate and sand. The different mortar mixtures are poured into parallelepiped-shaped moulds made of longitudinal wooden bars and divided into compartments by means of transversal partitions. Once filled with mortar, the moulds remain underwater between seven and eleven weeks. Thereafter, Zimmermann recovers the specimens and tests their flexural and tensile strengths. Among the tested mortar mixtures, the one made of one part trass from Brohl and 3/5 part hydraulic lime from Wallstädde, a

[72] *Ibid.*, p. 3.

[73] Clement-Louis Treussart, *Extrait d'un Mémorial sur les Mortiers Hydrauliques*, in, "Annales de Chimie et Physique," vol. XXVI, 1824, pp. 324–329 (Id., *Ueber den hydraulischen Mörtel*, in, "Polytechnisches Journal," vol. XV, 1824, pp. 341–345).

[74] "According to my tests," Zimmermann asserts, "no other cement can exceed the force of this kind of trass" ("Nach meinen Versuchen wird seine Wirksamkeit von keinem anderen Cement übertroffen") (Zimmermann, *Einige Bemerkungen über die Festigkeit*...cit., p. 14).

[75] See Clement-Louis Treussart, *Notiz des General Treussart über die Bereitung der künstlichen Puzzolanen oder Trasse*, in, "Polytechnische Journal," 1826, vol. XXI, pp. 40–47 (Id., *Note du M. le général Treussart sur la Fabrication des Pouzzolanes ou trass factices*, in, "Annales de Chimie et Physique," 1826, vol. XXXI, pp. 243–253); Pierre-Simon Girard, *Ueber einige natürliche und künstliche Puzzolanen*, in, "Polytechnische Journal," 1827, vol. XXIII, pp. 60–65, 1827, vol. XXV, pp. 404–409 (Id., *Note sur les Pouzzolanes naturelles et artificielles*, in, "Annales de Chimie et Physique," 1826, vol. XXXIII, pp. 197–204); Louis-Joseph Vicat, *Bemerkung über künstliche Puzzolanen*, in, "Polytechnische Journal," 1827, vol. XXIV, pp. 175–177 (Id., *Note sur les Pouzzolanes artificielles*, in, "Annales de Chimie et Physique," 1827, vol. XXIV, pp. 102–105); Pierre-Simon Girard, *Notiz über die natürlichen und künstlichen Puzzolanen*, in, "Polytechnische Journal," 1827, vol. XXV, pp. 409–420 (Id., *Note sur les Pouzzolanes artificielles*, in, "Annales de Chimie et Physique," 1827, vol. XXXV, pp. 140–154).

[76] "Die chemischen Untersuchungen, welche viele

village near Hamm, gives the optimal outcomes, and Zimmermann therefore adopts it to build the lock of Hamm.

Construction begins in the summer of 1828. Mortar and concrete are prepared manually – the former on wooden platforms, the latter in wooden boxes. Concrete is poured into the foundation pits using a vertical wooden pipe steered by workers standing on provisional carpentry. Zimmermann asserts that using pipes to pour concrete underwater is preferable when the foundation pit is not deep; otherwise he recommends using chests hanging from a pulley, similar to the ones that Bélidor describes in his *Architecture hydraulique*.

The realization of John's theory in Fuch's studies on lime and mortar

The nature and properties of artificial pozzolana, whether or not pozzolana, trass and brick powder play a chemical role in mortar, and why clay acts as cement only after having been burnt are critical issues in French technical debate during the second half of the 1820s. Several articles about these questions are written by Vicat, Treussart and Pierre-Simon Girard, an engineer of the *Corps de Ponts*. These articles are first published in the "Annales de Chimie et Physique" and then, as usual, translated into German and published in the "Polytechnisches Journal."[75] "The several chemical research studies aiming at discovering the origin of the properties of volcanic and artificial pozzolana that eminent chemists have carried out," writes Girard, "have not yet produced any acceptable theory about a phenomenon that has considerable importance for the arts."[76]

The Bavarian chemist Johann Nepomuk von Fuchs (1774–1856) brings the matter to a turning point. In the first half of the 1820s, he discovers much about the chemical structure and behaviour of silica through studies about porcelain.[77] He understands that nature always combines silica with other substances to form silicates, among which clay is the most common. This knowledge is helpful when he later researches hydraulic lime and mortar, on behalf of the director of the Bavarian *Ministerial-Bau-Section* Von Purgel, developing a theory in 1829.[78] Fuchs argues that to produce hydraulic lime, it is necessary to release silica from the other substances forming silicates, and he maintains that this is only feasible through a chemical reaction triggered by acids or fire. This explains why clay needs to be burnt to act as cement or as artificial pozzolana, while pozzolana and trass are considered by Fuchs as a kind of clay that has been burnt by nature.[79]

This is both the core and the most innovative point of Fuch's theory about lime ad mortar, whose practical implications are important as well. Fuchs in fact promotes the use of Bavarian marl to produce local hydraulic

ausgezeichnete Gelehrte anstellten, um die Ursache der Eigenschaften der vulkanischen und künstlichen Puzzolanen kennen zu lernen, haben bis auf diesen Tag noch auf keine annehmbare Theorie einer in den Künsten so häufigen Erscheinung geführt" (Girard, *Notizen über die natürlichen und künstlichen Puzzolane*... cit., p. 409) ("Les recherches chimiques auxquelles beaucoup de savants distingués se sont livrés, pour connaître la cause des propriétés dont jouissent les pouzzolanes volcaniques et artificielles, n'ont pas abouti jusqu'aujourd'hui à donner une théorie passable d'un phénomène pourtant si commun dans les arts") (Girard, *Note sur les Pouzzolanes naturelle et artificielles*...cit., p. 140). See also Treussart, *Ueber die Bereitung der künstlichen Puzzolanen*...cit.

[77] Johann Nepomuk Fuchs, *Ueber ein neues nutzbares Produkt aus Kieselerde und Kali*, in, Karl Wilhelm Gottlob Kastner, ed., *Archiv für die gesammte Naturlehre*, vol. V, Nürnberg, 1825, pp. 385–412.

[78] See Johann Nepomuk Fuchs, *Ueber Kalk und Mörtel*, in, "Beilage zum Kunst- und Gewerbe-Blatt des polytechnischen Vereins für das Königreich Bayern," XV, 1829, (coll. 1–56), col. 4.

[79] "The volcanic products that are known under the names of trass and pozzolana, and are considered good cements," he states, "are nothing else than burnt clay." ("Als gute Cämente sind längst sehr berühmt die unter den Namen Trass und Puzzolan, bekannten vulkanische Produkte; unter welche man sich aber doch auch nichts anderes worstellen darf, als gebrannte Thonsorten") (Johann Nepomuk Fuchs, *Ueber Kalk und Mörtel*...cit., col. 40).

lime, and he even maps places around Munich where it would be possible to quarry appropriate kinds of marl. He then encourages Bavarian master builders to use hydraulic lime produced from local marl, both for hydraulic works and above-ground masonries.

In 1829, Fuchs presents the upshot of his research at a special sitting of the *Polytechnischer Verein für das Königreich Bayern* (Bavarian Polytechnic Society), which attracts an audience wishing to hear the results of his studies.[80] The presentation sparks interest and, contemporaneously, the opposition of those who believe that the best hydraulic lime has already been discovered in France and England, and that Bavarians should acquire the principles of hydraulic lime manufacturing in these countries rather that invest in research. At that time, most people consider Roman cement from England as the best hydraulic lime available, and some even consider it impossible to produce a similar material in Bavaria because of the lack of coke, which is commonly used to burn Roman cement in England.[81] It is probably on the basis of such conviction that August von Cetto (1794–1879), a Bavarian diplomat at the court of London, sends a request for information about Roman cement to the engineer Brunel shortly before the rebuilding of the fortress of Ingolstadt begins in 1828.[82]

Despite the criticism from initial detractors, interest in Fuch's theory increases, as evidenced by the publication of three different editions of his research paper already in 1829.[83] The following year, Fuchs participates in a competition promoted by the Royal Holland Society of Sciences and Humanities with the aim of discovering "the characteristics that distinguish cement that hardens in water" and "the constituting principles and chemical combinations that act during their solidification."[84] Fuchs rapidly writes a report and submits it to the Holland Society in December 1830, making the first step toward the international recognition of his theory.

Concrete foundations in technical literature

Starting in the late 1820s, several pivotal articles addressing the topic of concrete hydraulic foundations are published in the "Journal für die Baukunst" and remain important references for over ten years. An early significant one is written by hydraulic engineering inspector Elsner, who depicts different types of concrete foundations observed since the beginning of the century along the Rhine, in northwest Germany and in Holland.[85] Shortly thereafter, the master builder Zimmermann writes an article about the concrete foundation that he built for the lock of Hamm, stating, furthermore, his convictions about hydraulic lime, trass and concrete.[86] Another noteworthy article comes

[80] See Anon., *Auszug aus den Protokollen der Sitzungen in den Monaten April, Mai und Juni 1829*, in, "Kunst- und Gewerbe-Blatt des polytechnischen Vereins für das Königreich Bayern," XV, 1829, n. 29, (coll. 389–392), col. 392.
[81] Anon., *Würdigung der Arbeit des Professors und Akademikers Fuchs über Kalk und Mörtel*, in, "Kunst- und Gewerbe-Blatt des polytechnischen Vereins für das Königreich Bayern," XVIII, 1832, (coll. 756–761), coll. 758–759.
[82] See Anon., *Hilfsmittel bei Bauführungen, Ueber den römischen Zement*, in, "Allgemeine Bauzeitung," II, 1837, pp. 390–392; also mentioned in, Otto Christian Friedrich Reinhold, *Sammlung practischer Erfahrungen*...cit., pp. 154–155.
[83] Fuchs, *Ueber Kalk und Mörtel*, in, *Die neuesten Forschungen im gebiete der technischen und ökonomischen Chemie*, vol. III, 1829, pp. 1–26, 132–162; Id., *Ueber Kalk und Mörtel*, in, "Beilage zum Kunst- und Gewerbe-Blatt des polytechnischen Vereins für das Königreich Bayern," cit.; Id., *Ueber Kalk und Mörtel*, Leipzig, Barth, 1829.
[84] "Quels sont le charactere, auxquels on reconna'itra les Cements, qui s'endourcissent sous l'eau? Quels en sont les principes constituants et quelle est la combinaison chimique,

24. Elsner, concrete foundations. Box to pour concrete; enbankment along the Rhine; pillar foundation; enbankments at the entrance of the port of Cologne along the River Rhine; lock along the *Grand Canal du Nord* (F. W. Elsner, *Über die Anwendung des Béton-Mörtels zum Fundamentiren unter Wasser*, in, "Journal für die Baukunst," 1829, vol. I).

from master builder Braun from Berlin who, moving from the example of the concrete foundation of the fountain on the Pfaueninsel, suggests using concrete, not only in hydraulic engineering but also to build the foundations of ordinary buildings on wet and unsteady soil.[87] "It is known," Braun maintains, "that master builders of the past and nowadays have used concrete to build foundations underwater, but they have rarely used it to stabilize bad foundation soils for considerable buildings. And yet, under certain circumstances, poured concrete could be used with advantage to do without expensive timber pilings, especially if it is mixed with a cementitious material."[88] Braun continues, "Instead of using natural stones, with low costs and even less effort, it is possible to build an artificial stone-mass that rapidly gains solidity, hardness, and becomes like a rocky base. A mixture of trass from Brohl upon the Rhine, lime, sand and gravel, besides small pieces of granite, stones or bricks are appropriate for such a purpose."[89]

In 1830, an anonymous master builder from Colberg in Thüringen writes an article dealing with foundations made of rubble and mortar, poured and rammed in pits, layer upon layer, straight on the foundation soil or even

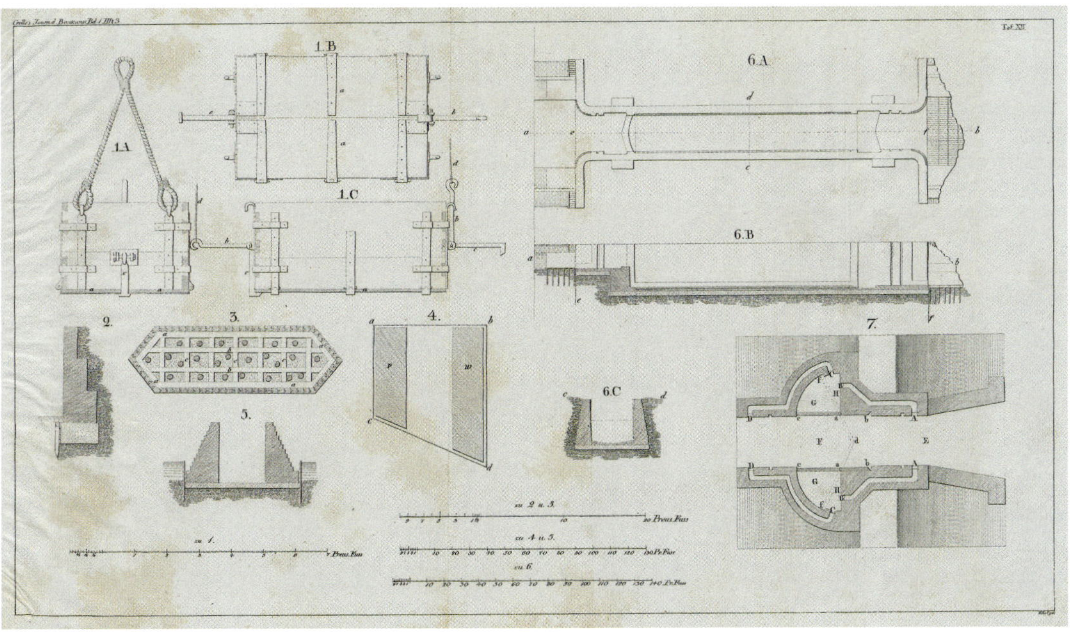

qui s'opere pendat leur solidification?" (Johann Nepomuck Fuchs, *Ueber die Eigenschaften, Bestandtheile und chemische Verbindung der hydraulischen Mörtel. Eine gekrönte Preisschrift vom Akademiker und Conservator Dr. J. N. Fuchs in München*, in, "Polytechnisches Journal," 1833, vol. XLIX, pp. 271–296).

[85] Elsner, *Über die Anwendung des Béton-Mörtels*...cit.

[86] Zimmermann, *Einige Bemerkungen über die Festigkeit*...cit.

[87] Braun, *Über Anwendung des Traß-Bétons*...cit., pp. 112–117.

[88] "Es ist bekannt, dass ältere und neuere Baumeister sich häufig des Bétons zu den Fundamenten der Wasserbauten bedienten; seltener scheint derselbe zur Befestigung eines schlechten Baugrundes, worauf Land-Gebäude von Bedeutung aufgeführt werden sollen, benutzt worden zu sein, und doch dürfte der, besonders mit einem cementartigen Materiale gemischte und zubereitete Béton, in Masse gelegt, in manchen von dem Baumeister nach den Umständen zu erachtenden Fällen nicht ohne bedeutenden Vortheil zu benutzen und besonders der kostspieligen Pfahlrosten vorzuziehen sein" (*ibid.*, p. 112).

[89] "Statt der Fels-Platte ist es nun möglich, durch Kunst, und sogar mit leichter Mühe, und nach Verhältniss mit geringen Kosten, an Ort und Stelle eine Steinmasse zu erzeugen, die sehr bald eine solche Härte und Dichtigkeit erhält, dass sich darauf fast wie auf einem Fels bauen lässt. Es eignet sich dazu der aus Trass von Brohl am Rhein, aus Kalk, Sand, Kies, kleinen Granit- oder Bruch und Ziegel- Stücken verfertigte Bèton ganz vorzüglich" (*ibid.*, pp. 112–113).

on timber piles, in case of extremely unsteady soil.[90] This kind of conglomerate, introduced as "gerammter Béton" (rammed concrete), is actually similar to Wiebeking's "pebbles and mortar," and may even derive from Bélidor's description of Milet de Monville's concrete. The use of the French term *demoiselle*, to better describe the pile driver used to ram rubble in mortar, and the French expression *en quinconce*, to explain the appropriate arrangement of timber piles in the soil, support the hypothesis of a French origin, even more so considering that Bélidor uses the term *demoiselle* and the expression *en quinconce* in the *Architecture hydraulique*. Unlike Milet de Monville, who uses pozzolana-based concrete, the German author gives up the use of trass, which is considered too expensive and not necessary for foundations out of water. Common slaked lime, or mixtures made of one part slaked lime and three parts sand are considered sufficient.[91] The unavailability of hydraulic binders in several German regions, along with the general adverse economic conditions, often urges master builders to do without cement, adapting foreign building techniques to local constraints.

Examples of French hydraulic works still play a role in German technical literature of the 1830s. The hydraulic engineer Otto Christian Friedrich Reinhold, for instance, mentions the 0.65-meter thick concrete bed foundation of the bridge over the Loire in Roanne in an article about the building of stone bridges, dating back to 1831.[92] And it is likely in reference to French examples that Friedrich Fyck, a building councillor from Hesse-Kassel, in his book about the construction of roads and bridges that is published in 1831, discusses the possibility of building a foundation by means of sunken caissons filled with concrete.[93] Far more critical, however, is Vicat's recollection of the concrete foundations of the bridge over the Dordogne that Johann Friedrich Wilhelm Dietlein (1787–1837), who teaches *Straßen-, Brücken-, Schleusen- und Kanalbau* at the Berlin *Bauakademie* since 1824, translates into German and publishes in "Journal für die Baukunst" in 1832.[94] The building of concrete foundations is also part of Dietlein's lectures at the *Bauakademie*, which are published between 1830 and 1832, in different issues of the "Journal für die Baukunst."[95] Dietlein describes concrete as "a mass made of stone-pieces and hydraulic mortar," and explains that hydraulic mortar "has been composed over centuries by mixing either lime and sand, or lime, sand and cement (pulverized volcanic products like pozzolana, brick powder, iron oxide, iron filings), or even just lime and cement in quite various ratios."[96] Dietlein knows John and Vicat's books, but he seems to be unaware of Fuchs' research about lime and mortar.

[90] See Anon., *Die Schwung-Ramme, ein Beitrag zum Grundbau*, in, "Journal für die Baukunst," vol. II, 1830, pp. 341–350.

[91] *Ibid.*, p. 344.

[92] See Otto Christian Friedrich Reinhold, *Beiträge zur Bestimmung der Höhe der Gewoelbsteine und Stärke der Widerlager und Mittelpfeiler grosser massiver Brücken; durch viele Beispiele erläutert und insbesondere für Practiker gesammelt und zusammengestellt*, in, "Journal für die Baukunst," 1831, (pp. 363–407), p. 399. The building of the bridge in Roanne dates back to 1789 (see *ibid.*).

[93] See Friedrich Fyck, *Die Verwaltung des Straße- und Brückenbaues mit Rücksicht auf möglichste Kosten-Ersparniß, Wohlthätigkeit für die ärmeren Volksklassen, uns Aufhebung der Frohdienste*, Cassel, Johann Christian Krieger, 1831, (pp. 57–67), p. 65.

[94] Louis-Joseph Vicat, *Bemerkungen beim Bau der Brücke über die Dordogne bei Souillac*, ed. Friedrich Wilhelm Dietlein, in, "Journal für die Baukunst," 1832, vol. V, pp. 73–86.

[95] Johann Friedrich Wilhelm Dietlein, *Grundzüge der Vorlesungen in der Königl. Bau-Akademie zu Berlin über Strassen- Brücken- Schleusen- Canal- Strom- Deich und Hafen-Bau*, in, "Journal für die Baukunst," 1830, vol. III, pp. 33–87, 327–358, 420–464, 1831, vol. IV, pp. 90–120, 302–341, 1832, vol. V, pp. 36–72, 155–203, 245–279, 315–385; for what concerns concrete, see especially vol. III, pp. 330, 341, 344–346, vol. IV, pp. 303, 311.

[96] "Béton ist eine aus Steinstücken und hydraulischem oder Wassermörtel (Mörtel, der unter Wasser erhärtet) zusammengesetzte Masse. Den hydraulischen Mörtel hat man schon seit sehr langer Zeit dadurch erhalten, dass man Kalk, entweder mit Sand allein, oder mit Sand und Cement (pulverisierten vulkanischen Produkte, wie Puzzolane, oder Ziegelmehl, Hammerschlag, Esienfeilspähne), oder auch mit Cement allein vermischte, unter ziemlich verschiedenen Verhältnisse der Menge des Kalkes, Sandes und Cements" (Dietlein, *Grundzüge der Vorlesungen*...cit., 1830, vol. III, p. 344).

25. Anon., foundation made of "rammed concrete," tools (Anon., *Die Schwung-Ramme, ein Beitrag zum Grundbau*, in, "Journal für die Baukunst," 1830, vol. II).

26. L. J. Vicat, Bridge of Souillac, 1812-1824. The construction of the concrete foundation of a pier, preparation of the pit (L. J. Vicat, *Bemerkungen beim Bau der Brücke über die Dordogne bei Souillac*, ed. J. F. W. Dietlein, in, "Jounal für die Baukunst," 1832, vol. V; drawings taken from B. Brisson, ed., *Recueil de 245 dessins ou feuilles de textes relatifs à l'art de l'ingénieur; extraits de la seconde collection terminée en 1825*, Paris, École Royale des Ponts et Chaussées, 1827).

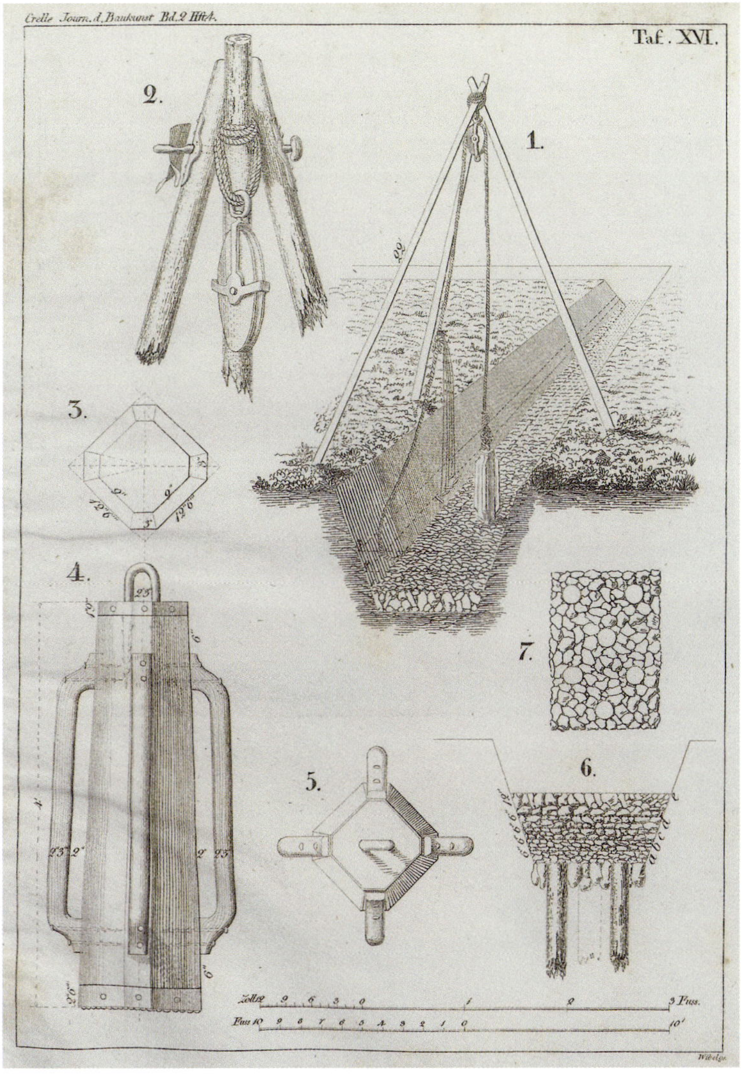

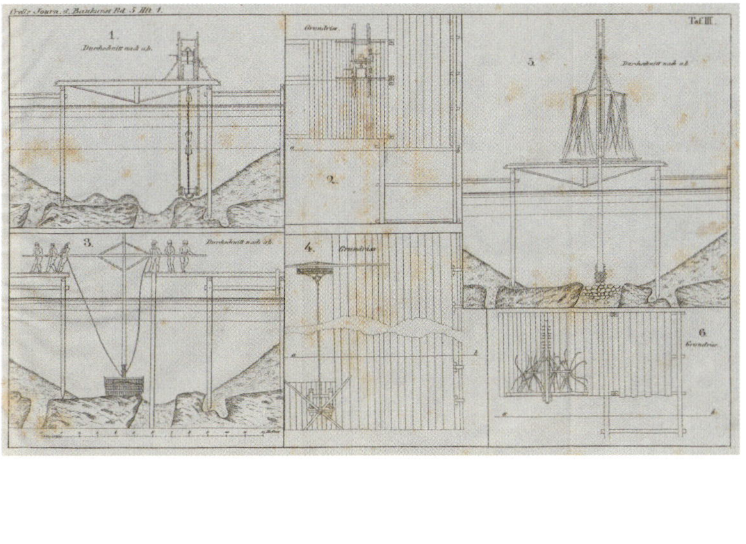

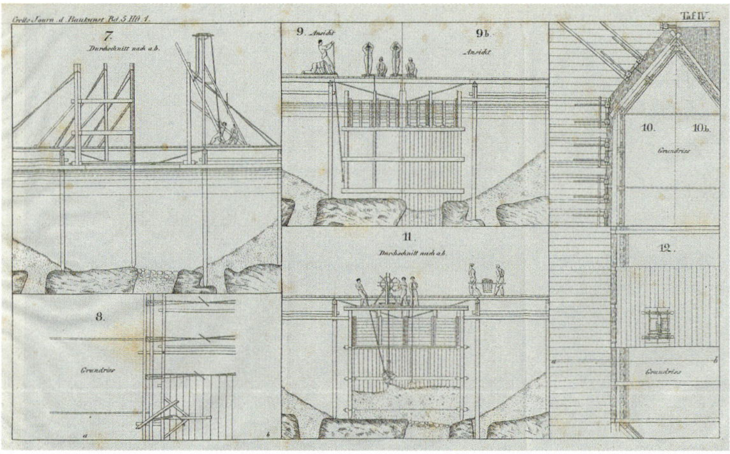

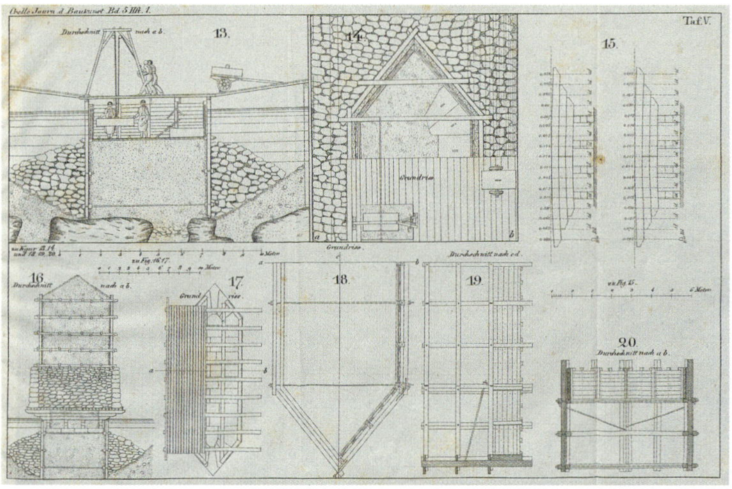

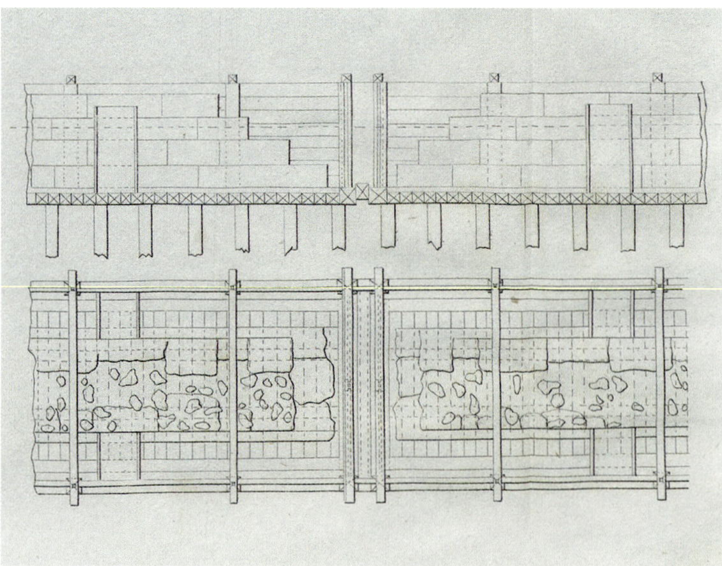

27-28. L. J. Vicat, Bridge of Souillac, 1812-1824. The construction of the concrete foundation of a pier (L. J. Vicat, *Bemerkungen beim Bau der Brücke über die Dordogne bei Souillac*, ed. J. F. W. Dietlein, in, "Jounal für die Baukunst," 1832, vol. V; drawings taken from B. Brisson, ed., *Recueil de 245 dessins ou feuilles de textes relatifs à l'art de l'ingénieur; extraits de la seconde collection terminée en 1825*, Paris, École Royale des Ponts et Chaussées, 1827).

27. Construction of the sheet pile enclosure and concrete pouring.

28. The completion of the foundation and the compressive strength test.

29. J. F. W. Dietlein, hydraulic foundation made of masonry in sinking caisson (J. F. W. Dietlein, *Grundzüge der Vorlesungen in der Königl. Bau-Akademie zu Berlin über Strassen- Brücken- Schleusen- Canal- Strom- Deich und Hafen-Bau*, in, "Journal für die Baukunst," 1830, vol. III).

This is probably the reason he does not acknowledge the origin of the hydraulic property, as evidenced by the following assertion: "Substances that endow mortar with the property of hardening underwater have not yet been discovered."[97] Instead, Dietlein has clearer ideas about the uses of concrete. As is typical at the time, he asserts that concrete is suitable to be poured underwater inside enclosures made of piles or sheet piles, straight on foundation soils or on timber pilings. This last case is thought to be particularly appropriate to build locks. To pour concrete underwater, Dietlein lists three kinds of tools: simple shovels, which should be used only when foundation pits are shallower than 4 or 5 feet; pipes, used if pits are between 5 and 10 feet deep; and chests hanging from a pulley for deeper pits. Dietlein also recommends using concrete to fill the spaces between sunken caissons in a row, and he even draws the sketch of an ashlar-faced wall filled with a kind of rubble masonry.

In the same year that Dietlein's lectures are published, the German translation of the huge French collection of lectures given by Joseph-Mathieu Sganzin at the *Ecole Polytechnique* in Paris is published, providing a further essential reference about concrete hydraulic constructions from the French building culture. A note on the text deserves a special attention, since it addresses the construction of the concrete foundation of the lock of Huningue by the French engineer Louis Alexis Beaudemoulin, which will later play an important role in the construction of similar locks in Bavaria in the second half of the 1830s.[98] Mention of Zimmermann's concrete foundation in Hamm is also made in the same note, showing that early German concrete construction begins to be considered as a reference, along with the French. The construction of stone-faced walls filled with concrete is furthermore described, evoking the ancient Greek *emplecton* and the Roman *opus incertum*.[99]

Accounts of how to produce concrete are also given in a number of other books pertaining to different domains, such as the treatise *Elemente der technischen Chemie*, written by the chemist and physician Ernst Ludwig Schubarth (1797–1868) and published in 1831, the book by Moritz Ferdinand Gaetzschmann (1806–1895) about the construction of walls for underground mines, also published in 1831, and the treatise about artificial stones and binders by Ludwig Friedrich Wolfram, dating back to 1833.[100]

[97] "Welche Bestandtheile eigentlich einem Wassermörtel die Eigenschaft geben, unter Wasser zu erhärten, ist noch genau ausgemittelt" (Dietlein, *Grundzüge der Vorlesungen*...cit., 1830, vol. III, pp. 344–345).

[98] Joseph-Mathieu Sganzin, *Lehrbuch des Straßen-, Brücken-, Kanal- und Hafen-Baukunde*, ed. [Hermann] F. Lehritter, G. H. Strauß, Regensburg, Friedrich Pustet, 1832, vol. I, (pp. 116–151), pp. 144–151, vol. II, pp. 52–56, 206–207, esp. n. 22.

[99] *Ibid.*, vol. I, pp. 160–162, vol. II, pp. 225–227.

[100] See Ernst Ludwig Schubarth, *Elemente der technische Chemie, zum Gebrauch beim Unterricht königl. Gewerbinstitut und den Provizial-Gewerbeschulen*, vol. I, Berlin, August Rücker, 1831, pp. 402–416 (1833^2, pp. 392–402); Moritz Ferdinand Gaetzschmann, *Anleitung zur Grubenmauerung*, Schneeberg, Carl Schumann, 1831, pp. 20–24; Ludwig Friedrich Wolfram, *Lehre on den künstlichen Bausteinen und Verbindungsstoffen, oder Lehre von getrockneten Lehmsteinen, von der Zubereitung, vom Brennen und von der Anwendung aller Ziegeleierzeugnisse vom Brennen des Kalkes und Gipses, vom Baue der Ziegel-, Kalk- und Gipsöfen, von den Mörteln aller Art, von den Kitten u. s. w., mit Rücksicht auf bedeutende Bau- und Kunstwerke der Alten und Neuern*, Stuttgart, Carl Hoffmann, Wien, Carl Gerold, 1833, pp. 53–56, 78–116.

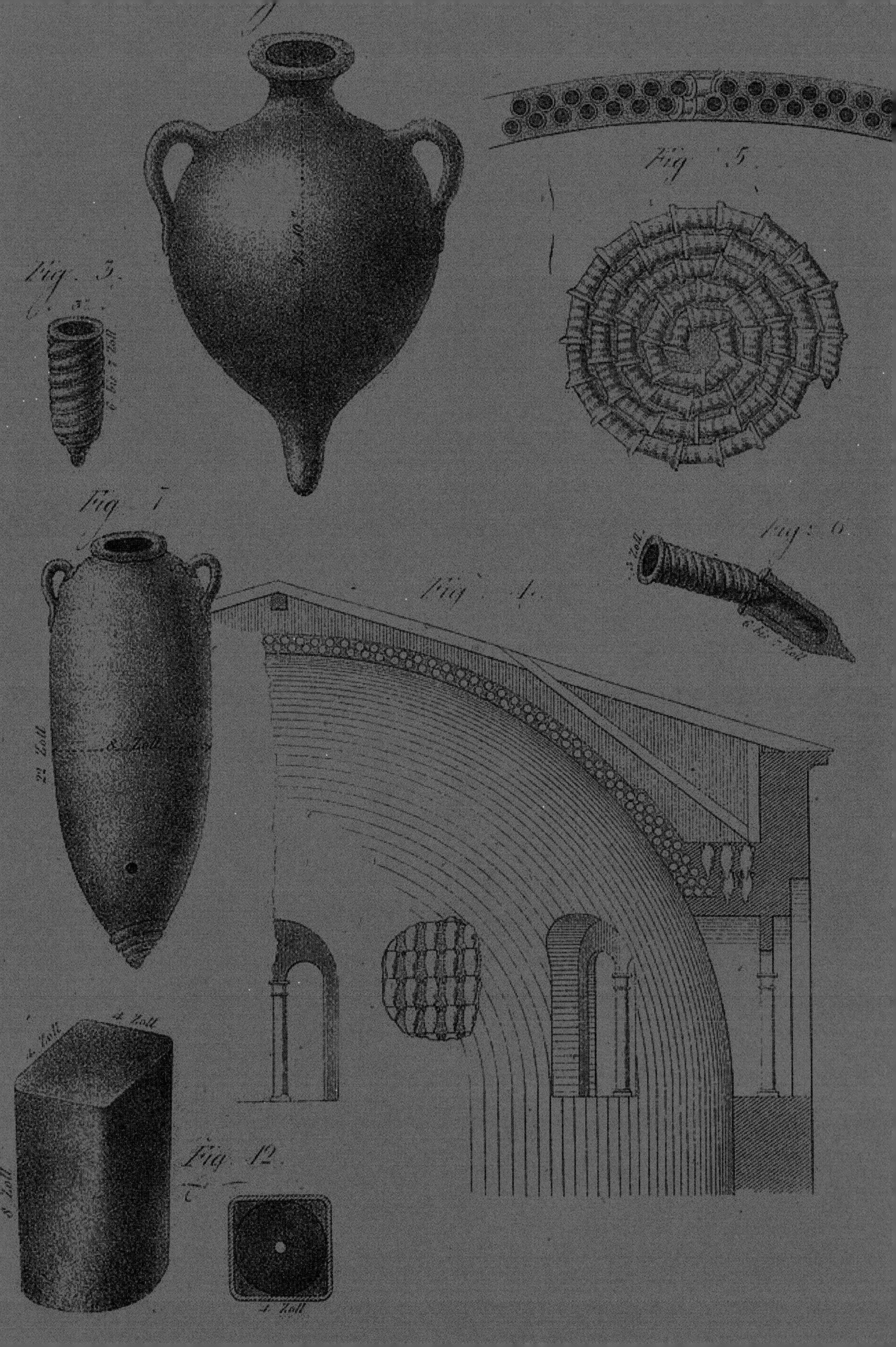

Fig. 3
Fig. 5
Fig. 7
Fig. 6
Fig. 4
22 Zoll
8 Zoll
4 Zoll
Fig. 12
8 Zoll
4 Zoll

chapter three
hydraulic lime and cement manufacturing

The Bavarian hydraulic lime, a *mineralischer Schatz*

Following the theoretical and practical achievements of the previous decade, the manufacturing of hydraulic binders flourishes in several German regions during the 1830s. It begins far away from the Rhine, in an area that goes from Friesland to Silesia, and southwards Bavaria, where trass is not easily available. In northeast and central regions, from the Frisian coast to Kassel, hydraulic binders are mainly produced on the initiative of private entrepreneurs imitating the manufacturing process of English Roman cement – in some cases, even using marl imported from England – and defined as cement. In southern regions, however, early hydraulic binders are produced as implementation of the German scientific studies about lime and mortar, and they are usually called hydraulic lime. What's more, in Bavaria, state institutions support the development of hydraulic lime manufacturing, with the aim of making available better-quality and more affordable mortars and plasters by exploiting local marl. The research about lime and mortar that Fuchs carries out on behalf of Von Purgel should also be seen in this light. Indeed, at that time a clear technical difference between natural hydraulic lime and cement burnt from marls has not yet been formulated. Cement is considered a very effective hydraulic lime and English Roman cement is thought to be the best hydraulic lime available on the market of building materials. Carl Wilhelm Dempp, a teacher at the *Königliche Baugewerbeschule* of Munich, writes succinctly on this topic. "The theory of binders," he states in 1838, "distinguishes common or aerial lime from the hydraulic one, which is also called cement."[1]

[1] "Die Baukunst unterscheidet in der Lehre von den Bindestoffen Beim Kalke - gewönlichen oder gemeinen, auch Luftkalk, und hydraulichen oder Wasserkalk, auch Zement genannt" (Carl Wilhelm Dempp, *Ueber die Benutzung des Mergelkalkes zu hydraulischem Kalk im Allgemeinen, und dessen Anwendung in Bayern insbesondere*, in, "Allgemeine Bauzeitung," III, 1838, (n° 49, pp. 442–443, n° 50, pp. 447–450, n° 51, pp. 454–457) n° 49, p. 442).

In the early 1830s, the main proponent of hydraulic lime production and use in Bavaria is the state civil engineer Friedrich Panzer (1794–1854), who, moving from Fuchs' theory, experiments and promotes the production and use of hydraulic lime burnt from local marl. Based in Wurzburg since 1826, Panzer is made responsible for the building of a city barn in 1830. He plans to build an underground masonry silo, which needs to be built with high-quality hydraulic mortar in order to preserve grain from moisture in the soil. In a preliminary survey into different kinds of existing barns and silos, Panzer takes into account the studies that French agronomist Charles Philibert de Lasteyrie carried out at the turn of the 1820s, proving the importance of using hydraulic mortar to build the masonry of grain silos.[2] De Lasteyrie even envisages building silos made of concrete by pouring mixtures of pebbles and hydraulic mortar into pits shaped as formworks for circular walls and vaults. Some German articles about techniques for storing grain, published between 1821 and 1824 in the "Polytechnisches Journal" and in the "Monatsblatt für Bauwesen und Landesverschönerung," report De Lasteyrie's building principles for silos.[3] The importance of using hydraulic lime is highlighted, but none of these articles deals with concrete silos, and Peter Ludwig Maréchaux (1764–1832), a naturalist based in Munich, who writes a long paper about the topic of grain storage, mistakenly asserts that De Lasteyrie proposes pouring concrete merely as filling material for stone-faced masonry.[4]

Panzer seizes on the importance of using hydraulic lime-based mortar to build a moisture proof masonry silo and, following Fuch's example, undertakes a survey of marl deposits in the valley of the River Main, between Bamberg and Aschaffenburg.[5] He then verifies the clay content of each kind of marl he finds and fires two samples: one in red burning coal for seventy-five minutes, the other in strong fire up to four hours.[6] The burnt samples are ground into powder and mixed with rainwater to form two different doughs, which are put into two glasses and covered with more water. Their hardening is checked first after 24 hours and then again after 14 days. In this way, Panzer establishes that lime produced from marl quarried close to the village of Randersacker, south of Wurzburg, is the best he had found. Consequently, he uses it to burn the hydraulic lime he needs to build the silo. Burnt marl is ground in a local chalk mill, since lime kilns, at that time, are not yet equipped with mills. Mortar is prepared by mixing one measure of lime, 2/3 of sand

[2] See Charles Philibert de Lasteyrie, *Des fosses propres à la conservation des Grains, et da la manière de les construire, avec différents moyens qui peuvent être-employés pour le même objet*, Paris, Imprimerie Royle, 1819; Id., *Rapport fait par M. de Lasteyrie, au nom du Comité d'agriculture, sur le silo ou fosses à conserver les grains, de M. de Lacroix, établi à Ivry, près Paris*, in, "Bulletin de la Société d'encouragement pour l'industrie nationale," XXI, 1822, pp. 286–288; Id., *Rapport fait par M. de Lasteyrie, au nom d'une Commission spéciale, sur les fosses à grains*, in, "Bulletin de la Société d'encouragement pour l'industrie nationale," XXII, 1823, pp. 241–249.

[3] Peter Ludwig Maréchaux, *Würdigung sämmtlicher bis jetzt bekannt gewordener Methoden, das Getreide, mehrere Jahre hindurch, ohne Nachtheil für dasselbe, aufzubewahren*, in, "Polytechnisches Journal," vol. V, 1821, (pp. 334–376), pp. 347–350; Anon., *Ueber die Silos oder Erdgruben zur Aufbewahrung des Getreides*, in, "Polytechnisches Journal," vol. IX, 1822, pp. 329–331; Anon., *Zweckmäßige Getreide-Magazine*, in, "Monatsblatt für Verbesserung des Landbauwesens und für zweckmäßige Verschönerung des Bayerischen Landes," II, 1822, pp. 9–13; Anon., *Miszellen. Ueber die Silos oder Erdgruben zur Aufbewahrung des Getreides*, in, "Polytechnisches Journal," vol. X, 1823, p. 123; Anon., *Ueber die Korngruben oder sogenannten Silos*, in, "Polytechnisches Journal," vol. XIII, 1824, pp. 255–259.

[4] See Marechaux, *Würdigung...* cit., p. 349.

[5] "The building of an airtight and moisture proof masonry silo," Panzer writes in 1830, "needs the use of hydraulic lime to bind stones, fill all gaps and plaster the interior and exterior surfaces of walls" ("Ein gemauertes Silo, welches aller Feuchtigkeit und der atmosphärischen Luft unzugänglich seyn soll, erfordet die Anwendung des hydraulischen Kalkes, sowohl als Bindungsmittel der Steine und zur Ausfüllung aller Zwischenräume, als auch als Ueberzug von innen und von außen") (Friedrich Panzer, *Abhandlung über die Aufbewahrung des Getreides in Silo's,* Würzburg, Carl Strecker, 1830, p. 18).

[6] At that time, lime makers gauge the degree of burning by the colour of the flame; see Valentin Biston, *Manuel théorique et pratique du chaufournier*, Paris, Roret, 1828, p. 55.

[7] Panzer, *Abhandlung über die Aufbewahrung*...cit.

and a little amount of fat lime. It is then used to bind the bricks and plaster the surfaces of the masonry.

During the construction of the silo, Panzer gathers the knowledge and expertise he is developing in a short pamphlet, where he particularly emphasizes the usefulness of hydraulic lime produced from Bavarian marls. The pamphlet is published in 1830 and, alongside Fuchs' paper, becomes a reference text on the production and use of natural hydraulic lime.[7]

Further crucial events for the development of Bavarian hydraulic lime happen during 1831. The *Polytechnischer Verein für das Königreich Bayern* promotes a competition aiming to award "the engineer or builder who, following Fuchs' theory and practical instructions about marls, discovers a kind of local hydraulic lime that is suitable to prepare good-quality mortar, suitable to protect façades exposed to the elements, to make wet walls and moist houses dry, to build water containers, grain silos and hydraulic works like sewer, basements etc."[8] A note on the announcement invites applicants "to take into account the interesting essay by the engineer Friedr. Panzer in Wurzburg."[9] This indicates the importance of Panzer's pamphlet as a source of knowledge about hydraulic lime. Only two competitors apply – the first is Panzer, and the second is Andreas Zenner, the owner of a lime kiln in Munich – but the initiative still demonstrates the rising confidence in the possibility of developing the Bavarian manufacturing of high-quality hydraulic binders, while opposition to Fuchs' theory decreases further. In the wake of such events, Panzer publishes a pioneering manual expressly focused on the production and use of Bavarian hydraulic lime in 1831.[10] "This essay," he writes in the preface, "aims at explaining the usefulness of hydraulic lime produced from marl, making people [...] more familiar with this mineral treasure that mountains shelter in bulk on both sides of the River Main, between Bamberg and the Spessart mountain range. Words from this book should come to private people who build new houses or ameliorate existing ones, and to artisans who more or less work with mortar, such as masons, stonemasons, painters, roofers, plasterers, fountain builders, lime makers, etc."[11] Beyond mere technical instructions, two concepts expressed in Panzer's book are particularly relevant in terms of the future manufacturing and use of hydraulic binders. The first concerns the necessity of theoretical and scientific knowledge, while the second addresses the suitability of hydraulic lime both for hydraulic works

[8] "Demjenigen Ingenieur, Bau- oder Maurermeister, welcher nach Anleitung der von Hrn. Hofrath und Akademiker Dr. Fuchs in München herausgegebenen Abhandlung über den Mörtel inländischen hydraulischen Kalk aufsucht, daraus hydraulischen Mörtel bereitet und anwendet, um feuchte Wände und Wohnungen troken, die Wetterseiten der Gebäude schützend und dauerhaft zu machen, dann Wasserbehälter, Wasserbauten überhaupt, Kloaken, Keller etc. herzustellen und trockene Silo's zur Aufbewahrung des Getreides zu erbauen: die dreifache goldene Medaille" (*Preisaufgaben des polytech. Vereins in Bayern*, in, "Polytechnisches Journal," vol. XL, 1831, (pp. 387–389), p. 388).

[9] "Man macht di Theilnehmer auch zugleich auf die interessante Abhandlung des k. b. Ingenieurs Hrn. Friedr. Panzer in Würzburg aufmerksam" (*ibid.*, p. 388).

[10] Friedrich Panzer, *Anleitung über die Bereitung des Mörtels aus hydraulischem Kalke in Beziehung auf die Auffindung des vorzüglich hierzu sich eignenden Mergels, so wie über das Verfahren, welches bei der Anwendung des hydraulischen Mörtels, sowohl bei Landgebäuden, als auch bei Bauten im Wasser zu beobachten ist*, Würzburg, Becker, 1831.

[11] "Der Zweck des gegenwärtige Aufsatzes besteht darin, den Nutzen des hydraulischen Kalkes zu zeigen und das Verfahren zu erläutern, wie man sich bei der Bereitung des Mörtels mit hydraulischem Kalke mit Benutzung des Mergels zu verhalten hat; ferner das Publikum, [...], näher mit diesem mineralischer Schatz, welche die Gebirge zu beyden Seiten des Mainflusses zwischen Bamberg und dem Spessart in unerschöpflicher Masse in ihren Schoos einschließen, vertraut zu machen. Daher sollen diese Worte zunächst an Private, welche neue Gebäude aufführen, oder an den bestehenden Verbesserungen vornehmen lassen, so wie an alle Handwerksleute, die mit der Anwendung des Mörtels in näherer, oder entfernterer Beziehung stehen, als Mauerer, Steinhauer, Tüncher, Dachdecker, Gypser, Brunnenmacher, Kalkbrenner u.s.w. gerichtet seyn" (Panzer, *Anleitung über die Bereitung des Mörtels aus hydraulischem Kalke in Beziehung auf die Auffindung des vorzüglich hierzu sich eignenden Mergels, so wie über das Verfahren, welches bei der Anwendung des hydraulischen Mörtels, sowohl bei Landgebäuden, als auch bei Bauten im Wasser zu beobachten ist*, 2nd. ed., Würzburg, München, Lindauer'sche Buchhandlung, 1832, p. 4).

and for ordinary masonries exposed to moisture and to the elements.[12] "Hydraulic lime," Panzer asserts, "was already known in Italy, Holland, England and France; but the lack of scientific knowledge has prevented a wider and general use of it. In recent times, the development of a theory, notably in chemistry and mineralogy, has shed a light on the deficient knowledge, which was based only on practical experiences, and we should be thankful to the eminent scholars who have discovered the chemical combination that allows mortar to harden underwater. Those who want to know more about this revolutionary discovery should read the dissertation by Dr. Fuchs [...], which I used as a guide."[13] German scholars and building masters should be aware of the suitability of hydraulic lime for hydraulic and ordinary works, since John, Von Prittwitz and Fuchs have already proven it. Nevertheless, in the early 1830s, the ancient belief in the superiority of aerial lime remains deep-rooted and influences most master builders. "The expression hydraulic lime should not mislead to think that this kind of lime is only effective in water," Panzer warns and reminds the reader that some kinds of hydraulic mortar composed with trass or brick powder, like the Loriot mortar, have already been successfully used out of water.[14]

Again in 1831, in addition to writing the manual about hydraulic lime, Panzer builds a stone masonry arch bridge in the village of Schönungen, using hydraulic lime-based mortar for the masonry, both underwater and above ground. "The external parts of the masonry," he relates in this regard, "are made of sandstone ashlars, whereas the internal parts are made of unrefined stones. The foundations stand entirely underwater, whereas the abutments, the vault and the wings only occasionally come into contact with water. The use of hydraulic lime has made this building particularly solid."[15]

Even to prepare the so-called *Reissboden* (drawing floor) – a flat and open-air place where all bridge stones are designed at real scale – Panzer builds a 30 square-foot floor made of bricks bonded with hydraulic-based mortar and covered with a 2-inch thick layer of the same kind of mortar. He probably conceives this kind of construction on the basis of an ancient Italian building technique, which he learnes from the architect Johann Gottfried Gutensohn (1792–1851) and then describes in the 1831 book about hydraulic lime.[16] Six months after the bridge has been completed and the drawing

[12] *Ibid.*, pp. 4–5, 23–29.

[13] "Der hydraulische Kalk war in Italien und in Holland, später in England und in Frankreich schon längst bekannt; es stand jedoch der weiteren Verbreitung und allgemeinen Anwendung desselben der Mangel an wissenschaftlicher Begründung entgegen. In neuester Zeit mußte die Unvollkommenheit bloßer Erfahrungen dem Lichte der Theorie, nämlich der Chemie und Mineralogie weichen. Großen Gelehrten, die deßhalb unseren innigsten Dank verdienen, gebührt das Verdienst, das chemische Verhalten der Substanzen des Mörtels ermitteln zu haben, wobei die Erhartung im Wasser vor sich geht. Wer sich näher mit dieser nützlichen, im Bauwesen Epoche machenden Erfindung, bekannt machen will muß die von Herrn Dr. Fuchs [...] verfaßte Schrift lesen, welche ich als leitende Norm benutzt habe" (*ibid.*, pp. 3–4).

[14] "Der Name des hydraulischen Kalkes darf nicht verführen, als leiste er seine Dienste los im Wasser" (*ibid.*, p. 5).

[15] "Die äußere Bekleidung bestehet aus gehauenen, großen Quadern von Sandsteinen, das Innere des Gemäures aus unzugerichteten Bruchsteinen. Die Fundamente der Widerlager des Gewölbes, so wie der flügel liegen ganz unter Wasser; die Widerlager und das Gewölbe selbst, die Flügel und dergl. sind dem Wasser abwechselnd ausgesetzt. Die Anwendung des hydraulischen Kalkes hat diesem Bauwerke eine große Festigkeit verschafft" (Friedrich Panzer, *An die Redaktion des Kunst- und Gewerbe-Blattes*, in, "Kunst- und Gewerbe-Blatt des polytechnischen Vereins für das Königreich Bayern," vol. X, 1832, coll. 111–144, col. 111).

[16] See Panzer, *Anleitung über die Bereitung des Mörtels...* cit., p. 27. Panzer mentions Gutensohn as source; see *ibid.*, p. 24. Between 1819 and 1823, Guthenson spends a long stay in Rom, where he uses to draw the monuments he observes; see Johann Gottfried

floor abandoned, Panzer is pleased to confirm that the hydraulic mortar layer is still in good condition.

Encouraged by Panzer, several builders and masons experiment with hydraulic lime-based mortar between the summer of 1831 and the winter of 1832, to build or restore masonry works of various sizes. Panzer collects reports about these experimental works and publishes them in the "Kunst- und Gewerbe-Blatt des polytechnischen Vereins für das Königreich Bayern" in 1832.[17] The reputation of the Bavarian hydraulic lime has improved so much that it is even compared with English Roman cement. "Until now," states an article from the "Polytechnisches Journal" in 1831, "people have considered hydraulic lime a rare natural product, and they have even believed that high-quality lime was only available in England. However, plenty of good hydraulic lime can also be found in Bavaria but we could not realize it, due to the lack of scientific knowledge, and we have been using, instead, expensive pozzolana from Rome, bitumen from Lobsan and trass from Andernach or from other places, leaving hydraulic lime unexploited."[18]

Indeed, the English way of producing cement is criticized for being empirical, and a scandal arises around the fact that, shortly after the publication of Fuchs' research, a civil servant was sent to learn the art of producing hydraulic lime in England, "where there was not a grain of theory, hydraulic lime was produced only from one kind of raw material, following a quite empirical method, and nobody was therefore able to provide any explanation based on general principles, making impossible the production of hydraulic lime under different conditions."[19] The civil servant sent to England is probably Christoph Maria Joseph Schmitz (1796–1866), inspector at the porcelain factory of Nymphenburg and member of several technical committees in Bavaria.[20] According to Dempp, Schmitz went on a research journey to England at the turn of 1831 to learn about cement production. Back in Munich, he is reported undertaking some successful experiments that later serve as references for the foundation of several cement kilns around Munich.[21] Nevertheless, Fuchs' and Panzer's contribution to the development of Bavarian hydraulic lime remain undeniably prevalent, and even officially recognized in 1832, as the Bavarian *Polytechnischer Verein* awards Panzer for the competition launched in 1831, and the Dutch Society of Science awards Fuchs for the competition

Gutensohn, *Denkmale der christlichen Religion oder Sammlung der ältesten christlichen Kirchen Roms vom 4ten bis zum 13ten Jahrhundert*, Roma, Johann Michael Knapp, vol. I, 1821, voll. II–III, 1824; Id. *Sammlung von Denkmalen und Verzierungen der Baukunst in Rom vom 16ten Jahrhundert*, Roma, Romanis, voll. I–III, 1826. The use of hydraulic lime to build floors, in some case even imitating natural stones, develops in the following decades.

[17] See Panzer, *An die Redaktion des Kunst- und Gewerbe-Blattes*...cit. A bridge is built in Veithhöchheim and a lock is built in Dühlstadt, while the lock of Schweinfurt and a canal near the Julius hospital in Wurzburg are restored. In addition, several masons test hydraulic lime to plaster walls in private houses.

[18] "Der hydraulische Kalk, den man sich bis jetzt nur als ein seltnes Naturproduct dachte, der z. B. nur in England von der erforderlichen Güte gefunden würde, der aber nun in ganz Bayern, von jeder Qualität und im Ueberflusse vorhanden ist, den wir aber bisher, aus Mangel einer gründlichen Theorie, nicht kannten und unberührt ließen, während man Puzzolane von Rom, Traß von Andernach und anderen Orten, Lobsaner Erdpechkitt u. mit großen Kosten statt des einfachen hydraulischen Mörtels verwendete" (Anon., *Ueber Aufbewahrung des Getreides in Silo's und die Benuzung des hydraulischen Mörtels zur Erbauung derselben*, in, "Polytechnisches Journal," vol. XXXIX, 1831, (pp. 427–433), p. 428).

[19] "Wo man damals hierüber noch kaum den Schatten von einer Theorie hatte, und nur mit einem gewissen Materiale auf ganz empirischem Wege hydraulischen Kalk darzustellen mußte, und daher gar nicht im Stande war, eine, auf allgemeine Grundzüge sich stützende Anleitung zu geben, ihn unter verschiedenen Umständen zu bereiten, wie sie hier schon hatten" (Anon., *Würdigung der Arbeit des Professors und Akademikers Fuchs*...cit., col. 759).

[20] About Christoph Maria Joseph Schmidtz, see Anon., *Nekrolog. Christoph Schmitz*, in, "Kunst- und Gewerbe-Blatt des polytechnischen Vereins für das Königreich Bayern," vol. XLIV, 1866, coll. 442–448, esp. coll. 445–446.

[21] See Carl Wilhelm Dempp, *Ueber die Benutzung des Mergelkalkes*...cit., p. 443. Dempp mentions the following cement producers: Gasteiger, Rechel, Schmidt and Höß, Stießberger und Zenner, who is probably the one who applies for the competition promoted in 1831 by the Bavarian *Polytechnischer Verein*.

dating back to 1830. In the wake of this acknowledgement, the Bavarian *Polytechnischer Verein* also awards Fuchs a special double gold medal for having contributed to the scientific development of the country.[22]

In April 1832, Panzer moves from Würzburg to Bamberg, where he studies the so-called *schwarzer Kalk* (black lime), a local lean lime, which is burnt from marl quarried in villages close to Bamberg and has been known for centuries for being particularly solid and durable.[23] Panzer observes mortar joints containing black lime in several important buildings in Bamberg, such as the Cathedral, the gate of the Carmelite monastery and the city walls. He realizes that mortar joints are sometimes better preserved than sandstone. Still, Panzer claims black lime-based mortar was used to plaster the façades of the town hall building, before Johann Anwander (1715–1770) painted them with frescos in 1755, and to bind bricks and tiles of several humble buildings in Bamberg and in the surroundings. Panzer himself verifies the solidity of the brick-masonry walls assembled with black lime-based mortar when he has to demolish the ones in an old stable. The walls prove to be so solid that Panzer let them break into big blocks, which he then uses as artificial stones to build the walls of a new stable.[24]

Panzer analyses the chemical composition of the Bamberg black lime and discovers that it contains clay, which he considers the origin of solidity, in accordance with Fuchs' theory. "The solidity of this lean lime," he asserts, "is due to chemical features, it namely contains between 5 and 14 percent clay."[25] To develop the local manufacturing of hydraulic lime, Panzer asks the royal home office for the construction of a mill, which he then designs himself. It is a typical animal-powered machine, with an inclined wheel moved by horses or oxen, putting in motion the millstones through a machine made of shafts and cogwheels. The mill is built up in the workshop of the *Königliche Bauinspektion* (public office for building) and is provided with a stamp mill to crush burnt marls roughly before they are ground into a powder. According to Panzer's initial plan, burnt marls should have been crushed by hand, but the construction of a mechanical stamp mill is later considered more advantageous, since this could be connected to the grind mill in a way that both machines would be powered by the same animals. On a request from Panzer, the mill is entrusted to Karl Zelger, an independent architect from Bamberg, with the aim of integrating the private sector into the business and spreading the awareness of the advantages that hydraulic lime provides.[26] This seems to be quite unusual at that time. "Being a civil servant," Panzer observes, "I was allowed neither to sell hydraulic lime to private citizens, nor to distribute it to communities, but this was necessary in order to spread the word about this material."[27] At first, the mill is only used occasionally and is powered by daily workers instead of animals, since wages are considered more affordable than the regular feeding of animals.[28] In the absence of containers expressly con-

[22] Anon., *Preisvertheilung des polytechnischen Vereins*, in, "Kunst- und Gewerbe-Blatt des polytechnischen Vereins für das Königreich Bayern," XVIII, 1832, (coll. 646–649), col. 649.

[23] See Friedrich Panzer, *Ueber das Vorkommen des hydraulischen Kalkes in der Keuperformation, über die Dauerhaftigkeit absolute und rückwirkende Festigkeit desselben und der Steine, als Beitrag zu dem Bau des Kanales zur Verbindung der Donau mit dem Main*, München, Lindauer'sche Buchhandlung, 1836, p. 3. Panzer mentions the following villages: Tütschengereuth, Triesenbach, Oberhaid and Burgebrach.

[24] See *ibid.*, pp. 3–5.

[25] "Die Festigkeit dieses mageren Kalkes ist lediglich seinen chemischen Eigenschaften zuzuschreiben; es befindet sich nämlich in demselben eine Quantität Thon, welche zwischen 5 und 14 Prozent variiert" (*ibid.*, p. 5).

[26] Karl Zelger is probably the architect who attends the Second Meeting of German Architect and Engineers, which takes place in Bamberg from September 8 to 10, 1843. In this regard, see *Verzeichniß der Theilnehmer an der zweiten Versammlung der deutschen Architekten und Ingenieure zur Bamberg vom 8. bis 10. September 1843*, in, "Zeitschrift

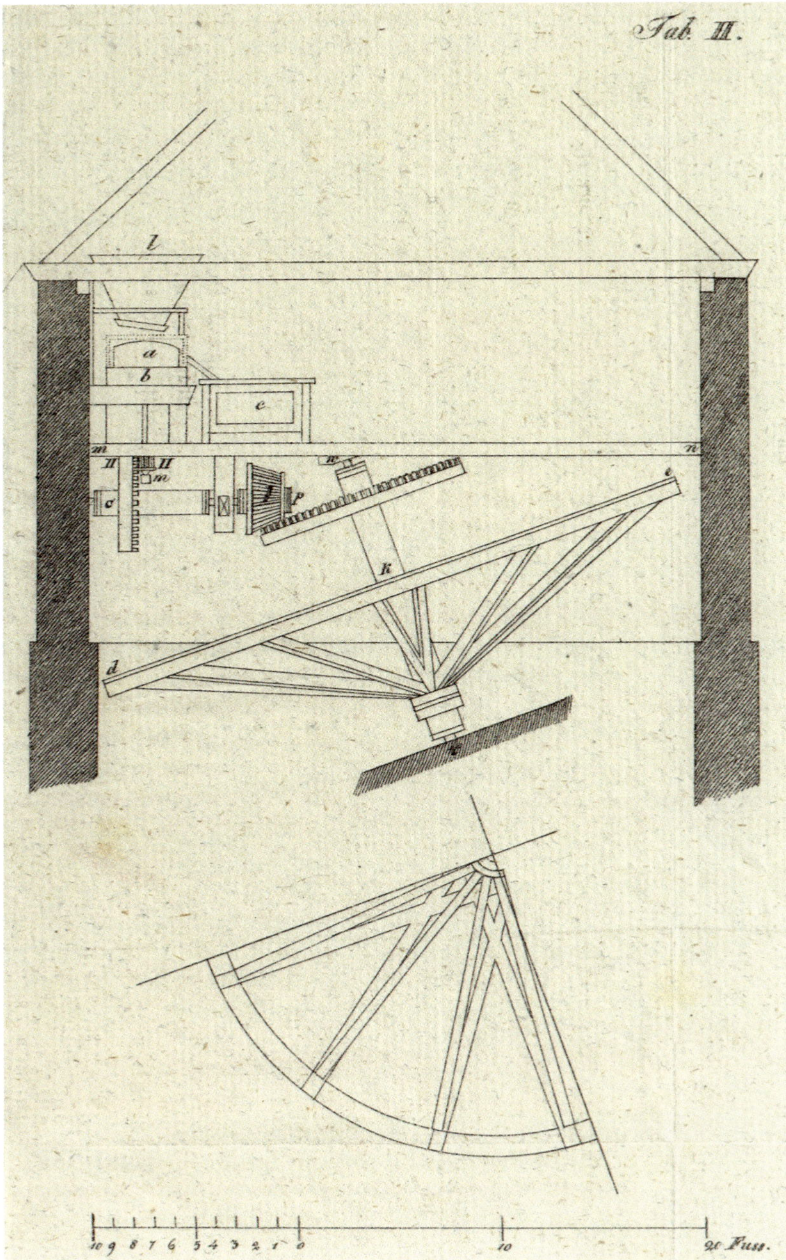

30. F. Panzer, project for the Bamberg mill to grind hydraulic lime, 1832 (F. Panzer, *Ueber das Vorkommen des hydraulischen Kalkes in der Keuperformation, über die Dauerhaftigkeit absolute und rückwirkende Festigkeit desselben und der Steine, als Beitrag zu dem Bau des Kanales zur Verbindung der Donau mit dem Main*, München, Lindauer'sche Buchhandlung, 1836).

für praktische Baukunst," III, 1843, pp. 228–229).

[27] "Meine Stellung als Königlicher Baubeamter ließ nicht zu den Absatz des hydraulischen Kalkes an Private, oder Gemeinden zu besorgen, und doch war dieses nothwendig, um dieses nützliche Material mehr bekannt zu machen" (Panzer, *Ueber das Vorkommen*...cit., p. 7).

[28] *Ibid.*, p. 23.

31. G. A. Böckler, drawing of a mill powered by a man (G. A. Böckler, *Theatrum Machinarum Novum*, Nürnberg, Paulus Fürstum, 1661).

ceived for lime, the lime, once ground, is packaged in barrels previously used for salt (and washed), and it is delivered with instructions for use.

In 1833, hydraulic lime from Zelger's mill is used for restoration work on the Bamberg Cathedral, in which Panzer is personally involved. Together with the architect Carl Alexander von Heideloff (1789–1865), who has been appointed responsible of the restoration in 1831, he develops a plan to recover the exterior parts of the building, with special regard for the seriously damaged, western towers.[29] The plan is achieved in April 1833 and works start in summer. Panzer is the only supervisor from August to December, as Von Heideloff is occupied with other restoration work in Nuremberg. He uses hydraulic lime-based mortar to repair and, in some cases, even to rebuild several architectural elements such as the cornices of windows and the bases and capitals of columns on the facades of the towers. "The cathedral western towers," he later relates, "which are about 700 years old, are built with a kind of keuper sandstone that is little cohesive; columns were so weathered that they were on the point of collapse; the bases, the capitals and all other ledges were ruined to the point of being unrecognisable, as if they were eroded stones. All damaged ornaments, ledges, etc., were remade affordably, and this western part of the cathedral is now safe for centuries to come."[30]

Between February 1834 and April 1835, a short advertisement written by Zelger to promote hydraulic lime from the Bamberg mill is published in at least three different local newspapers.[31] Zelger lists several kinds of works for which hydraulic lime is considered appropriate. "One can build," he writes, "dry houses, watertight basins, imperishable canal systems made of clay pipes laid in mortar, durable exhaust channels, watertight and dry basements, watertight and snow-resistant roofs made of tiles and hydraulic lime-based mortar, durable plaster on façades exposed to the elements, dry floor in breweries, distilleries and son on."[32] Furthermore, he asserts that hydraulic lime is appropriate "to produce relief decorations in plaster and ornaments that, as proved by solid experience, perfectly withstand the elements."[33] These claims probably refer to the restoration work carried out at the Bamberg Cathedral in 1833. They demonstrate that a certain interest in the production of decorations and objects using hydraulic mortar is growing in Bavaria. As a matter of fact, some objects made of hydraulic mortar, along with some others made of brick bond and plastered with hydraulic mortar, are exhibited at the *Industrieausstellung* (industrial exhibition) that takes place in Munich in 1834.

"It is with intense joy that we observe," states an official exhibition report, "several products made of hydraulic lime. They inspire great expectations and deserve encouragement by the state and by the public. The iron trader J. G. Schmidt from Munich exhibits cement powder and several objects

[29] See Achim Hubel, *Die beiden Restaurationen des Bamberger Domes. Zur Geschichte der Denkmalpflege im frühen 19. Jahrhundert*, in, "Berichte des Historischen Vereins Bamberg," vol. CXXI, 1985, (pp. 45–90), pp. 78–81.

[30] "Die westlichen Thürme des Domes, welche nahe an 700 Jahren alt seyn mögen, sind aus Keupersandsteine von geringer Kohärenz erbaut; die Säulen waren so sehr verwittert, daß sie den Einsturz drohten, und ihre Basen und Kapitäle, so wie die übrigen Gesimse bis zur Unkenntlichkeit vom Steinfraße zerstört. Mit verhältnissmäßig wenig Kosten wurden alle beschädigten Ornamenten, Gesimsglieder, u. s. w. wieder hergestellt, so, daß dieser gegen Westen gekehrte Theil des Domes auf Jahrhunderte hinaus gesichert ist" (Panzer, *Ueber das Vorkommen*...cit., p. 6).

[31] Karl Zelger, *Hydraulischer Kalk*, in, "Korrespondenten von und für Deutschland," 1834, n° 52; in, "Königlich Bayerisches Intelligenzblatt für den Ober-Main-Kreis," 1834, n° 27, p. 257; and in, "Tagblatt der Stadt Bamberg," 1835, n° 117, p. 468.

[32] "Man baut hiermit trockene Häuser, wasserdichte Bassins, unvergängliche Wasserleitungen, wobei thönerne Röhren in hydraulischen Mörtel gelegt werden, dauerhafte Abzugskanäle, trockne, wasserdichte Keller, haltbare, von Regen und Schnee undurchdringliche Dächer, wenn nur die Ziegel gehörig

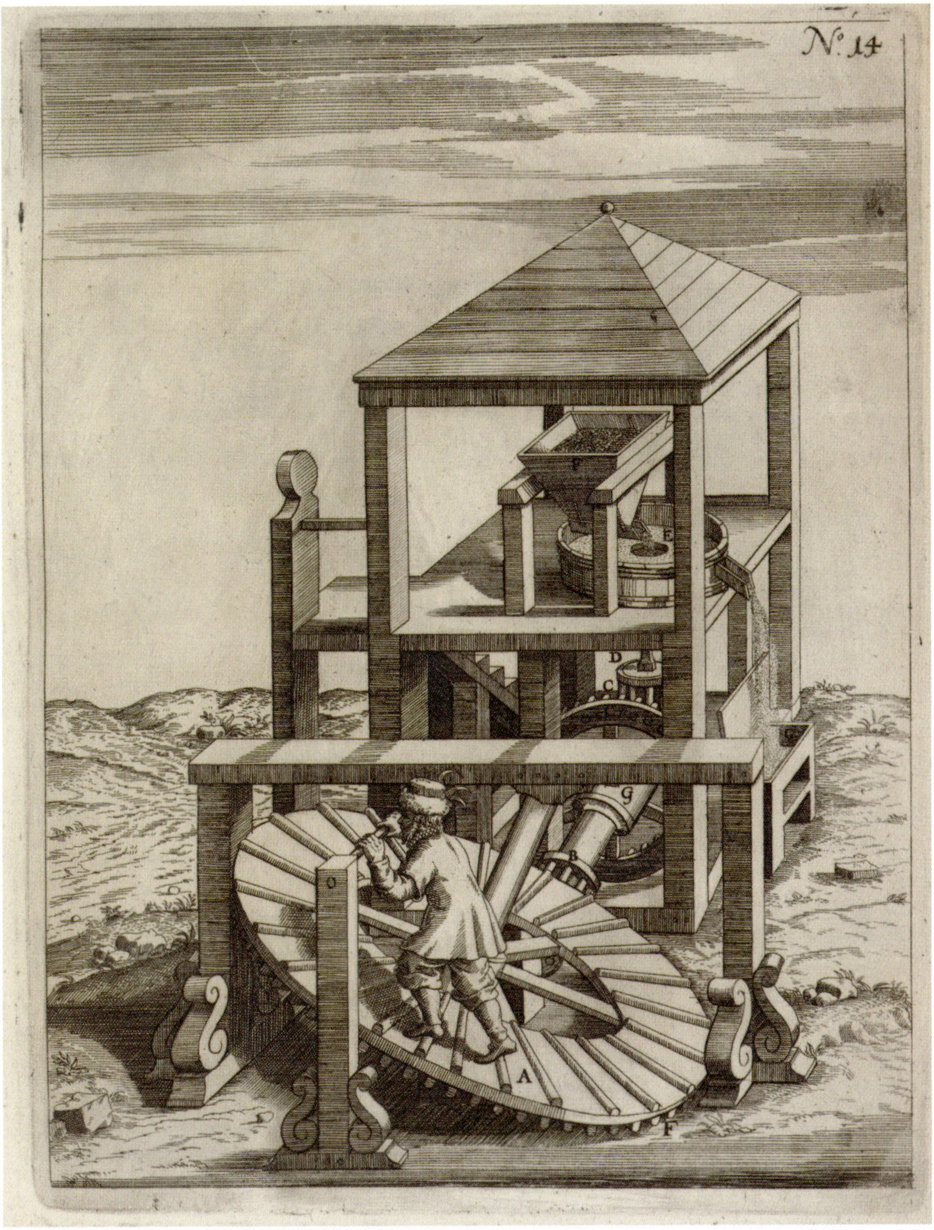

damit eingespeißt werden fertig dauerhaften Bewurf an den Wetterseiten, trockne Böden in Bräuhäusern, Brantweinbrennereien, Färbereien u.s.w." (*ibid.*).

[33] "Selbst plastische Verzierungen lassen sich in dem Bewurfe der Gebäude mit hydraulischem Kalke herstellen und eine feststehende Erfahrung hat beweisen, daß solche Ornamente dem Einflusse der Witterung vollkommen widerstehen" (*ibid.*).

32. F. Panzer, drawings of contraptions for tensile and compressive strength tests (F. Panzer, *Ueber das Vorkommen des hydraulischen Kalkes in der Keuperformation, über die Dauerhaftigkeit absolute und rückwirkende Festigkeit desselben und der Steine, als Beitrag zu dem Bau des Kanales zur Verbindung der Donau mit dem Main*, München, Lindauer'sche Buchhandlung, 1836).

made of cement, among which a water container is especially interesting. G. Reihl from Munich has produced four grindstones made of hydraulic lime and quartz-sand, which probably still permits improvements. Finally, Mr. J. Karlinger from Miesbach has sent several busts made of hydraulic lime."[34]

Even on this occasion, Fuchs is praised for his scientific contribution of having made the proper exploitation of hydraulic lime and mortar possible. "Thanks be to the eminent court councillor and academic Dr. Fuchs," asserts the same report, "for the scientific basis of knowledge about hydraulic lime and, even more, for the development of it in our homeland."[35] The objects exhibited in Munich in 1834 seem to anticipate the huge variety of objects of the same kind exhibited at several industrial expositions in the second half of the 19th century.

During the development of these events, between 1832 and 1834, Panzer collects different kinds of marl around Bamberg, which he first analyses and then burns into lime to produce hydraulic mortar specimens, in preparation for compression and tensile strength tests he intends to undertake. He prepares different cylindrical specimens set to harden underwater until January 1835, when the tests take place. An original contraption conducts the tensile strength tests. It consists of two symmetric parts, each made of two curved stirrups joined through two metallic plates with a half–circular cut on one side. The whole is constructed such that, when the two symmetric parts are brought near each other, the metallic plates form an empty circle, in which is fixed a cylindrical specimen provided with a groove on the curved surface. A couple of stirrups hang from a hook, while the others hold a suspended load, which is increased until the specimen cracks. The installation for the compression strength test recalls the one conceived by Émiland-Marie Gauthey to test the stones that Jacques-Germain Soufflot intended to use for the dome pillars of the Church of Saint-Geneviève in Paris.[36] It consists of a lever inserted into a wall on one side that holds a hanging weight on the other side. This weight compresses a specimen put on a pole close to the wall. To evaluate the real compression strength of the specimens, Panzer develops a calculation that takes into account the weight of the lever and a multiplier effect from the distance between the specimens and the hanging weight.[37] Using the two described installations, Panzer also tests some local natural stones, bricks, and between 800 and 350 years-old mortar samples taken from the Bamberg Cathedral, from the Carmelite Church and from the Church of Saint Elizabeth. Through these tests, Panzer realizes that burnt bricks and hydraulic lime-based mortar are a good combination for building solid masonries that are supposed to be even more resistant than some kinds of natural stones commonly used in Bamberg.

Around 1835, Bavarian hydraulic lime appears among the teaching topics at the *Landwirthschaft- und Gewerbe-Schule* in Aschaffenburg. In 1836 the

[34] "Mit inniger Freude erblikten wir bei der Ausstellung von 4 Ausstellern mehrere Fabrikate aus hydraulischem Kalke, die noch zu großen Erwartungen berechtigen, und jede mögliche Aufmunterung vom Staat und Publicum verdienen. Eisenhändler J. G. Schmidt von München stellte nicht bloß Cementpulver aus, sondern auch mehrere Fabrikate aus solchem, worunter sich namentlich ein Wasserbehälter sehr vorteilhaft auszeichnete. G. Reihl in München hatte 4 Schleifsteine aus hydraulischem Kalke und Quarzsand verfertigt, die vielleicht noch einstige Vervollkommnung zulassen. Hr. J. Karlinger in Miesbach endlich sandte mehrere Büsten aus hydraulischem Kalke verfertigt" (Anon., *Die im November 1834 zu München gehaltene Industrieausstellung*, in, "Polytechnisches Journal," LIV, 1834, (pp. 393–427) p. 410). Joseph Karlinger has several kilns in Miesbach and burns hydraulic lime using coal that he quarries in a site close to his plants; he later develops and patents a method to produce pipes made of poured hydraulic lime-based mortar. On this matter, see Anon., *Beyträge zur Geschichte des Bergbaues auf Braun- und Steinkohle im Königreich Bayern*, in, "Kunst- und Gewerbeblatt des Polytechnischen Vereins

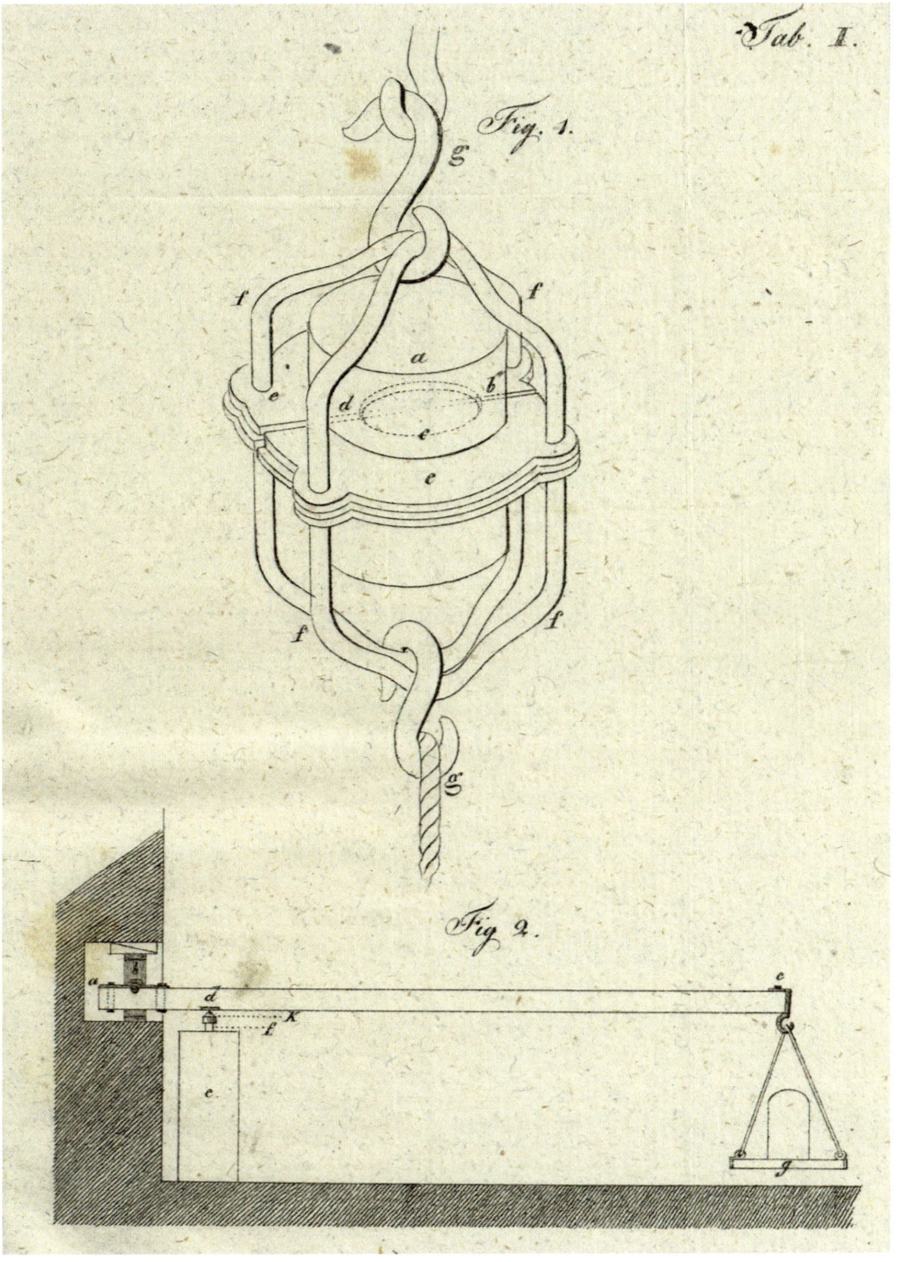

für das Königreich Bayern," vol. XXVI, 1840, (coll. 236–268), coll. 263–264; Anon., *Vaterländische Notizen und Briefe*, in, "Bayerische National-Zeitung," 1838, n° 192, p. 780.

[35] "Dank sey hier herzlich wiederholt unserem hochverdienten Hofrathe und Akademiker Dr. Fuchs, dem wir die wissenschaftliche Begründung der Lehre vom hydraulischen Kalke verdanken, und der sich auch um die immer weiter greifende Anwendung desselben in unserem Vaterlande so hochverdient gemacht" (Anon., *Die im November 1834 zu München*...cit., pp. 410–411).

[36] See Émiland–Marie Gauthey, *Traité de la Constructions des Ponts*, ed. Claude-Louis-Marie-Henry Navier, vol. I, Paris, Firmin Didot frères, 1832, p. 199, pl. XIII.

[37] See Panzer *Ueber das Vorkommen*...cit., pp. 24–38.

dean of the school, naturalist Martin Balduin Kittel (1798–1885), writes and publishes a pamphlet in which, with reference to Fuchs and Panzer, he traces all essential knowledge of that time about hydraulic lime. He furthermore describes Panzer as "the first one in Bavaria who really seizes and develops this discovery in the domain of his activity," and whose "commitment made it possible to establish natural cement plants in Würzburg, Schweinfurt and Bamberg."[38]

Cement manufacturing in Northwest and central Germany, from Hamburg to Kassel

Around 1829, two Hamburg entrepreneurs, Shipmann and Hester, start on cement manufacturing. They import marls from England, which are burnt in a common lime kiln and ground in a mill.[39] In 1831, Hamburg hydraulic engineer Reinhard Woltmann (1757–1837) writes a glowing report about Shipmann's and Hester's cement. "In hydraulic engineering," Woltmann asserts, "Hamburg cement has the same good properties as the best English Roman cement, and both cements are better than trass or *tarras*, since mortar prepared using this cement, and as little water as possible, remains cohesive underwater, gains hardness over days and weeks, and becomes solid like a stone, making it almost impossible to scrape it with a knife. This kind of mortar hardens in a few hours and can be used inside and outside buildings, in moist and in dry places. Instead, trass–based mortar falls apart if it is poured into water immediately after having been prepared, and it is not effective if it is used in dry places. It hardens like a stone only in moist places, in foundations, in basements and in retaining walls."[40] Woltmann also recommends using cement-based mortar to build "concrete walls, made of every kind of rubble." This is quite a pioneering idea at that time, as concrete is rarely used and is essentially prepared with trass.[41]

In the early 1830s, further cement kilns are founded in the Frisian villages of Carolinensiel and Emden, in Hamelin south of Hannover, in Barbis near the Harz Mountains, in Lerbeck near Minden, and in Kassel.[42] The owner of the cement kiln in Emden is [Burchard Wilhelm] Rodewyk, who defines the

[38] "Er war der Erste in Bayern, welcher die Entdeckung richtig aufgriff und im Bereiche seines Wirkungskreises mit Erfolg in das Leben führte. Auf seine Bemühung entstanden in Würzburg, Schweinfurt und Bamberg Niederlagen von hergerichtetem natürlichem Cämente, welcher nichts zu wünschen übrig läßt" (Martin Balduin Kittel, *Der hydraulische oder Cäment-Kalk aus der Umgegend Aschaffenburgs und dessen Benutzung*, Aschaffenburg, M. I Wailandt's Wittib und Sohn, 1836, p. 3).

[39] See Reinhold, *Sammlung practischer Erfahrungen*...cit., pp. 94–96; J. Cohn, *Ueber die Wichtigkeit der Cemente: in Beziehung auf gesunde Wohnungen, dauerhafte Wasser- und Prachtbauten, Kunststein, Anlegung von Silo's, so wie auf Fabrik-Industrie; nebst Vorschlägen zur Gründung fester Steinufer als Bedämmung gegen Stromgefahr, so wie zur Nachforschung auf inländische Puzzolane oder Trass*, Breslau, Druck der Stadt- und Universitäts-Buchdruckerei von Grass, Barth & Comp., 1855, p. 10.

[40] "Dieser Hamburgische Cement hat zum Wasserbau gleich gute Qualität mit dem besten Englischen Patent–Cement, und beide übertreffen den Trass oder Tarras darin, dass sie, zum Mörtel mit so wenig Wasser als möglich zubereitet und unverzüglich ins Wasser geworfen, ihre Cohärenz behalten, selbige schnell zunehmend vermehren und in einigen Tagen und Wochen dergestalt versteinern, dass man mit dem Messer kaum noch etwas davon abschaben kann; auch ist dieser Mörtel überall, an feuchten und trocknen Orten, unter und über der Erde, im Innern wie im Aeussern der Gebäude, vortrefflich, und erhärtet in einigen Stunden; der Trass-Mörtel hingegen, frisch ins Wasser geworfen, fällt auseinander, ist auch in freier Luft nicht dauerhaft; nur an feuchten Orten, in Fundamenten und Kellern, Futter- und Vorsetzmauern u.s. w. erhärtet er allmälig zu Stein" (Reinhard Woltmann, *Nachricht über den Cement aus der Fabrik der hiesigen Bürger Schipmann et Hester*, in Reinhold, *Sammlung practischer Erfahrungen*...cit., p. 94).

[41] "Béton-Gemäuer von allerlei Steinbrocken" (*ibid.*).

[42] See Reinhold, *Sammlung practischer Erfahrungen*...

cement he produces as "Römischer Cement," since, he explains, "it is in every respect similar to the product that is fabricated in England."[43] Rodewyk also highlights the relation between hydraulic lime and cement. "Cement has the same properties," he writes, "as a high-quality hydraulic lime, that is to say, a kind of lime that turns into a mush once mixed with water and with sand in some cases, hardens underwater and above ground, becomes like a good sandstone and gains hardness over time."[44] Rodewyk, as usual, suggests using cement to compose hydraulic mortar for masonry exposed to water or moisture, and to produce mortar-based artificial stones. "Cement mixed with sand and water," he explains, "can be used to fabricate objects of whatever form, instead of using cut stones."[45]

The cement manufacturing in Kassel starts on the initiative of Ernst Friedrich Martin Koch (1786–1860), a gunpowder producer, who undertakes the production of building materials after the end of the Napoleonic wars, in view of increasing post war building activities.[46] Around 1817, Koch builds a stamp mill to crush and grind chalk, which he hopes to sell for the restoration works of the Kassel city castle, the so-called *Kattenburg*, which a fire had damaged during the period of the French occupation. As the restoration work on the *Kattenburg* stops in 1821, following the death of Prince Wilhelm I, who had promoted it, Koch assumes other manufacturing activities. In 1823, he rents a mill for fertilizers and, shortly after, he joins the *Ziegel- und Braunkohlwerk Mönchenberg*, a company that exploits local coalmines and burns bricks from local lime deposits. The brothers Johann Werner (1782–1850) and Carl Anton Henschel (1780–1861) own the company. Together with their father Georg Christian Carl (1759–1835), the two brothers also own a foundry, for which Carl Anton later conceives a dome whose construction plays an important role in the spread of cement use in Kassel.[47]

Around 1825, mining engineer Adolph Schwarzenberg (1799–1864) discovers marl deposits in *Mönchenberg*, close to the clay deposits that are exploited for the brickyard. He announces this discovery in the "Landwirthschaftliche Zeitung für Kurhessen" in 1825, but Mönchenberg marl remains unexploited

cit., p. 89; Fr. von Reden, *Bericht über die von dem Gewerbe-Vereine für das Königreich Hannover in den Monaten Mai und Juni 1835 veranstaltete erste Ausstellung Inländischer Gewerblicher Eszeusnisse*, in, "Mittheilungen des Gewerbe-Vereins für das Königreich Hannover," 1835 (coll. 527-529); Karin Kraus, *Hydraulische Bindemittel im 19. Jahrhundert auf dem Gebiet der heutigen Bundesländer Hessen, Rheinland-Pfalz, Saarland und Thüringen*, in, "Berichte des Instituts für Steinkonservierung e. V.," n° 43, 2012). Mr. Wode is the owner of the factory in Barbis, while the factory in Carolinensyhl belongs to a certain Mr. Timmen.

[43] "Derselbe wird deshalb Römischer Cement genannt, weil er in jeder Beziehung demjenigen Fabricate gleich kommt, welches in England bereitet und in den Handel gebraucht wird" ([Burchard Wilhelm] Rodewyk, *Ueber die Anwendbarkeit und die Art des Gebrauches des sogenannten Römischen Cementes*, in Reinhold, *Sammlung practischer Erfahrungen...* cit., (pp. 90–93), p. 90).

[44] "Die Eigenschaften des Cements sind die eines hydraulischen Kalks der trefflichsten Art, das heisst, eines Kalks, welcher, mit Wasser zu einem dicken Brei angerührt, sowohl ohne als auch mit Sandzusatz unter Wasser nicht weniger schnell wie an der Luft, zu einer steinartigen Masse erstarret, sehr bald die Härte eines guten Sandsteins gewinnt, älter werdend fortwährend an Härte zunimmt" (Rodewyk, *Ueber die Anwendbarkeit*...cit., p. 90).

[45] "Endlich kann man aus dem Cements, mit Sand und Wasser gemengt, steinartige Körper von jeder beliebigen Form bilden, welche die Stelle von behauenen Steinen vertreten" (Rodewyk, *Ueber die Anwendbarkeit*...cit., p. 91).

[46] See Albrecht Hoffmann, *Ernst Koch (1786-1860). Ein Pionier der hessischen Zementindustrie*, in, *Jahrbuch 2010 Landkreis Kassel*, Kassel, 2010, pp. 69–74; Harm-Hinrich Brandt, *Die Industrie- und Handelskammer Kassel und ihre Vorlaeufer, 1763–1963*, Kassel, Industrie- und Handelskammer Kassel, 1963, pp. 78–79.

[47] Johann Werner Henschel (1782–1850) founds the *Ziegel- und Braunkohlwerk Mönchenberg* around 1820 and his brother Carl Anton joins the company in 1827. In this respect, see Sigfried Lotze, *Beiträge zu einer Geschichte hessischer Unternehmer im frühen 19. Jahrhundert. Die Familienbeziehungen niederhessischer Textil- und Farbenfabrikanten*, in, "Zeitschrift des Vereins für hessische Geschichte und Landkunde," vol. XCVI, 1991, (pp. 233–254), pp. 235, 249, n. 12.
About Carl Anton Henschel, see Conrad Matschoß, *Grosse Ingenieure. Lebensschreibungen aus der Geschichte der Technik*, München, J. F. Lehmanns, 1954, pp. 161–73; Brandt, *Die Industrie- und Handelskammer Kassel*... cit., pp. 75–77; Kurt Ewald, *Henschel, Carl Anton*, in, *Neue Deutsche Biographie*, vol. VIII, Berlin, Hartmann Heske, 1969, pp. 553–554.

until the fall of 1832, when Koch decides to move from chalk to cement production, after having received a positive assessment by Carl Anton Henschel and by Johann Conrad Bromeis (1788–1855), who is the head of the *kurhessische Bauverwaltung* (Hessian Building Office).[48] In 1833, Koch writes an explanatory leaflet about cement and sends a copy of it to all Hessian Building Offices.

Almost contemporary with the beginning of Koch's cement production is the foundation of the cement plant in Hamelin, on the initiative of engineer Georg Dietrich Wendelstadt (1790–1860) and banker Adolph Meyer (1807–1866). The plant is built on the western side of the River Weser. At first it only consists of a single burning chamber kiln and a steam-powered mill.[49] The production starts in 1833, following the release of a decade-long exclusive license of King Wilhelm IV.[50] On June 1833, Meyer writes a letter to the local land magistrate, soliciting the ban on the exportation of raw stones suitable for producing cement, to reserve them for the cement plant. Clearly, the production of cement using local raw materials is running full speed at that time.[51] The factory grows during the 1830s. It is provided with a second and more modern kiln, with separate combustion chamber.[52] In 1835, Wendelstadt und Meyer participate in the first *Ausstellung inländischer gewerblicher Erzeugnisse* (Exhibition of inland manufacturing products) organized by the *Gewerbe-Vereine für das Königreich Hannover* (Society for Manufacturing and Trade in the Kingdom Hannover), where they are awarded a silver medal for the quality of the cement they produce and for having contributed to

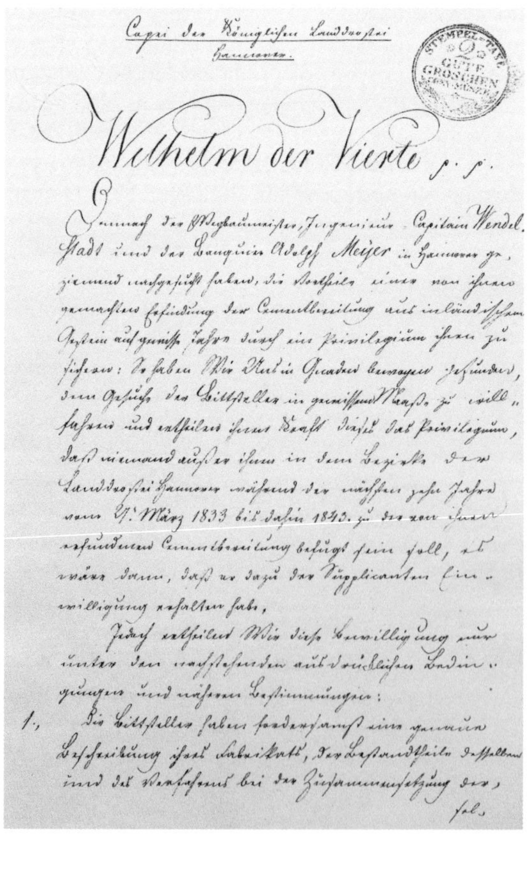
Wilhelm der Vierte

33. Weldelstadt and Meyer's licence for the manufacturing of cement, 1833 (NLH, Hann. 80, n. 15903, n. fol.).

[48] See Hoffmann, *Ernst Koch...* cit., p. 70.

[49] See Silke Schulte, *Geschichte der Stadt Hameln*, in, Niedersächsisches Ministerium für Inneres, ed., *Niedersachsenbuch 2009. Rattenfängerstadt Hameln*, Hameln, C. W. Niemeyer, 2009, (pp. 24–32), p. 29; Friedrich Wilhelm Otto Ludwig Freiherr von Reden, *Amtlicher Bericht über die siebzehnte Versammlung der deutschen Naturforscher und Aerzte zu Pyrmont im September 1839*, in, Lorenz Okenfuß, ed., *Isis oder Encyclopädische Zeitung von Oken*, 1840, (coll. 928–930), col. 929.

[50] See *Königligles Privilegium zu Cementbereitung* (NLH, Hann. 80, n° 15903). See also Von Reden, *Bericht über die von dem Gewerbe-Vereine für das Königreich Hannover in den Monaten Mai und Juni 1835 veranstaltete erste Ausstellung*...cit., col. 528; Id. *Das Königreich Hannover statistisch beschrieben, zunächst in Beziehung auf Landwirtschaft, Gewerbe und Handel*, Hannover, Hahn'sche Buchhandlung, vol. I, 1839, p. 464; Id., *Amtlicher Bericht über die siebzehnte Versammlung der deutschen Naturforscher und Aerzte zu*

the development of a promising manufacturing branch. "Cement (N°1644) by Wendelstadt und Meyer," declares the exhibition's report, "has generally proved to be particularly good; it is equal to the English cement but more affordable. For this reason and for having opened the way to a new manufacturing branch, Wendelstadt and Meyer are awarded a silver medal."[53]

Brunkhorst und Westphalen's lime kiln in Buxtehude begins to burn cement from English marls in 1834, in addition to grinding trass from stones imported from Holland and fat lime from shells and Rüdersdorf limestones.[54] The production grows in the following years and the plant is provided with a second mill at the turn of the 1840s. In 1842, the year of the great fire of Hamburg, after which huge quantities of building materials are needed to rebuild a great part of the city, the plant is producing 60 barrels of cement a day.[55]

Further production of hydraulic lime and the persistent fascination with English Roman cement in Prussia

At the turn of the 1830s, Prussia seems torn between the persisting ambition to reproduce English Roman cement and the actual development of hydraulic lime manufacturing. In the early 1830s, after the pioneering experiences that arise around the Werbellinsee, hydraulic lime begins to be produced in the town of Oppeln, Silesia, exploiting marl discovered in a place called Moritzberg.[56] Lime from Oppeln is moderately hydraulic and is slaked by pouring water on burnt

Pyrmont...cit; Ernst Meyer Hermann, *"Leicht wie ein Gedanke". Erste Große deutsche Kettenbrücke: 1836/39 bei Hameln erbaut*, in, "Die Weser," XLIV, 1970, n. 1, (pp. 48–49), p. 48; Hort Knobe, *Hamelner Wasserbauwerke an der Weser. Die Geschichte der Schleusen und Wehre, der Münsterbrücke und des Hafens*, Bielefeld, Verlag für Regionalgeschichte, 2003, p. 119.

51 See Adolph Meyer, letter to the *Königlicher Großbritannischer Hannoverscher Landdrost zu Hannover*, June 27, 1833 (NLH, Hann. 74, Hannover, Hameln n° 3102).

52 See Von Reden, *Amtlicher Bericht über die siebzehnte Versammlung der deutschen Naturforscher und Aerzte zu Pyrmont*...cit.

53 "Der von HH. Wendelstadt und Meyer auf die Ausstellung gebracht Cement (N° 1644) hat sich, nach ziemlich allgemeinen Urtheile, als besonders gut bewährt; er ist dem Englischen vollkommen gleich zu stellen, und sein Preis sehr mäßig. Deshalb, so wie in Betracht, daß jene Herren dem Königreiche einen neuen nicht unwichtigen Industrie-Zweig zugewandt haben, ist denselben die silberne Medaille zuerkannt" (Von Reden, *Bericht über die von dem Gewerbe-Vereine für das Königreich Hannover in den Monaten Mai und Juni 1835*...cit., col. 529).

54 See "Fragen, welche von Seiten der auswärtigen Mitglieder des Gewerbe-Vereins den einzelnen Besitzern der Fabriken und von Gewerbe-Anstalten zur Beantwortung vorgelegen sind," 1834 (Stadtarchiv Buxtehude, Rep. HG 28,28, n. fol.); "Königliches Ministerium des Innern. Resolution," August 9, 1835 (Stadtarchiv Buxtehude, Rep. HG. 29,18, n. fol.); Anon., *Zusammenstellung der Nachrichten über die seit 1838 im Königreiche Hannover neu entstanden oder wesentlich vergrößerten und verbesserten Fabriken*, in, "Mittheilungen des Gewerbe-Vereins für das Königreich Hannover, 1842-1843," 1843, (coll. 325–408), coll. 399–400; Reinhold, *Sammlung practischer Erfahrungen...* cit., p. 89; Heike Linderkamp, "Die Industrialisierung in Buxtehude. Anfänge und Entwicklung bis zum Beginn des 20. Jahrhunderts," Buxtehude, Buxtehude Stadtarchiv, 1992. Tuff stones are exported from Rhineland to Holland via the Rhine and then imported via the River Weser or via Hamburg to Niedersachsen, due to the lack of railways and proper waterways between Rhineland and Niedersachsen.

55 Anon., *Zusammenstellung der Nachrichten über die seit 1838 im Königreiche Hannover neu entstanden oder wesentlich vergrößerten und verbesserten Fabriken*...cit., col. 400.

56 Today Opole, in Poland.

stones laid on sieves, so that slaked lime falls through nets as a powder, which is then usually mixed with trass or brick powder. During the 1830s, lime from Oppeln is used to build the masonries of several hydraulic works in Oblau, Breslau and Brieg, and it is known even in Berlin.[57]

While the development of local hydraulic lime manufacturing is in progress, the Prussian *Gewerbverein* promotes a new competition in 1833 to urge, once more, the development of inland hydraulic cement with properties similar to the ones of English Roman cement.[58] This competition has similarities with the ones promoted in 1822 and 1829, except that its reference to Roman cement is openly declared. Shortly thereafter, the Prussian *Ober-Bau-Deputation* organizes a series of tests in Berlin between 1834 and 1835 aiming to shed light on the effectiveness of the principal materials available in Prussia for producing hydraulic mortar, namely trass from Cologne, hydraulic lime from Joachimsthal, hydraulic lime from Oppeln and Roman cement from England.[59] Four different kinds of hydraulic mortar are prepared using the aforementioned materials as essential components. Trass is mixed with lime from Rüdersdorf in equal measures; lime from Oppeln is mixed with 1 and ¼ quartz-sand and ¼ brick powder; Roman cement is mixed in equal measures with quartz-sand. Each kind of mortar is used to bind samples of yellow bricks from the village of Glindow and red bricks from Rathenow, which are then put into wooden boxes so that, between the brick samples, two different free compartments remain and are filled with mortar and crushed stones (sandstone, granite and limestone), forming a sort of coarse concrete. Once filled, the four boxes are placed underwater in the River Spree, where they remain for one and one-half years. In addition, samples of the four different kinds of mortar are laid on a wall exposed to moisture, where they remain as long as the boxes stay underwater. On September 1835, the samples are checked and the four different kinds of mortar are assessed for their use both underwater and as plaster.[60]

The constant search for raw materials suitable for producing hydraulic lime and cement in Prussia produces further positive consequences in the second half of the 1830s, and results in the foundation of some new lime and cement kilns. One is founded in 1836 in Tarnowitz, north of Oppeln, while a further kind of limestone appropriate for hydraulic lime is discovered in Schimntiz, south of Oppeln.[61] In 1837, a kind of lime that slakes very slowly is produced with limestone located on a small hill near Rüdersdorf, called Krienberg.[62] Although chemical analyses prove that this limestone contains the right substances for cement, manufacturing does not start in earnest due

[57] Feller, *Bekanntmachingen vermischten Inhalts. Baumaterialien-Lieferung*, in, "Oeffentlicher Anzeiger als Beilage des Amtsblatts der königlchen Regierung zu Oppeln," 1834, (pp. 618–619), p. 619; Carl Hoffmann, *Notizen über den Kalk vom Moritzberge bei Oppeln, und dessen Verwendung zum Mörtel bei Wasserwerken zu Breslau*, in, "Notiz-Blatt des Architekten-Vereins zu Berlin," 1839, pp. 18–19.

[58] See *Preisaufgaben für die Jahren 1833–1834. Erste Preisaufgabe betreffend die Anfertigung eines hydraulischen Cements aus inländischen Materialien von gleicher Güte als der Englische Roman Cement*, in, "Verhandlungen des Vereins zur Beförderung des Gewerbfleißes in Preußen," XII, 1833, pp. 30–31.

[59] See Anon., *Auszug aus den Protokollen der Versammlung des Vereins in den Monaten Novemebr und Dezember d. I. J.*, in, "Verhandlungen des Vereins zur Beförderung des Gewerbfleißes in Preußen," XIV, 1835, (pp. 277–280), pp. 279–280; Brunsch, *Die historische Verwendung...* cit., pp. 61–66.

[60] Mortar made of trass and lime from Rüdersdorf appears as the best binder underwater, followed by mortars composed with Joachimsthal lime, Roman cement and Oppeln lime. The latter seems, instead, to be the best binder for plastering wet walls, followed by Roman cement and Joachimsthal lime. On this matter, see Brunsch, *Die historische Verwendung...* cit., pp. 61–66.

[61] See Cohn, *Ueber die Wichtigkeit...*cit., p. 10, n. 1; Hoffmann, *Notizen über den Kalk vom Moritzberge...* cit., p. 19. Tarnowitz cement is used for the building of the fortresses of Neisse and Posen.

[62] See Anon., *Notiz über den in Moabit angefertigten Cement: nach einer Mittheilung der Herren Haslinger und Schondorff*, in, "Notiz-Blatt des Architekten-Vereins zu Berlin," I, 1847, pp. 12–13.

to the lack of proper installations. After having sparked the interest of the *Stadtbaurath* Langerhans in 1839, limestone from Krienberg is once again considered in 1841, as the entrepreneurs Haslinger and Schondorff, who build a kiln in Moabit based on English models, perform new tests comparing Krienberg cement with Roman cement imported from England. The two products appear similar; Krienberg cement hardens a little slower than Roman cement, but it is considered good enough to be mixed with twice as much sand as Roman cement, making significant savings possible. Haslinger and Schondorf start manufacturing cement in Moabit and succeed in selling it for the construction of important new buildings in Berlin and in the vicinity, such as the Berlin Cathedral, the *Neues Museum* by Friedrich August Stüler (1800–1865) and the *Schloß Babelsberg* in Potsdam.[63]

Mortar, plaster and concrete, experimental uses of Kassel and Hamelin cement, and the attempts to build continuous masonry elements

The earliest applications of Koch's cement in Kassel are disappointing. They are actually quite simple and mainly concern restorations of roofs and walls, but the lack of expertise causes several failures. The architect Johann Daniel Wilhelm Eduard Engelhard (1788–1856) later reports on this false start, which also involves some of his own construction. "People who had trusted cement," he relates, "were so disappointed that a general discontent about it arose. Many failed reparations of walls, roofs, etc., were to be seen in Kassel. I myself used cement to whitewash the wet façade of a building, but, less than two years later, cement fell off. I cladded a Corinthian pediment with stone plates fixed and plastered by means of cement, but, during the same summer I did it, parts of the cladding-work cracked, causing danger to pedestrians. And this happened despite having chosen workers with care, instructing them and overseeing their work. As a consequence, nobody wanted to hear about cement any longer and some people complained openly about having spent considerable amounts of money in vain."[64]

Probably in response to early failures, Koch manages to get a positive assessment in an article dealing with hydraulic lime, which is published in 1833 in the "Annalen der Pharmacie," a periodical edited by the chemist Justus von Liebig (1803–1873). "Mr. Koch's cement," the text goes, "has all the properties that a binder should have. It forms real artificial clinking stones, a sort of regenerated limestone, and it has the same quality as the celebrated Roman cement [...]. In addition to the huge masonry works underwater for which Koch's cement has already been successfully employed, we have observed big

[63] "Der Krienberg-Cement ist übrigens in Berlin schon angewendet worden: Beim neuen Dom, dem neuen Museum, beim Normalkrankenhaus, der neuen Ulanen-Kaserne, dem Zellen-Gefängniss, dem Artillerie-Laboratorium, dem Königl. Stadtgericht, dem Friedrich-Wilhelms-Hospital, bei der Städtischen Gasanstalt; ferner beim Schlossbau zu Babelsberg, zu verschiedenen Bauten der Berlin-Hamburger Eisenbahn, und noch mehreren grossen Privatbauten" (*ibid.*, pp. 12–13).

[64] "Das Publicum fasste nun schon einiges Zutrauen zu dem Gegenstande, fand sich aber doch vielfältig in seinen Erwartungen getäuscht, so dass die Cementconstructionen fast zum öffentlichen Ärgerniss wurden. Man sah an vielen Gebäuden in Cassel misslungene Ausbesserungen an Mauern, Wänden, Dächern u. s. w. Auch ich hatte an einem Gebäude, mit einer Plateform, die feuchte Frontmauer mit Cement tünchen lassen, musste aber sehen, dass die Tünche im zweiten Jahre wieder abfiel. Die Steinplattenbedeckung eines Ziemlich grossen Frontons corinthischer Ordnung hatte ich mit Cement überziehn und dichten lassen, und in demselben Sommer löseten sich ganze Stücke derselben ab und droheten auf die Vorübergehenden zu fallen; gleichwohl hatte ich die Werkleute zu diesen beiden Arbeiten sorgfältig gewählt und sie mit ängstlicher Genauigkeit dazu angewiesen und überwacht. So kam es denn, das hier Niemand mehr etwas von dem Cement wissen wollte und Mancher eine beträchtliche vergebliche Ausgabe zu beklagen hatte" (Johann Daniel Wilhelm Eduard Engelhard, *Über den Casseler Cement*, in, "Journal für die Baukunst," vol. XX, 1844, (pp. 165–188, 189–198), pp. 166–167).

water containers made of bricks put on edge and bonded with this cement, which have been now full of water for a long time and remain watertight, as if they were carved from single pieces of stone. When we have tried to break masonry assembled with this cement, cracks arose in the stones earlier than in the cement joints; the cement mass had formed a sort of continuum with the stones."[65] Such enthusiastic words are probably not entirely reliable. In any case, they directly contradict Engelhard's memories, and Engelhard is far from being a detractor of Koch's cement. To the contrary, he has high expectations for it. Having observed applications in pozzolana-based mortar during a research journey to Italy in 1809, Engelhard hopes that modern cement could replace all traditional cementitious mixtures.[66] "If this material proves to be good," he reasons, "a gap in our current way of building could be filled, since mixtures of lime and brick powder are very defective, and trass and pozzolana are hardly, if at all, available in Germany."[67] Engelhard is particularly fascinated by the Roman way of building aqueducts made of bricks, tuff and a mixture of gravel and mortar, and by what remains of plaster and cement coverings on the side columns of the Temple of Fortuna Virilis in Rome, where he furthermore observes the use of pozzolana-based mortar for the rebuilding of the pavement of Ponte Sisto. "First a roughly one-foot high basement of tuff stone was laid in pozzolana and lime," he relates, "then, this basement was covered with 8 inches wide and long cobbles, placed very close to each other in a second bed of lime and pozzolana, to form the real pavement."[68]

In 1835, with the aim of improving the reputation of the cement he produces, Koch writes a pamphlet of instructions for applying cement. Three years later, the pamphlet will be published as a book, serving as one of the earliest examples of German manuals about cement. In 1836, Koch asks the *Hessian Handel- und Gewerbeverein* (Hessian Society for Manufacturing and Trade) for an official inspection of the cement-based whitewashing, plaster and mortar applied during the construction of Schwarzenberg's private house between 1834 and 1835. A committee is set up for that purpose, gathering the major local experts about technical matters, namely Engelhard, Arzenberg, the Henschel brothers and Heinrich Bluff (1805–1878), who is the editor of the "Gewerbeblätter für Kurhessen," a periodical edited by the *Hessian Handel- und Gewerbeverein*. The inspection produces satisfactory results, and

[65] "In der That leistet das Cement des Herrn Koch Alles, was man nur von einem Kitt verlangen kann. Es bildet einen wahren künstlichen klingenden Stein, einen regenerirten Kalkstein, und kommt in seinen Eigenschaften vollkommen mit dem berühmten romischen Cement überein […]. Aufser den gröfseren Mauerungen unter Wasser, bei denen das Koch'sche Cement angewendet war und sich vollkommen bewährt erwiesen bat, haben wir z. B. grofse Wasserbehälter gesehen, welche aus auf die schmale Kante gestellten und mit diesem Cement verbundenen Ziegeln (Backsteinen) zusammengesetzt waren, die, schon seit langer Zeit mit Wasser angefüllt, so wenig die geringste Menge Wassers durchliefsen, wie ein aus einem ganzen Stein gehauener Trog. Als wir versuchten, ein, mit diesem Cement verbundenes Mauerwerk zu zerschlagen, so entstanden die Bruche eher in den Steinen selbst, als in den Cementfugen; die Masse des Cementes hatte mit den Steinen gleichsam ein Continuum gebildet" (Anon., *Ueber den hydraulischen Kalk*, in, "Annalen der Pharmacie," vol. V, 1833, (pp. 241–246), pp. 244–245).

[66] See Johann Daniel Wilhelm Eduard Engelhard, *Instruction für junge Architekten zu Reisen in Italien*, in, "Journal für die Baukunst," vol. XI, 1837, pp. 1–24, 155–174, 203–238, 303–333, esp. pp. 308–309, 312.

[67] "Bewährt sich dieses Material wirklich, so wurde eine bedeutende Lücke in unserem hiesigen Constructionssystem ausgefüllt sein. Die Mischung von Kalk und Ziegelmehl ist sehr unvollkommen, und Traß und Puzzolane haben wir in Deutschland wenig oder gar nicht" (Engelhard, *Über den Casseler Cement*…cit., p. 167).

[68] "Zuerst wurde über dem dazu erbauten Planum ein durchlaufendes Strafsenfundament von Tufsteinen in Puzzolana und Kalk gemauert; es mochten diese Lagen zusammen einen Fufs hoch sein; über diesem wurde sodann das eigentliche Steinpflaster, bestehend aus regulär-quadratischen, übereck stehenden, etwa 8 Zoll breiten und langen Steinen gesetzt, ebenfalls in Kalk und Puzzolana, aber sehr dicht aneinander gefügt" (Engelhard, *Instruction*…cit., p. 308).

[69] "Selbst bei Bauten an solchen Stellen mit größtem Vortheil verwendet werden kann, wo es entweder ganz im Trocken steht, oder abwechselnd den Einflüssen feuchter und trockner Witterung ausgesetzt ist" (Adolph Schwarzenberg, Carl Anton Henschel, Johann Werner Henschel, Johann Heinrich Buff, *Gutachten die Brauchbarkeit des*

it assessed that Koch's cement is suitable not only for hydraulic works, "but also for works to realize in places that are dry or alternatively dry and moist, as long as cement is carefully handled and mixed with an appropriate amount of sand, according to the kind of work to do."[69]

When Engelhard builds his own villa between 1837 and 1839, he makes wide use of cement. He builds with typical masonry in moist places, but he also carries out interesting construction experiments.[70] Firmly convinced of the perfect cohesion between stones and cement mortar, he builds bond-beams made of stone plates and cement mortar inside the bearing walls, at the base of each floor. He describes such construction as "cohesive stone-chains through the walls."[71] To improve the cohesion between bricks and cement mortar, he even creates special porous bricks made of clay and coal slivers, which he orders at the *Mönchenberg* brickyard and then uses to build the window corners.[72]

But the most interesting experiments that Engelhard carries out during the construction of his house probably concern two roof gardens. Engelhard admired flat roofs during his journey to Italy and wants to build similar ones in his own house while attempting to improve their impermeability. "Such constructions are realized in southern lands more frequently than in our country," he writes in this regard, "but this does not necessary mean that they are always well built. Quite the contrary, disadvantages of flat roofs are the same in northern and in southern lands, where moisture is even more unfavourable for health than in the North, but people do not care about it. I have seen many such roofs, several of which could be preserved for many years if carefully built and maintained; thus, I could not resist building something similar, while trying to overcome the imperfections I observed."[73] Engelhard first builds a flat roof by pouring a 1/3 inch thick layer of mortar composed of cement and sand on a load-bearing floor made of wedge-shaped stone plates, which form flat vaults between timber beams. The whole is then covered with clay and vegetable soil to arrange a garden. Apart from the garden arrangement, this kind of construction is similar to certain British floors of the late 18th century, which are made of flat brick vaults spanning timber beams, with the whole covered by sand and brick tiles.[74] The second flat roof that Engelhard builds consists of grooved timber beams, transversal iron-sticks and crushed bricks embedded between the sticks. Once this is

Koch'schen Cements betreffend. Protokoll 18ten Juni 1836, in, "Gewerbeblätter für Kurhessen," I, 1836, (pp. 137–39), p. 139).

[70] At that time, Engelhard's villa is known as the Engelsburg, because of its impressive size; see Ruth Stummann-Bowert, *Philippine Engelhard, geborene Gatter. Ein Bürgerliches Frauenleben zwischen Aufklärung und Empfindsamkeit*, in, Trauden Weber-Reich, ed., *"Des Kennenlernens werth". Bedeutende Frauen Göttingens*, Göttingen, Wallstein, 2002, (pp. 27–52), p. 49. Engelhard also reports having successfully built a container for effluent made of bricks and cement mortar in a public cowshed, at the insistence of a local mason, and despite previous failures; see Engelhard, *Über den Casseler Cement*...cit., p. 167.

[71] "Zusammenhängenden Steingurt durch jedes Stockwerk" (Engelhard, *Über den Casseler Cement*...cit., p. 175)

[72] It is worth mentioning that, basing on such strong confidence in the perfect cohesion between bricks and cement mortar, in 1837, a bearing wall at the ground floor of a house in Kassel was cut over a length of 13 feet, and replaced with an arch made of bricks and cement mortar; see Ernst Friedrich Martin Koch, *Gesammelte Erfahrungen über die Verarbeitung u. die verschiedenen Anwendungen des Cemente aus den Cementfabriken von Ernst Koch in Hessen-Kssel und Hanau*, Kassel, Leipizig, Theodor Fischer, 1838, pp. 17–18.

[73] "So Etwas freilich in südlichen Ländern öfter vorkommt, als hier; aber man muss nicht glauben, dass die Terrassen dort immer vollkommen gelungen wären; eher das Gegentheil. Man beachtet nur die Nachtheile davon dort nicht so sehr; wiewohl man im Grunde ebensowohl Ursach dazu hätte, als im Norden; denn Feuchtigkeit ist im Süden der Gesundheit noch nachtheiliger, als bei uns. Ich hatte viele dergleichen Terrassen gesehen, und mehrere derselben in Bau und Besserung Jahre lang erhalten; so dass ich dem Wunsche nicht widerstehen konnte, Etwas der Art hier machen zu lassen, und zu versuchen, die Nachtheile und Unvollkommenheit derer, die ich gesehen hatte, dabei zu vermeiden" (Engelhard, *Über den Casseler Cement*...cit., p. 182).

[74] See Bill Addis, *Iron in floor structures: the development of fireproof floors*, in, Gargiani, ed., *L'architrave*... cit., (pp. 492–498), p. 495, fig. 2.

achieved, cement mortar is poured over the entire construction. Together with the crushed bricks, it forms a sort of coarse concrete between the beams, while, in the upper part, a homogeneous and watertight base of cement mortar supports the garden. This complex floor structure seems to merge the building technique of floors made of timber beams and clay rolls (*Lehmpatzen*) with the building technique of the Italian flat roofs, having a concrete casting on boards fixed on beams. Furthermore, it also serves as a predecessor of the floors made of iron beams and concrete filling, whose construction develops in Germany in the 1870s. Despite the efforts to improve the waterproofness, Engelhard's floors are permeable because of cracks in the cement mortar. One year after they are finished, the suspended gardens must be dismantled and the floor structures are covered with tiles and limestone plates laid in cement mortar.

In addition to Engelhard's constructions, the dome of the new Henschel foundry in Kassel, which is conceived by Carl Anton Henschel and realized between 1837 and 1839, after a fire destroyed the previous building in September of 1836, deserves special attention for being a masterful example of construction exploiting the rapid and strong cohesion between cement mortar and bricks. Henschel designs a circular building made of brick masonry, covered with a dome and provided with two side wings on the back.[75] For the construction of the dome, he envisages using hollow clay pots and cement-based mortar, without resorting to any centring, but rather relying on the rapid setting of cement and on its strong cohesion with bricks. Describing Henschel's plans for the dome, Koch maintains that such vaults "become like single masses and therefore demand less care around the abutment."[76] Henschel's idea of building a masonry vault without any centring recalls the construction of the Thames tunnel, especially given that Henschel met Brunel during a journey to England between 1833 and 1834, just as Brunel was testing the cohesion of bricks and cement mortar through the experimental construction of two impressive half arches.[77] A reference for the use of hollow clay pots could be found, instead, in Schinkel's *Vorlägeblätter für Maurer* of 1834, which illustrates and describes fireproof ancient vaults made of hollow clay pots bonded with pozzolana based-mortar along with more recent ones made of pots and gypsum-based mortar, as they are usually built in Paris.[78] Henschel conceives special hollow clay-pots provided with grooves on the external circular surface in order to improve adhesion between bricks and cement mortar, and he orders such clay-pots in different diameter sizes, since the section of the dome masonry work is intended to decrease as building proceeds. Pots are laid tangentially to the curve; double curved hooks provided with a load on the lower side keep pots together during the mortar setting, while stirrups hold the mortar applied on the intrados until it set. In the absence of centring, the layout of the curve is shown by different length cables hanging from a metallic bar fixed on the top of the high pole in the centre of the building.[79]

[75] See Jos Tomlow, *Die Kuppel des Gießhauses der Firma Henschel in Kassel. Eine frühe Anwendung des Entwurfsverfahrens mit Hängemodellen*, in, "Architectura," vol. XXIII, 1993, pp. 151–171.

[76] "Da sich das Gewolbe fast zu einer einzigen Masse vereingt, so braucht man weniger auf die Stärke der Widerlagen zu sehen" (Koch, *Gesammelte Erfahrungen*...cit., p. 20).

[77] It is a construction consisting of a pier supporting an half arch on a side, and a quarter of arch on the other sides, both made of bricks, cement mortar and some iron bars in mortar, and built without centring; see Francis, *The Cement Industry*... cit., p. 48.

[78] See Schinkel, *Grundlage der praktischen Baukunst*...cit., p. 13, pl. XVIII.

[79] See Koch, *Gesammelte Erfahrungen*...cit., pp. 19–20.

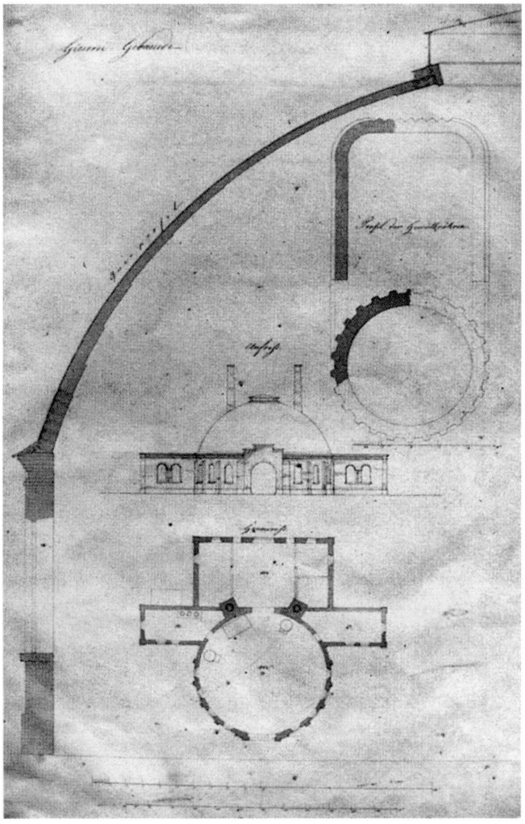

34. C. A. Henschel, foundry, Kassel, 1837–1839. First project, 1837 (courtesy of the *Museumslandschaft Hessen Kassel*)

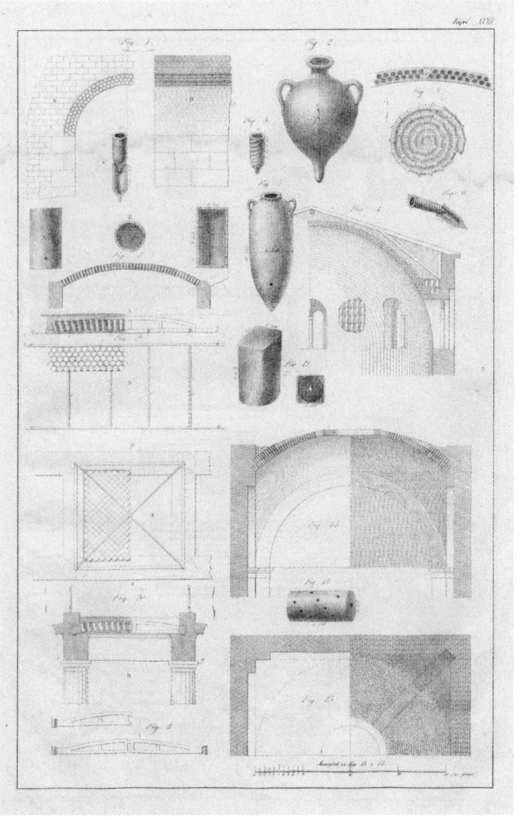

35. K. F. Schinkel, drawings of vaults made of hollow clay pots and different kinds of clay pots (Karl Friedrich Schinkel, *Grundlage der praktischen Baukunst*, vol. I, *Mauerwerkkunst*, Berlin, Schenk & Gerstäcker, 1834).

36. C. A. Henschel, provisional tools to hold hollow bricks and plaster during mortar setting (E. F. M. Koch, *Gesammelte Erfahrungen über die Verarbeitung u. die verschiedenen Anwendungen des Cemente aus den Cementfabriken von Ernst Koch in Hessen-Kassel und Hanau*, Kassel, Leipzig, Theodor Fischer, 1838.

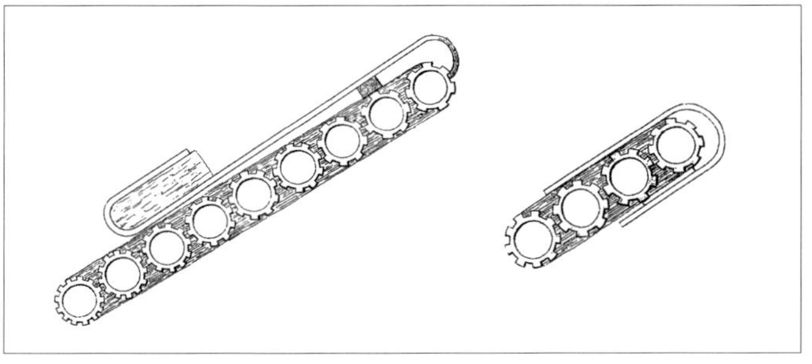

Despite all expectations, the dome collapses while still under construction.[80] Henschel therefore plans a new construction made of six alternate rings of bricks and hollow clay pots bonded with cement mortar. Unlike the previous ones, the hollow pots now have smooth surfaces and are laid perpendicularly to the tangents. The lowest ring is made of ordinary bricks and forms a sort of solid base for the further rings, which are made of hollow pots. The third ring is made of pots filled with cement mortar, probably to form a sort of solid belt approximately in the middle of the dome, and the fifth one is made of pots with half the diameter of the other pots. The last ring forms a crowning of two concentric circular rows of ordinary size hollow pots that support a lantern serving as a chimney.[81]

In the wake of its restored reputation, Koch's cement begins to be used even outside Kassel, and Koch founds a second cement kiln in early 1836, in Hanau, southern Hessen, partnering with the entrepreneur Friedrich Deines (1802–1876).[82] In 1837, the Thuringian *Baukondukteur* Lünzner, reports having experimented with Koch's cement on a request from the Erfurt *Gewerbverein*, when building four bridges over the River Lache, along the road from Weißensee to Kindelbrück. Lünzner describes having used pure cement mortar for the underwater masonry and mortar containing ½ part sand to each part cement for the above-ground masonry. Workers used winnowing fans to throw water against the above-ground masonry to improve the mortar hardening, a practice that later becomes customary. Lünzner expresses satisfaction with Koch's cement and even suggests using it to compose concrete for hydraulic foundations.

While Koch's cement is experimented with variously in Hessen and Thüringen, Wendelstadt and Meyer's cement is used for some small and large-scale applications in northwest Germany. In 1837, Carl Johann Elbers (1768–1845), the owner of a dye-works situated near Hagen in Rhineland, needs to repair a wooden container for bleaching powder.[83] He asks the director of the *Gewerbeschule* in Hagen, Goerthe, for technical information about the matter.[84] Goerthe suggests that he use Wendelstadt and Meyer's cement to plaster the internal surface of the container after driving iron nails with large heads to better fix the plaster work to the wood. Instead of following Goerthe's advice, Elbers builds a completely new container made of bricks assembled and plastered with cement-based mortar. Elbers uses mortar made of two equal measures of cement and sand to bind bricks, and pure cement-based mortar to plaster the internal surface of the container. The pure cement mortar shrinks too much and cracks, and Elbers needs to plaster the container twice, taking care to repair each crack that arises while the mortar is hardening. Once this is achieved, Elbers builds a container for oil using the same technique, and both constructions gain a certain distinction when Goerthe reports on them at the seventeenth assembly of the *Deutsche Naturforscher und Aertzte zu Pyrmont* (German Naturalists and Physicians in Pyrmont), which takes place in 1839. "The suitability of hydraulic lime for building

[80] See Koch, *Gesammelte Erfahrungen*...cit., p. 20; Gotthilf Heinrich Ludwig Hagen, *Ueber Form und Stärke gewölbter Bogen*, Berlin, Ernst und Korn, 1862, p. 25.

[81] Based on this project, the construction of the dome is successfully achieved in 1839.

[82] See Hoffmann, *Ernst Koch*... cit., p. 70.

[83] See Anon., *Amtlicher Bericht über die siebzehnte Versammlung*...cit., col. 938; August Zoller, *Stuttgart und seine Umgebungen*, Stuttgart, Köhler, 1841, p. 258. Elbers founds the dye-works near Hagen in 1822. It is a so-called *Türkischrotfärberei* (from *Türkischrot*, which means dark red). This kind of plant produces red wool clothes with yellow and withe patterns, called Bandanos. English clothes long dominates the German market; however, starting in the 1840s, German competition becomes significant.

[84] Beuth founds the *Gewerbeschule* in Hagen in 1824.

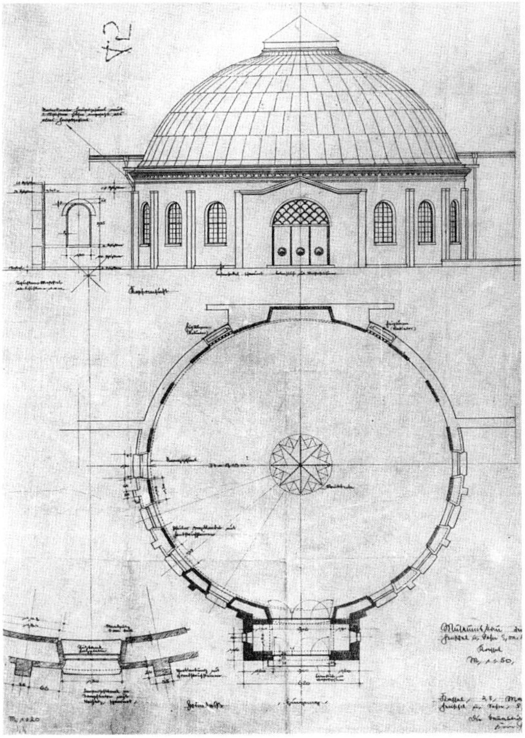

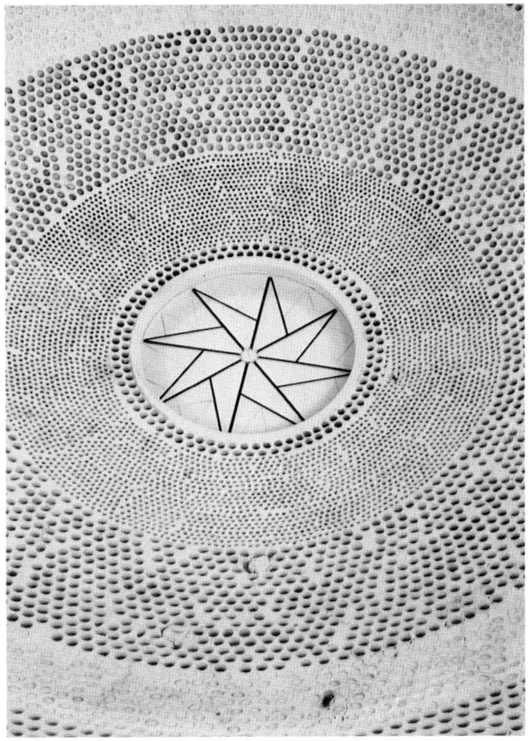

37-38. C. A. Henschel, foundry, Kassel, 1837–1839.

37. Second project, [1838] (courtesy of the *Henschel Museum*, Kassel).

38. View of the dome from inside, photo, n.d. (courtesy of the *Henschel Museum*, Kassel).

39. J. H. Martens, view of the Heschel factory with the foundry in the background, engraving, 1837 (*Denkschrift zum 100 jährigen Bestehen der Maschinen- und Lokomotivfabrik Henschel & Sohn, 1810-1910*, Cassel, 1910).

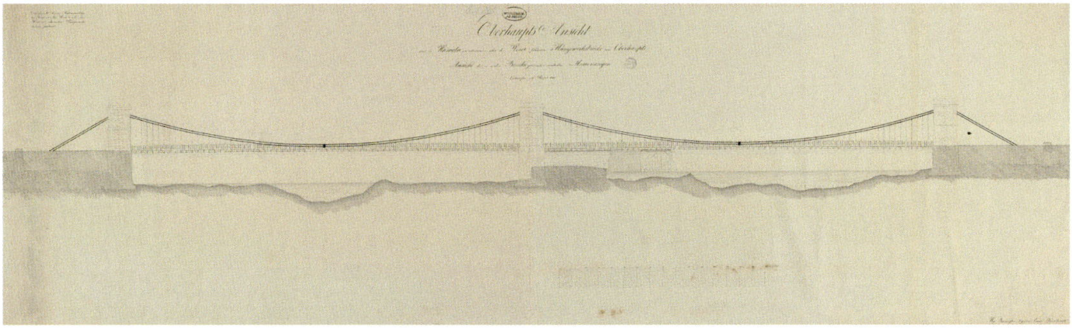

water containers is well known, and attempts at using hydraulic lime-based masses to manufacture water pipes have been already carried out in France," Goerthe asserts, "but hydraulic lime has not yet been used either for barrels, notably for those intended to contain bleaching liquids, or for oil containers." He continues, "Cement from the Weser, as well as trass from Andernach, Roman-Cement, and other similar materials can be recommended exactly for this kind of uses, seeing that such tools are more affordable when they are built using hydraulic lime than sandstone."[85] Aforementioned mining engineer Schwarzenberg, from Kassel, participates in the same assembly and discusses "lime, trass and other kinds of cement," mentioning successful cases of natural cement used to build "containers for brine, alum lye, salt, sodium chlorate and potassium chlorate."[86]

Wendelstadt and Meyer's cement is implemented more impressively in the building of the bridge over the River Weser in Hamelin. The establishment of Wendelstadt and Meyer's cement plant along the Weser is probably connected to the construction of this bridge, seeing that Wendelstadt is charged with the design of it in 1831. He conceives a suspension bridge

[85] "Die Benutzung des hydraulischen Kalks zu Wasserbehältern ist allgemein bekannt, selbst die Anfertigung von Wasserleitungsröhren aus dieser Masse ist in Frankreich versucht worden; aber Küpen, besonders zur Anfertigung von Bleichflüssigkeiten, und Oelbehälter möchten wohl früher nicht daraus dargestellt worden seyn, doch dürfte sich hierzu und zu ähnlichen Zwecken der Wesercement, eben so aber auch der Andernacher Traß, der Roman-Cement und andere diesen sich gleich verhaltende Stoffe sehr empfehlen, um so mehr, da die aus dem hydraulischen Kalk gefertigten Geräthe billiger zu stehen kommen, als solche aus Sandstein" (Anon., *Amtlicher Bericht über die siebzehnte Versammlung*...cit., coll. 939–940).

[86] "Bergrath Schwarzenberg fügt Bemerkungen über verschiedene Kalk-, Traß-, und Cementarten hinzu und führte Beyspiele an, wo der natürliche Cement mit Glück zu Gefäßen verwendet worden ist, in denen Salzlaugen, Alaunlaugen, Chlornatrium, behandelt werden sollten" (Anon., *Amtlicher Bericht über die siebzehnte Versammlung*... cit., coll. 940).

[87] See Thomas Telford, *An historical and descriptive account of the suspension bridge constructed over the Menai Strait in North Wales: with a brief notice of the Conway bridge*, London, Ibotson and Palmer, 1828, p. 24.

[88] For an exaustive account about the construction of the Hamelin Bridge, see Anon., "Zusammenstellung wichtiger Erscheinungen und Ergebnisse in Beziehung auf die bei der Hamelner Kettenbrückenbau gemachten Erfahrungen," [n.d.], (NLH, Hann. 109, n° 1069). See also Michael Mende, *"Not only a Matter of Taste but ... of the Laws of Mechanics." The Adoption of British Models in Nineteenth-Century Continental Suspension Bridge Design*, in, "Journal of Design History," vol. VI, 1993, (pp. 77–95), pp. 84–93.

40. G. D. Wendelstadt, the Suspension Bridge of Hamelin, 1833–1838. Elevation, 1833 (NLA Hannover – Plankammer, Kartensammlung, n° 13 b Hameln, pl. 3k, 4pg).

41. G. R. D. Osterwald, J. G. F. Poppel, the Suspension Bridge of Hamelin, engraving, [1850] (personal collection).

made of two stone masonry abutments and a central pier supporting two chain-suspended road decks. During the planning phase, Wendelstadt studies similar previous works and meets specialists who know about cement. An important reference is the Menai Suspension Bridge by Thomas Telford, the abutments of which are made of stone-faced walls filled with rubble masonry; the mortar is partly composed with hydraulic lime and the exterior joints are sealed with Roman cement mortar.[87] Furthermore, around 1833, Wendelstadt tests metal chains in the Henschel foundry in Kassel, in the presence of Carl Anton, who has just returned from England where he met Brunel, as previously mentioned.

The project for the bridge is accomplished in 1833, but construction does not begin before 1836, after the plans have been approved. The foundations stand straight on the soil and consist of stone masonry bonded with pure cement mortar.[88] The free spaces between ashlars and sheet piles around the foundation pits are filled with crushed stone and cement mortar, which can be seen as a sort of rudimentary coarse concrete. The masonry work of the abutments and of the pier consists of stone-faced walls bonded with pure cement mortar and a filling made of unrefined stones and inferior-quality mortar called *Cementstein-Asche*. This is made of slags from the cement production, namely bad burnt cement stones and ash ground into powder, and it is also used to build a 57 feet long retaining wall close to the central pier.

The foundations of the abutments for the chain anchorages on both sides of the bridge need special attention, since they consist of beds of concrete poured on a clayish and almost unsteady soil. The concrete is made of crushed stones, slaked lime and *Cementstein-Asche*. In addition to the construction of the major bearing parts of the bridge, cement-based mortar is used to build the drawing floor, similarly to what Panzer had done to build the small bridge in Schönungen.

chapter four
concrete foundations, walls, artificial stones

Concrete foundations in regions along the Rhine, the persisting use of trass

In the wake of the theoretical and practical advancements of the first third of the century, the use of concrete to build hydraulic foundations gradually spreads starting in the mid-1830s, against the backdrop of a favourable scenario. Manufacturing, mining activities and enhanced trade are boosted, inter alias, by the construction of the early railways and by the institution of the *Zollverein* (Customs Union). As a consequence, several new hydraulic works need to be built, namely foundations for docks, bridges and locks, in order to realize or ameliorate roads, waterways and the aforementioned railways. Such works need to be built with greater solidity than their predecessors, sometimes despite difficult orographic features and, when possible, rapidly and without resorting to arduous and expensive provisional works such as cofferdams. Considering these constraints, the use of concrete appears in several cases as an appropriate solution, since, as Zimmermann has already highlighted, concrete can be poured directly underwater, in enclosures delimited only by sheet piles, or by poles rammed close to each other. Furthermore, the use of concrete is supported by the improved knowledge and expertise about hydraulic binders and about pebbles and mortar, which, in some cases, have already been acknowledged as concrete.

Since the manufacturing of artificial hydraulic binders has yet to spread across Germany, the use of concrete develops mainly in regions where natural hydraulic binders are available at an affordable cost. The composition of concrete therefore differs from region to region, according to the local availability of binders. Trass-based concrete is mainly used in regions along the Rhine, where a number of locks, bridges and river docks need to be built in the mid-1830s, following the development of local mining activities. In 1834, the old *Papiermühleschleuse* (lock at the paper mill) in Werden on the River Ruhr, which is a right tributary of the Rhine, is replaced by a new construction standing on a trass-based concrete foundation. This construction serves as a model for the building of further locks along the Ruhr between the second

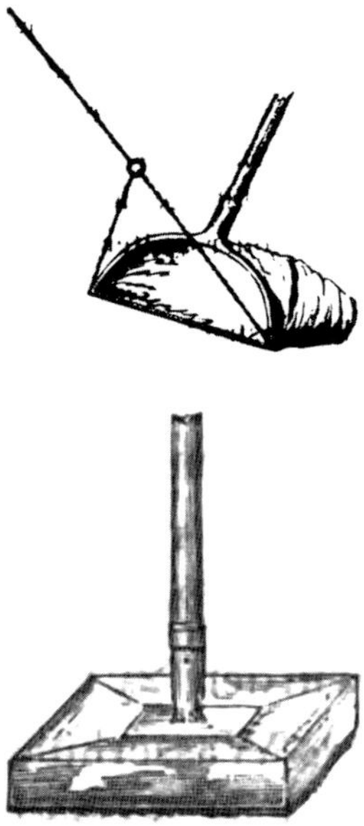

42. C. Hoffmann, drawings of tools used to build the foundations of the locks along the River Ruhr, 1834–1845, approx. A shovel to dredge, a concrete tamper (C. Hoffmann, *Beton-Fundirungen bei den Wasserbauten an der Ruhr*, in, "Notiz-Blatt des Architekten-Vereins zu Berlin," vol. I, 1847).

half of the 1830s and the end of the 1840s.[1] The foundations of these locks consist essentially of 88 feet long, 18 feet large and 3½ feet thick beds of concrete. Slaked lime, trass and different kinds of crushed stones, namely quartz gravel, sandstones and bricks are mixed together to produce concrete. Lime is taken from different lime kilns along the Rhine and from the nearby towns of Velbert and Ratingen, whereas trass is taken from Brohl and Andernach, and transported first along the Rhine, then along the Ruhr up to the building sites. Each measure of trass is mixed with about ½ measure of slaked lime to prepare mortar, which is then mixed with twice as much aggregate. The foundation site is first dug up and shored with sheet piles along the minor sides, whereas poles are rammed 4 feet apart from each other along the major sides, and the spaces between the poles are filled with boards and strips. Once this is achieved, mud is swept away from the bottom of the pit using a special shovel provided with a long handle and a rope, then, concrete is poured down using boxes similar to those that Bélidor describes. Three layers of concrete are poured on each other and tamped with a flat stone fixed at a long pole. About eight days later, when the concrete has set and hardened a little, three further provisional timber shores are built on the concrete bed, two along the major sides and one along the short side upstream, at a certain distance from the previous shores. They form three formworks to be filled with concrete, in order to build three 10 feet high, permanent retaining walls along the most stressed sides of the lock chamber. This last one is then built out of ordinary stone and brick masonry, standing partially on the concrete bed and partially on the retaining walls.

The building of the locks along the Ruhr also deserves special attention for the use of mortar and concrete mixers powered by steam engines. The concrete mixers consist of eight-sided wooden drums hinged on standing structures. They are provided with built-in rotating iron shafts, so that drums rotate and mix the mortar and aggregate together.[2] Mortar mixers, instead, are similar to pug mills; a typical one essentially consists of an immobile horizontal drum containing a central rotating shaft provided with several knives and connected to the steam engine. The drum can either be made of wood or sheet iron; it is fixed on two poles and has a longitudinal opening on the upper part to charge lime and trass. When the steam engine works, the shaft rotates and the several knives mix lime and trass, which come out as mortar through a trap door on the bottom of the drum.[3] The fact that the drum lies horizontally is special, as pug mills of that time generally stand and, consequently, have a vertical rotating shaft. Writing about the building of the Ruhr locks, Gotthilf Heinrich Ludwig Hagen (1797–1884), a notable hydraulic engineer, highlights this feature and refers to mortar pug mills used in France, namely the ones employed to build the *Pont de Neuilly*, the *Canal de Saint–Martin* and the harbours of Lorient and Toulon.[4] Apart from the vertical position, one of the pug mills used to build the *Canal de Saint–Martin*

[1] See Carl Hoffmann, *Beton-Fundirungen bei den Wasserbauten an der Ruhr*, in, "Notiz-Blatt des Architekten-Vereins zu Berlin," I, 1847, pp. 5–8; Gotthilf Heinrich Ludwig Hagen, *Handbuch der Wasserbaukunst*, vol. I, *Die Quellen*, Königsberg, Gebrüder Bornträger, 1841, p. 769; Id., *Handbuch der Wasserbaukunst*, vol. II, *Die Ströme und Kanäle*, Königsberg, Gebrüder Bornträger, 1852, p. 20; Id., *Brunnen, Wasserleitung und Fundirungen*, Berlin, Ernst & Korn, 1870, p. 311. Further locks along the Ruhr are built in Ruhrort (completed in 1841), Duisburg (the lock along the Ruhr canal and the lock of the harbour, completed in 1844), Dahlhausen (completed in 1844) and Mühlheim (completed in 1845); see, in this regard, Hoffmann, *Beton-Fundirungen*...cit., p. 5).

[2] Hagen, *Handbuch der Wasserbaukunst*, vol. I, cit., p. 777; Id., *Brunnen, Wasserleitung*...cit., p. 323.

[3] Hagen, *Handbuch der Wasserbaukunst*, vol. I, cit., pp. 768–769; Id., *Brunnen, Wasserleitung*...cit., p. 311.

[4] Hagen, *Handbuch der Wasserbaukunst*, vol. I, cit., pp. 769–770.

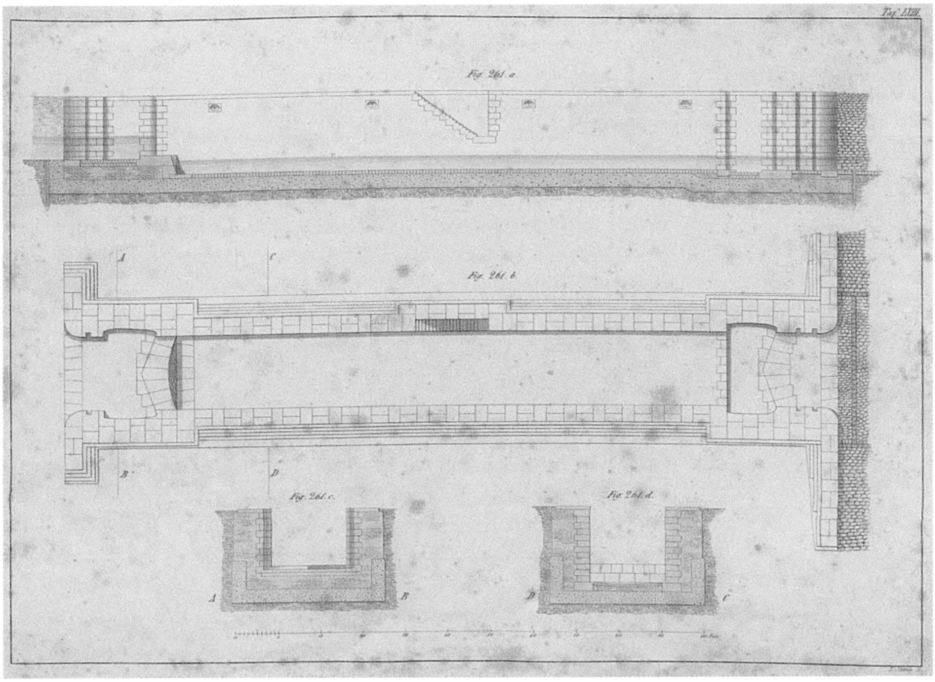

43. G. H. L. Hagen, drawings of a lock along the River Ruhr, approx. 1834–1845. Sections, plan (G. H. L. Hagen, *Handbuch der Wasserbaukunst*, vol. II, *Die Ströme und Kanäle*, Königsberg, Gebrüder Bornträger, 1852).

44. C. Hoffmann, drawings concerning the production of concrete and construction of a concrete foundation for a lock along the River Ruhr, approx. 1834–1845. A drum to mix, and a box to pour concrete; sections and plan of the pit during the pouring process (C. Hoffmann, *Beton-Fundirungen bei den Wasserbauten an der Ruhr*, in, "Notiz-Blatt des Architekten-Vereins zu Berlin," vol. I, 1847).

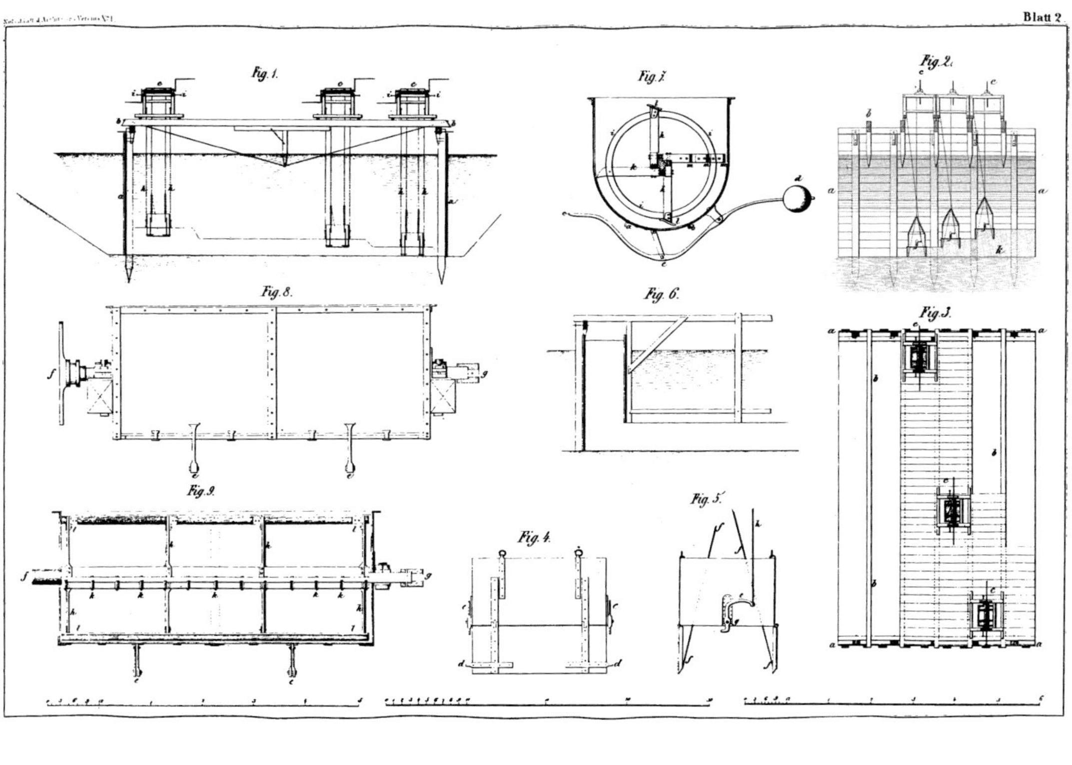

45. F. Bastenaire Daudenart, a British clay cutting tool (F. Bastenaire-Daudenart, *Die Kunst weisses Steingut anzufertigen mit durchsichtiger Glasur nach Art der Franzosen und Engländer anzufertigen: nebst einer Abhandlung über die im Musselfeure eingeschmolzenen Steingutfarben*, ed. Georg Frick, Ilmenau, Bernhard Friedrich Voigt, 1832).

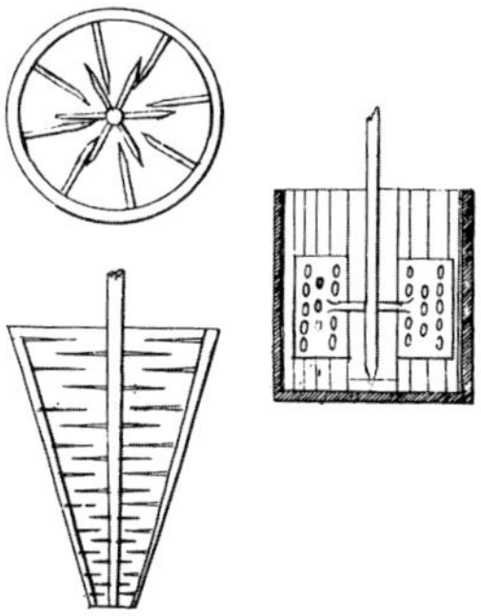

shows important similarities with the pug mills used along the Ruhr. A careful description of it is given in 1826 by the engineer of the *Corps des Ponts* Édouard de Villiers du Terrage. The machine "consists of an iron framed barrel," De Villiers relates, "which is larger at the top and tighter at the bottom, and of a vertical iron shaft provided with knives and rakes that turn horizontally inside the barrel at different levels."[5] De Villiers defines this pug mill *machine à mortier anglaise*, hinting at an English origin, which is actually the so-called "clay cutting tool" used in British porcelain factories.[6] Mortar mixers used in France are human or animal-powered machines, unlike the German ones used on the building sites of the Ruhr locks, and this is probably why the former stand vertically and the latter lie horizontally. A horse moves on a horizontal plan and can easily push a beam fixed on a vertical shaft, whereas steam engine flywheels normally stand vertically and can rotate a horizontal shaft by means of a simple chain more easily than a vertical shaft. The pug mills used along the Ruhr could therefore be seen as machines adjusted to the new steam technology, whose adoption is favoured by the local availability of coal.

At the same time that the River Ruhr begins to be improved as waterway, a railway line is planned to link the towns of Düsseldorf and Elberfeld, through a mining region stretching along the River Wupper, a further tributary of the Rhine. Built between 1838 and 1841, the railway requires the construction of a huge bridge across the Wupper, close to the village of Sonnborn.[7] Friedrich Eduard Salomon Wiebe (1804–1892), an engineer who is already entrusted with the building of the entire railway, designs the bridge. Wiebe studied at the Berlin *Bauakademie* at the time that Dietlein is teaching *Straßen-, Brücken-, Schleusen- und Kanalbau* and, once promoted master builder in 1836, he travels through Belgium, France and England.[8] It is probably through these experiences that he learns about concrete and, when designing the bridge of Sonnborn, he chooses to build three of the five piers on concrete foundations. The foundation of the central pier is provided with timber piles, whereas the two lateral ones stand straight on the soil, which is made of gravel and clay. The concrete is produced using local trass, and it is poured into the foundation pits only after they have been dried out; this shows evidence of the lack of expertise about concrete building at that time.[9]

Further examples of concrete foundations are in the town of Mannheim, situated at the confluence of the Rivers Rhine and Neckar. A new port is built there between 1835 and 1840, along the artificial canal that connects the two rivers bypassing the natural mouth considered unfit for navigation. The quays of the port are built on foundations made of timber piles, frameworks

[5] "Elle se compose d'un tonneau cerclé en fer, et un peu plus étroit en haut qu'en bas, [...] et d'un arbre vertical en fer, armé de couteaux et de rateaux [sic]; tournant horizontalement à différentes hauteurs dans ce tonneau" (Édouard de Villiers du Terrage, *Description du Kanal de Saint-Denis et du Canal Saint-Martin*, Paris, Carilian-Goeury, 1826, p. 37).

[6] See F. Bastenaire Daudenart, *Die Kunst weisses Steingut anzufertigen mit durchsichtiger Glasur nach Art der Franzosen und Engländer anzufertigen: nebst einer Abhandlung über die im Musselfeure eingeschmolzenen Steingutfarben*, ed. Georg Frick, Ilmenau, Bernhard Friedrich Voigt, 1832, pp. 54–55 (Id., *L'Art de fabriquer la Faience blanche, recouverte d'un émail transparent à l'instar français & anglais suivi d'un traité de la peinture à revebère*, Paris, Anselin, 1830). A further description and an illustration of the "clay cutting tool" is also to find in, Johann Wilhelm David Korth, Johann Georg Krünitz's *Oekonomische Encyklopädie oder allgemeines System der Staats- Stadt- Haus- und Landwirthschaft*, vol. CLXXII, Berlin, Pauli, 1839, entry "Steingut," pp. 342–390, esp. pp. 361–362 and pl. I, fig. 8974.

[7] See [Eduard] Römer, *Die Düsseldorf-Elberfelder Eisenbahn*, in, "Allgemeine Bauzeitung," VIII, 1843, (pp. 60–77), pp. 62–64.

[8] See Hermann Julius Meyer, ed., *Meyers Konversations-Lexikon: eine Eycyklopädie des allgemeinen Wissens*, Leipzig, Wien, Bibliographisches Institut, vol. XVI, 1890, entry "Wiebe," p. 592.

[9] "In order to pour the concrete layer as much carefully as possible, water was pumped out during this work" ("Um die Betonschicht möglichst sorgfältig schütten zu können, wurde das Wasser während dieser Arbeit ausgepumpt") Römer, *Die Düsseldorf-Elberfelder Eisenbahn*...cit., p. 63.

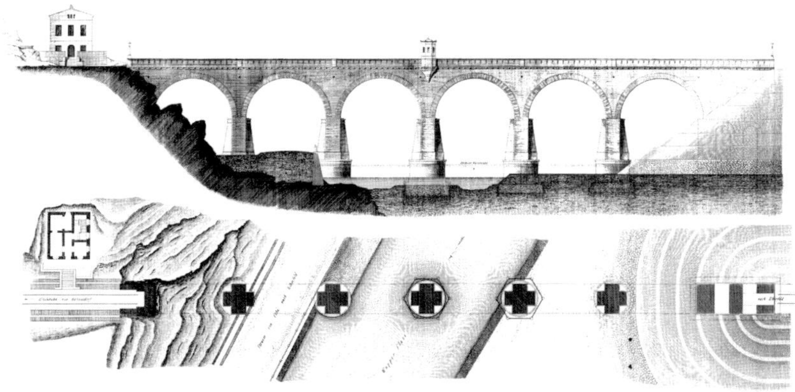

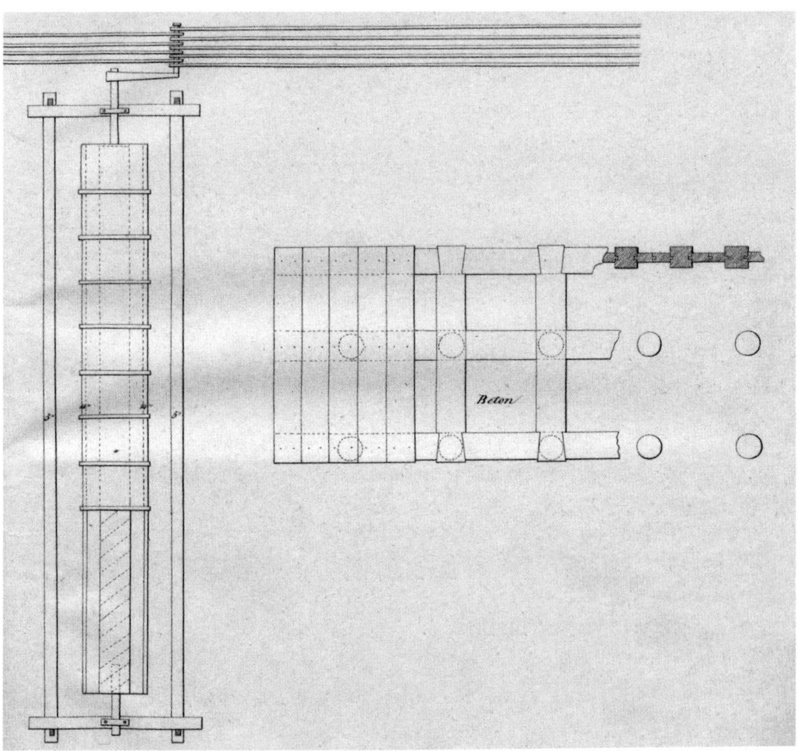

46-47. F. E. S. Wiebe, the Bridge of Sonnborn (1838-1841).

46. Elevation and cross sections of the concrete foundations; plan ([E.] Römer, *Die Düsseldorf-Elberfelder Eisenbahn*, in, "Allgemeine Bauzeitung," VIII, 1843).

47. View of the bridge in the early 20th century (Stephani, *Umbau der Eisenbahnbrücke über die Wupper in Elberfeld-Sonnborn*, in, "Zentralblatt der Bauverwaltung," XXXIV, 1914, n° 101).

48. Anon., Port of Mannheim 1835–1840. View from above of the screw pump and a stretch of the foundation for a quai (Anon., *Technische Notizen auf einer Reise nach dem Unter-Rhein, im Herbst 1835*, in, "Zeitschrift über das gesammte Bauwesen," I, 1836).

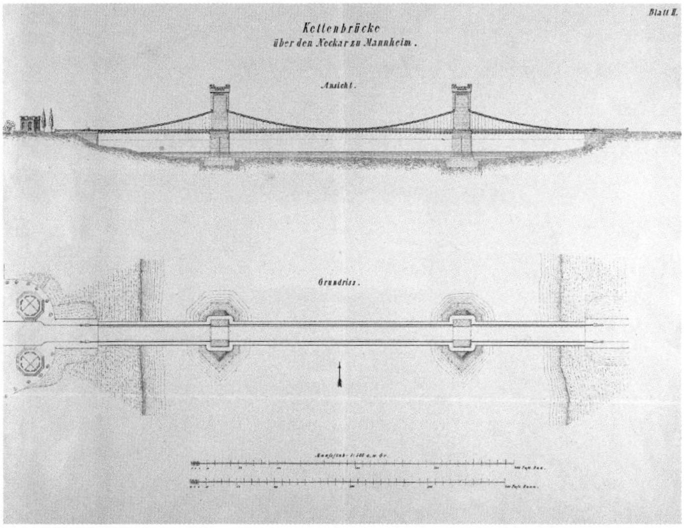

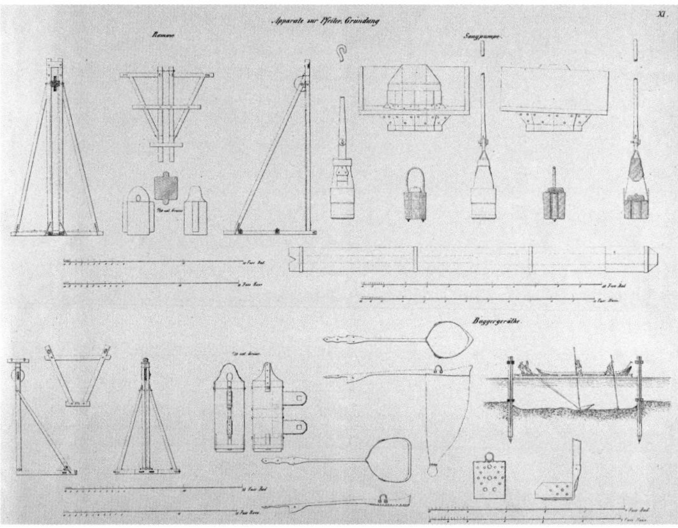

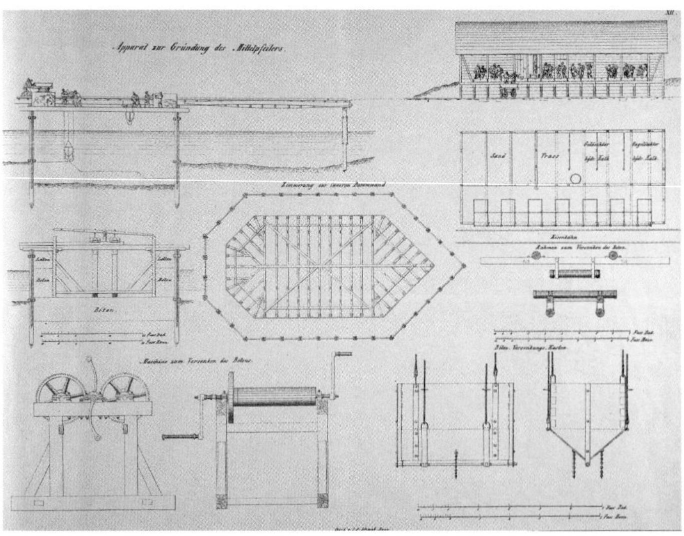

49-51. G. D. Wendelstadt, Suspension Bridge of Mannheim, 1842-1845.

49. Longitudinal section, plan (NLA Hannover – Plankammer, Karten Deutschland n° 73 SD/1 pg, pl. 3).

50. Tools to dig the pit and drawing of the digging process (NLA Hannover – Plankammer, Karten Deutschland n° 73 SD/1 pg, pl. 13).

51. Production of concrete, construction of the concrete foundation of a pier; crane and box to pour concrete (NLA Hannover – Plankammer, Karten Deutschland n° 73 SD/1 pg, pl. 12).

and trass-based concrete poured into the free fields of the frameworks, among the piles heads, similar to certain foundations by Wiebeking. As with the construction of the bridge in Sonnborn, water is drained from the foundation pits before concrete is poured; an impressive Archimedes screw powered by about twenty men, pulling and pushing beams connected to the screw shaft, is set up for this purpose.[10]

During the building of this port, the city council of Mannheim decides to build a suspension bridge over the Neckar and entrusts Wendelstadt to plan it in 1839, shortly after the opening of the Hamelin Bridge. Wendelstadt submits the plans in 1840 and the bridge is built between 1842 and 1845, with the collaboration of Adolph Funk (1819–1889), an architect from Hannover, and Georg Lüttich, an engineer who participated in the construction of the Hamelin Bridge. The piers of the bridge in Mannheim stand on 7 feet high beds of concrete, which is poured straight onto the soil inside enclosures made of sheet piles dried out by means of lift pumps. A closed cofferdam is built all around the top of each concrete bed to form a further watertight enclosure where workers build the early masonry layers of the piers. The lowest parts of these cofferdams (about 7 feet high) are made of concrete, whereas the upper parts (about 5 feet high) are made of clay. In a provisional cabin on the riverside, concrete is produced by mixing slaked lime, trass and aggregate. Once prepared, it is deposited into wagons on rails and transported via a provisional bridge from the riverside to the pouring sites. It is then poured into the foundation pits by means of boxes, each one hanging from a crane by eight ropes, with two trap doors that workers open by releasing two chains and four ropes. Each crane is provided with two rollers and a gearshift; it can be moved on rails, which, in turn, lie on a movable bridge, so that concrete can be easily and uniformly poured all over the bottom of the pit. The use of local trass to build the foundation of the Mannheim Bridge, despite the fact that Wendelstadt is a cement maker, demonstrates the importance of the local availability of trass to the composition of concrete. This is further empha-

52. W. Hablitscheck, Suspension Bridge of Mannheim, engraving, appr. 1850 (personal collection).

[10] See Anon., *Technische Notizen auf einer Reise nach dem Unter-Rhein, im Herbst 1835*, in, "Zeitschrift über das gesammte Bauwesen," I, 1836, (pp. 101–107), p. 101.

sized by two constructions in Mühlheim am Rhein: the trass-based concrete foundations of the suspension bridge that is built between 1842 and 1844 - known as the *Schloßbrücke* - and the trass-based concrete foundation of the lock along the Ruhr that is also built between 1843 and 1844, following the model of the lock at the paper mill in Werden.[11]

Concrete foundations in Bavaria and Pomerania, the exploitation of local hydraulic lime

The building of the canal to connect the Rivers Danube and Main, commonly known as *Ludwig-Kanal*, becomes a significant occasion for using Bavarian hydraulic lime to produce mortar and concrete, in light of the advice from Fuchs and Panzer. Heinrich Joseph Alois Freiherr von Pechmann (1774–1861), an eminent Bavarian engineer, designs the canal between 1828 and 1832.[12] It stretches from Kelheim to Bamberg, partially exploiting the Rivers Altmühl and Regnitz, and involving the building of ninety-eight masonry locks, several bridges, culverts and embankments.[13]

In 1836, shortly before works start, some arguments arise about how to waterproof the bed of the canal.[14] Von Pechmann intends to exploit the principle of filtering cloudy water through sand. He plans to fill up the canal, stretch after stretch, with water containing clay, which is supposed to filter through the natural sandy bed of the canal, until it is completely clay-logged and, therefore, watertight. "Cloudy water," Von Pechmann reasons, "becomes clear after flowing through sand, which keeps back all particles that make water cloudy; in this way, sand becomes waterproof."[15] To prove his assertion, Von Pechmann mentions the water purification plant in Paris, which periodically replaces sand filters because they get saturated. In addition, he mentions the canals and basins in the park of Nymphenburg near Munich, which are built in sandy soil without resorting to any special waterproofing system.[16] To test his claim, in March 1836, Von Pechmann digs a 4 feet deep

[11] See Hoffmann, *Beton-Fundirungen*...cit., p. 6.
[12] About the conception and the building of the *Ludwig-Kanal* see Heinrich Joseph Alois von Pechmann, *Entwurf für den Kanal zur Verbindung der Donau mit dem Main*, München, 1832; Carl Theodor von Kleinschrod, *Die Kanal-Verbindung des Rheins mit der Donau: unter Benützung amtlicher Quellen beleuchtet*, München, G. Frank'sche Buchhandlung, 1834; Anon., *Auszüge aus den Protokollen der zweiten Generalversammlung der Aektiengesellschaft für den Ludwig-Donau-Main-Kanal*, voll. I–XII, 1836–1848, esp. vol. II, 1837, pp. 31–49, vol. III, 1838, p. 20, vol. IV, 1839, pp. 47–49, vol. V, 1840, p. 24; Heinrich Joseph Alois von Pechmann, *Der Ludwig-Kanal. Eine kurze Beschreibung dieses Kanal's und die Ausführung desselben*, München, Joseph Lindauer'sche Buchhandlung, 1846; Friedrich Schultheis, Alexander Marx, *Der Ludwig-Kanal- Seine Entstehung und Bedeutung als Handels-Strasse*, Nürnberg, n.p., 1847; Heinrich Joseph Alois Freiherr von Pechmann, *Der Ludwigskanal. Kurze Geschichte seines Baues und seiner noch bestehenden Mängel, sowie die Mittel, sie zu entfernen und zu verbessern und den Kanal zu seiner Vollkommenheit zu erheben*, Nürnberg, Friedrich Korn'schen Buchhandlung, 1854; G. Dulk, *Notizen über den Ludwigs-(Main-Donau-) Kanal*, in, "Zeitschrift für Bauwesen," XII, 1862, coll. 227–238. Building materials are quarried along the course of the waterway, including limestone to produce hydraulic lime; see Schultheis, *Der Ludwig-Kanal- Seine Entstehung und Bedeutung*... cit., pp. 42, 53.
[13] The course of the River Altmühl is developed between Kelheim and Dietfurt, whereas the course of the River Regnitz is used between Bamberg to the confluence with the Main. Von Pechmann initially plans to build ninety-four locks, but, during the building, the number of locks needs to be raised to ninety-eight.
[14] The bill for the building of the *Ludwig-Kanal* is released in 1834. One year later, the *Aektiengesellschaft* für den *Ludwig-Donau-Main-Kanal* is founded, and early excavation works begins in 1836, near Nurnberg and Bamberg.
[15] "Die Natur hat dem Sande die Eigenschaft gegeben, trübes Wasser rein durch sich hindurch laufen zu lassen, und die Körper, die es trübes machen, in seinen Zwischenräumen zurück zu halten. Dadurch allein macht die Natur den Sand wasserdicht" (Heinrich Joseph Alois von Pechmann, *Über die Mittel, welche angewendet werden, um die in Sandboden gegraben Theile des Ludwigs-Kanals wasserdicht zu machen*, in, "Journal für die Baukunst," (XV, 1841, pp. 160–179; XXII, 1845, pp. 1–14), XV, 1841, p. 162). About the waterproofing of the *Ludwig-Kanal*, see also Id., *Ueber die angewendeten Mittel, um die in Sand gegrabenen Theile des Ludwigskanales wasserdicht zu machen*, in, "Allgemeine Bauzietung," V, 1840, pp. 375–382, (this articles is translated into French, Id., *Moyens employés pour rendre étanche le Canal Louis au droit des biefs creusés dans le sable*, in, "Annales des Ponts et Chausses," 1841, pp. 18–34); Id., *Der Ludwig-Kanal. Eine kurze Beschreibung*...cit., pp. 38–42; Id., *Der Ludwigskanal. Kurze Geschichte seines Baues*...cit., 1854, pp. 53–67.
[16] Von Pechmann, *Über die Mittel, welche angewendet werden, um die in Sandboden gegraben Theile des Ludwigs-Kanals wasserdicht zu machen*... cit., XV, 1841, p. 163; Id., *Über die angewendeten Mittel, um die in Sand gegrabenen Theile des LudwigsKanales wasserdicht*... cit., p. 375.

53-55. H. J. A. von Pechmann, *Ludwig-Kanal*, 1836–1846.

53. Plan, engraving (F. Schultheis, A. Marx, *Der Ludwig-Kanal- Seine Entstehung und Bedeutung als Handels-Strasse*, Nürnberg, 1847).

54-55. Views of the locks in Berching and Forchheim, engravings (F. Schultheis, A. Marx, *Der Ludwig-Kanal- Seine Entstehung und Bedeutung als Handels-Strasse*, Nürnberg, 1847).

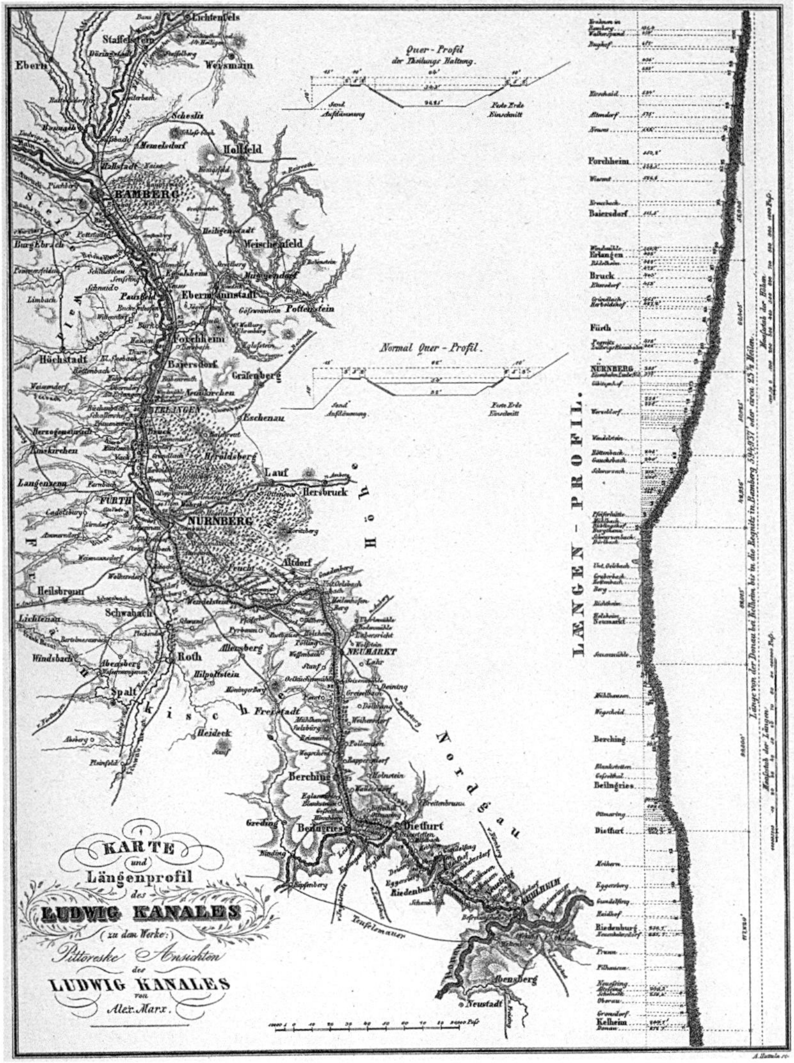

56-58. H. J. A. von Pechmann, *Ludwig-Kanal*, 1836–1846.

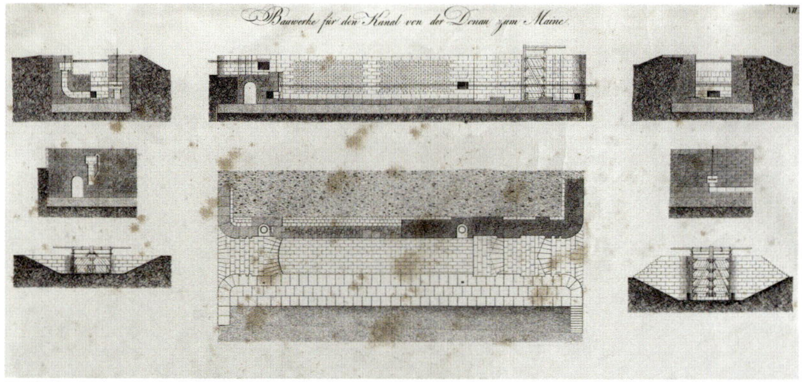

56. Design of a lock, plan and sections (H. J. A. von Pechmann, *Atlas zu dem Entwurfe für den Kanal von Donau an den Main*, [1832]).

and 6 square feet wide pit in the sandy soil of a garden in Nurnberg; he fills it up with running water coming through an appropriate pipe from a near basin. Eight days later, he verifies that the sandy soil has become watertight. However, such evidence does not sufficiently convince the *Kanalbaudirektion*, the body that manages the building of the canal. In an atmosphere of increasing perplexity, Panzer, who, meanwhile, has been appointed responsible for the building of the canal stretch between Forcheim and Bamberg, intervenes and proposes pouring a mortar made of hydraulic lime and water onto the canal bed to make it waterproof. Other engineers propose, instead, to either tamp clay onto the canal bed or cover it with stone plates joined by means of hydraulic mortar or moss. Von Pechmann agrees to test these different techniques, including the one proposed by Panzer, on a canal stretch near Nurnberg. Panzer's mortar hardens quite rapidly, but several cracks make it ineffective. Despite that, Panzer does not give up on the idea of exploiting hydraulic lime to waterproof the canal; he makes a second attempt using a mixture of slaked lean lime, hydraulic lime, sand and gravel, which can be considered a sort of concrete. This attempt also fails and Von Pechmann complains that the mixture used by Panzer "could be scraped off with the sole of a boot."[17]

Panzer's proposals are set aside but further occasions to use Bavarian hydraulic lime-based concrete arise when construction of the locks begins. Von Pechmann originally plans to build locks on masonry foundations, resorting to timber pilings only in the presence of significantly unsteady soil. He even claims to have illustrated a lock standing on a timber piling foundation in the atlas containing the initial plans for the building of the canal.[18] An atlas still in existence, whose dating remains controversial but should predate 1835, shows, however, a stone and brick masonry lock standing on a 2 feet high layer of unexplained material, which is represented using a pattern that hints at a sort of

[17] "Von Froste ausgelöset, daß ich, so wie jeder, der es versuchte, sie mit der Sohle des Stiefels abschaben konnte" (Von Pechmann, *Der Ludwigskanal. Kurze Geschichte seines Baues*...cit., p. 58). Despite these essential failures, Panzer attaches great importance to the use of hydraulic lime for the building of the *Ludwig-Kanal*. He describes his attempts in a specific chapter joined to his 1836 book about hydraulic lime; see Friedrich Panzer, *Mittel gegen das Versickern des Wassers in Sandboden*, in, *Ueber das Vorkommen*... cit., pp. 38–40, also published in, "Allgemeine Bauzeitung," III, 1838, p. 402). Seeing the disappointing results of the tests, Von Pechmann finally waterproofs the canal bed filtering water mixed with clay and, over some stretches, with mud; see Von Pechmann, *Der Ludwigskanal. Kurze Geschichte seines Baues*...cit., pp.58–59).

[18] See Von Pechmann, *Der Ludwigskanal. Kurze Geschichte seines Baues*...cit., 1854, p. 68.

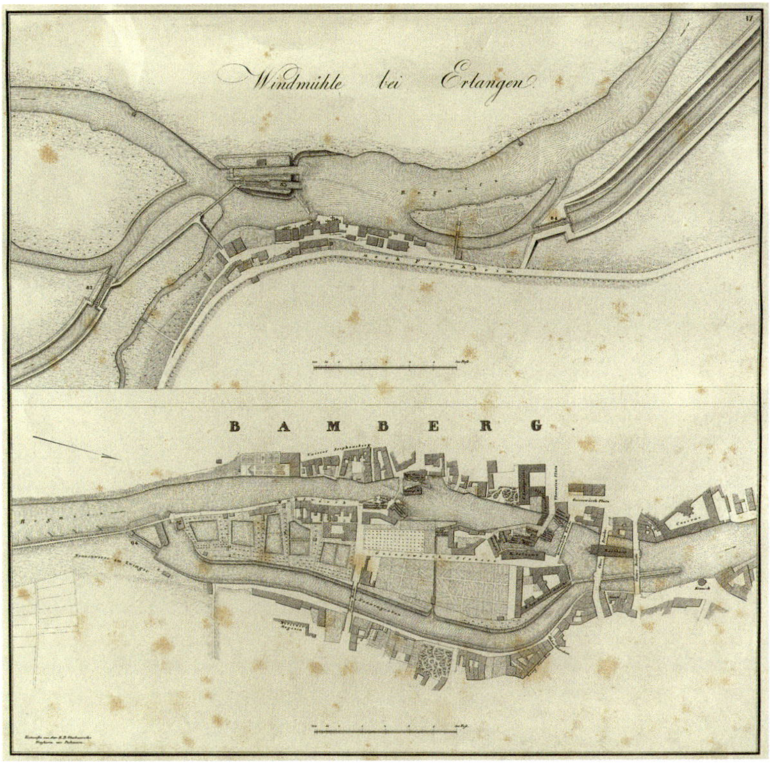

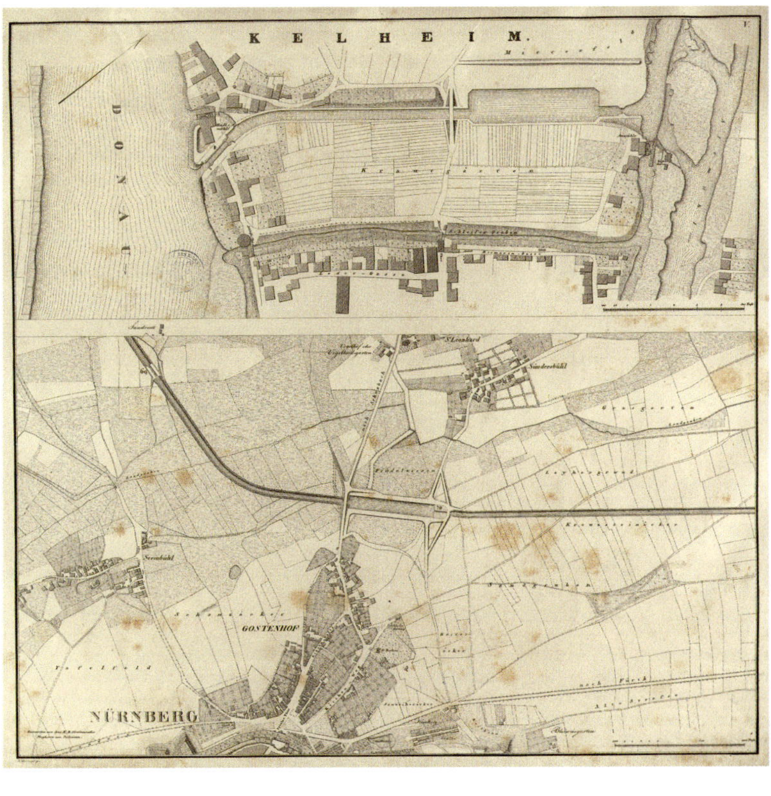

57-58. Plans of the locks in Erlangen, Bamberg (Nonnegraben), Kehlheim and Nuremberg (H. J. A. von Pechmann, *Atlas zu dem Entwurfe für den Kanal von Donau an den Main*, [1832]).

59-60. H. J. A. von Pechmann, *Ludwig-Kanal*, 1836–1846. Views of the locks in Kehlheim and Bamberg (Nonnegraben), engravings (F. Schultheis, A. Marx, *Der Ludwig-Kanal- Seine Entstehung und Bedeutung als Handels-Strasse*, Nürnberg, 1847).

conglomerate.[19] The same pattern is used to draw the covering of artificial canal beds on navigable bridges.[20] Despite plans, and despite the fact that the ninety-eight locks of the canal are "to be built on very different kinds of soil, from the most solid made of rocks to some kinds of marshy peat soil," timber piles are not used, and most locks are "built on between 3 and 4 feet thick beds of masonry laid all over the foundation sites and bound with hydraulic mortar."[21]

Difficulties arise in the building of four locks situated close to the Rivers Danube, Altmühl and Regnitz, which provoke strong flooding and make drainage systems ineffective.[22] The building of cofferdams is therefore considered necessary and Von Pechmann takes inspiration from the lock of Huningue, which the aforementioned French engineer Beaudemoulin has built in 1825 to connect the "Rhone-Rhine Canal" to the Rhine.[23] Beaudemoulin originally plans to build a masonry lock standing on a concrete bed foundation, inside a pit surrounded by ordinary cofferdams made of clay rammed between timber walls. However, an unexpected landslide along one of the major sides of the pit during the excavation works reduces the available space and drives Beaudemoulin to conceive an innovative construction consisting of a watertight encasement composed of a concrete bed as a base, two rows of sheet piles along the minor sides and two trapezoidal concrete cofferdams along the major sides, with sloping faces towards the interior of the encasement. Once the water has

[19] See Heinrich Joseph Alois von Pechmann, *Atlas zu dem Entwurfe für den Kanal von Donau an den Main*, n.p., n.d. Most libraries date this atlas back to 1832, as it should have been joined to the book *Entwurf für den Kanal zur Verbindung der Donau mit dem Main* published in 1832. As a matter of fact, the *Description des principales collections que renferme l'établissement géographique de Bruxelles, fondé par Ph. Vandermaele*, contained in *Dictionnaire des hommes de lettres, des savans, et des artistes de la Belgique* also lists an *Atlas zu dem Entwurfe für den Kanal von Donau an den Main* dating 1832; see, in this regard, Ph. Vandermaelen, *Dictionnaire des hommes de lettres, des savans, et des artistes de la Belgique: présentant l'énumération de leurs principaux ouvrages; suivi de la description des principales collections que renferme l'Établissement géographique de Bruxelles*, Bruxelles, Etablissement géographique fondé par Ph. Vandermaelen, 1837, p. 245). However, for Von Pechmann's Atlas, the *Landesarchiv Baden-Würtemberg* only provides the general indication of "antecedent 1835," see Atlas *zu dem Entwurfe für den Kanal von Donau an den Main*, Landesarchiv Baden-Würtemberg, GA 105 n° 842, pp. 1–10.

[20] Initially 13 navigable bridges are planned, but only 10 are actually built; see Rotraut Trapp, *Die Brücken des Ludwig-Donau-Main-Kanals*, in, "Frankenland. Zeitschrift für fränkische Landeskunde und Kulturpflege," 2003, (pp. 287–306), p. 289.

[21] "Die 98 Schleusen des Ludwigs-Kanals mußten, wie bei der Länge desselben von 23 deutschen Meilen zu erwarten war, auf sehr verschiedenartigen Boden gegründet werden. Er wechselt von festem Felsengrund bis zu beinahe sumpfartigem Torfboden. Dennoch wurde bei keiner einzigen Schleuse ein Rost angewendet. Sie sind alle auf eine 3 bis 4' hohe, über die ganze Grundfläche der Schleuse sich ausdehnende Schicht von Mauerwerk gegründet, das mit gutem, hydraulischem Mörtel ausgeführt ist" (Von Pechmann, *Der Ludwig-Kanal. Eine kurze Beschreibung*...cit., p. 43).

[22] "As we began to dig the pits for these locks, floods from the nearby Regnitz and Danube were so strong that it was impossible to drain out water" ("Bei 4 dieser Schleusen war, als man die Baugrube auszugraben anfing, der Zudrang des Wassers in dieselbe aus der nahen Regnitz und Donau so groß, daß man es unmöglich fand, die Baugrube durch Schopfwerkzeuge leer zu erhalten") (Von Pechmann, *Der Ludwig-Kanal. Eine kurze Beschreibung*...cit., p. 44).

[23] See Louis Alexis Beaudemoulin, *Recherches théoriques et pratiques sur la fondation par immersion des ouvrages hydrauliques et particulièrement des écluses*, Paris, Carilian-Goeury, 1829, esp. pp. 6–32.

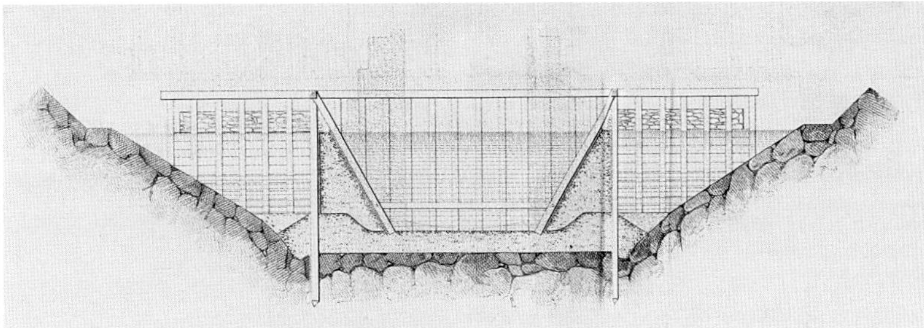

61. L. A. Beaudemoulin, lock in Huningue, 1825. Concrete foundation, cross section (L. A. Beaudemoulin, *Recherches théoriques et pratiques sur la fondation par immersion des ouvrages hydrauliques et particulièrement des écluses*, Paris, Carilian-Goeury, 1829).

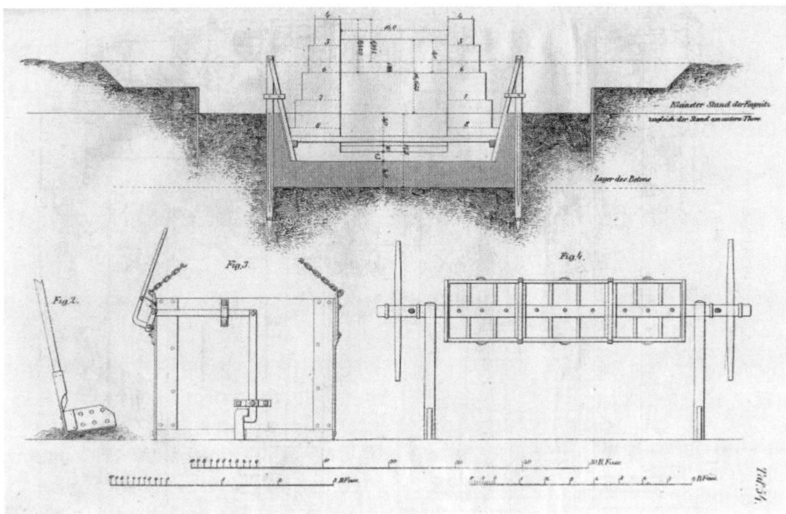

62-63. H. J. A. von Pechmann, *Ludwig-Kanal*, 1836–1846.

62. Lock at the Bughof, Bamberg, cross section, tools to build the foundation - a shovel to dredge, a drum to mix, and a box to pour concrete (W. Winckelmann, *Ueber die Fundirung der Schleuse Nr. 93 des Ludwigs-Kanals, an der Ausmündung desselben in die Regnitz bei Bamberg*, in, "Zeitschrift für praktische Baukunst," 1843).

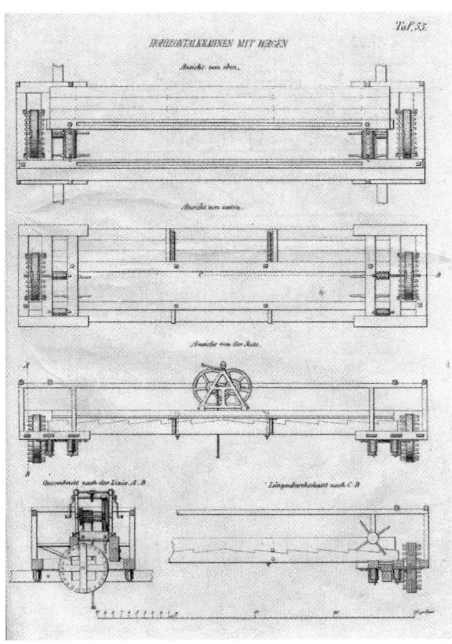

63. Bridge crane to pour concrete, elevations, view from above and sections (W. Winckelmann, *Ueber die Fundirung der Schleuse Nr. 93 des Ludwigs-Kanals, an der Ausmündung desselben in die Regnitz bei Bamberg*, in, "Zeitschrift für praktische Baukunst," 1843).

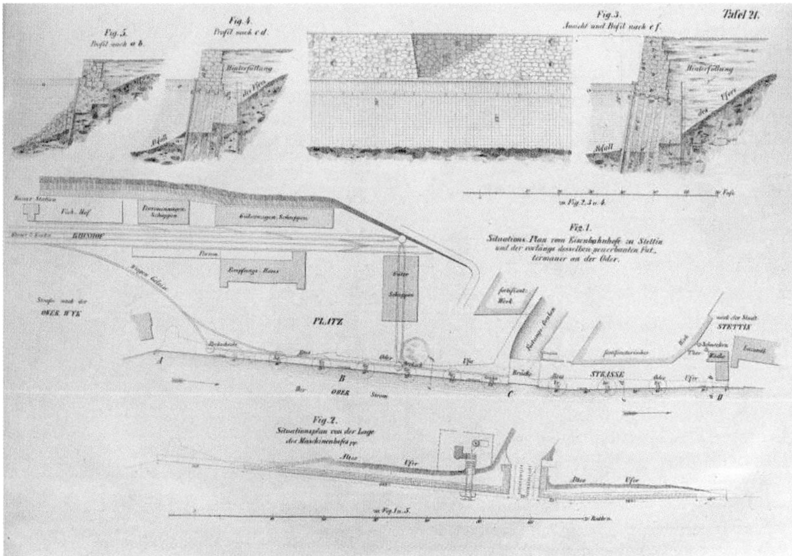

64-65. F. Neuhaus, L. Schwan, dock along the River Oder, Stettin, 1842–1844.

64. Cross sections, elevation, plan of the site (L. Schwan, *Beschreibung des Baues einer Kaimauer zu Stettin an der Oder vorlängs des dortigen Eisenbahnhofes, insbesondere des dabei in Anwendung gebrachten Gründungsverfahrens mittelst Beton aus Wolliner Mergelkalk*, in, "Zeitschrift für Praktische Baukunst," VIII, 1848).

65. Drawings of the installations to produce mortar and concrete (L. Schwan, *Beschreibung des Baues einer Kaimauer zu Stettin an der Oder vorlängs des dortigen Eisenbahnhofes, insbesondere des dabei in Anwendung gebrachten Gründungsverfahrens mittelst Beton aus Wolliner Mergelkalk*, in, "Zeitschrift für Praktische Baukunst," VIII, 1848).

been drained, the walls of the lock chamber are partially built on the sloping sides of the concrete cofferdams, which become an essential part of the entire construction, instead of just being provisional. Beaudemoulin himself writes of "a concrete encasement, which becomes part of the lock walls, after having acted as cofferdams."[24] Moving from this example, Von Pechmann develops the idea of building similar watertight encasements, each one made of a 3½ feet thick bearing bed of concrete and four concrete cofferdams – two trapezoidal ones along the major sides, similar to those that Beaudemoulin had built, and two minor ones with rectangular transversal sections along the short sides, which are intended to be demolished later to build up the lock gates. The concrete is mixed in manually-powered horizontal drums and is poured into the foundation pits by means of boxes hanging from movable manual bridge cranes. The concrete caissons remain full of water throughout winter and, once the concrete has hardened and the water drained out, masons build the walls of the lock chamber partially on the trapezoidal cofferdams. Beyond the reference to Beaudemoulin's lock, it is worth noting that the practice of building the walls of the lock chamber partially on the underground retaining concrete wall is already observable, although just *in nuce*, in the aforementioned locks along the Ruhr.

The four locks on concrete foundations are built between 1836 and 1841. Two are situated in Bamberg (one at the so-called Bughof, south of the town, and one at the Nonnengraben, in the centre of town), one in Kelheim, and the forth probably in Erlangen, because of its proximity to the River Regnitz – although Von Pechman does not explain this in detail. The building of the lock at the Bughof starts first and it therefore acts as a sort of trial

[24] "Un encaissement de béton qui, après avoir fait fonction de batardeaux, entrerait ensuite dans le massif de la maçonnerie des bajoyers" (Beaudemoulin, *Recherches théoriques et pratiques*...cit., p. 12).

[25] See W. Winckelmann, *Ueber die Fundirung der Schleuse Nr. 93 des Ludwigs-Kanals, an der Ausmündung desselben in die Regnitz bei Bamberg*, in, "Zeitschrift für praktische Baukunst," 1843, (pp. 231–235), p. 231.

[26] "Die beiden Seitenwände dieser Kästen konnten auf Grund ihrer erreichten Festigkeit für die Seitenmauern der Schleusenkammer mit verwendet werden. Die an den beiden Schmalseiten von Ober-

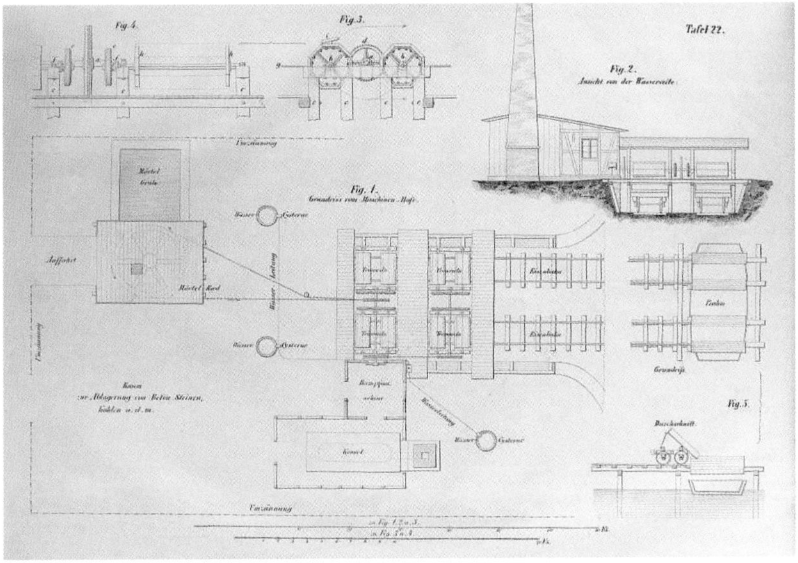

construction, as it also provides the opportunity to experiment with three different concrete mixtures. The first is made of fine milled hydraulic lime and coarse-grained brick powder, the second is made of hydraulic lime and sand, and the third is a mixture of black lime from Bamberg, sand and brick powder.[25] This last mixture is the most affordable and it also proves to be as good as the first two. Von Pechmann mentions the hardness of the concrete. "The two longitudinal cofferdams," he writes in this regard, "were so strong that they could be partially used as sidewalls of the chamber, whereas the short cofferdams upstream and downstream were demolished and the debris was used to build the wing-walls."[26]

A further, albeit minor, occasion to use concrete along the *Ludwig-Kanal* presents itself in the building of several timber framework road-bridges. These rest on masonry abutments made of unrefined cut stones and featuring 40 cm high concrete platforms at the top, which act as bearing decks for the timber beams.[27]

While the building of the *Ludwig-Kanal* in Bavaria is finishing, the engineer Friedrich Neuhaus (1797–1876) and the master builder Schwan construct an impressive concrete foundation, partially on timber piles, for a 1.309 feet long quay, which is built between 1842 and 1844 along the River Oder in Stettin.[28] The quay is part of the general construction plan of the Stettin railway station and is intended to be the juncture between land and water transports, serving as a mooring for ships and as platform for rails.

This construction project is characterized by a number of difficulties. Due to the tightness of the lot, the quay needs to be built in water, rather than on the riverside, so as to create later some land-surface by burying a strip of river. Moreover, the consistency of the soil changes along the riverbank, from

und Unterhaupt befindlichen Wände dieser Kästen wurden ausgebrochen und diese Bruchstücke zum Bau der Flügelmauern verwendet" (Von Pechmann, *Der Ludwigskanal. Kurze Geschichte seines Baues*...cit., p. 69).

27 Trapp, *Die Brücken des Ludwig-Donau-Main-Kanals*... cit., p. 298.

28 See L. Schwan, *Beschreibung des Baues einer Kaimauer zu Stettin an der Oder vorlängs des dortigen Eisenbahnhofes, insbesondere des dabei in Anwendung gebrachten Gründungsverfahrens mittelst Beton aus Wolliner Mergelkalk*, in, " Zeitschrift für Praktische Baukunst," VIII, 1848, coll. 145–162. About Fridrich Neuhaus, see Anon., *Friedrich Neuhaus - Ein vergessener Eisenbahnpionier*, in, "Mitteilungen des Vereins für die Geschichte Berlins," LXXVIX, 1983, pp. 109–113.

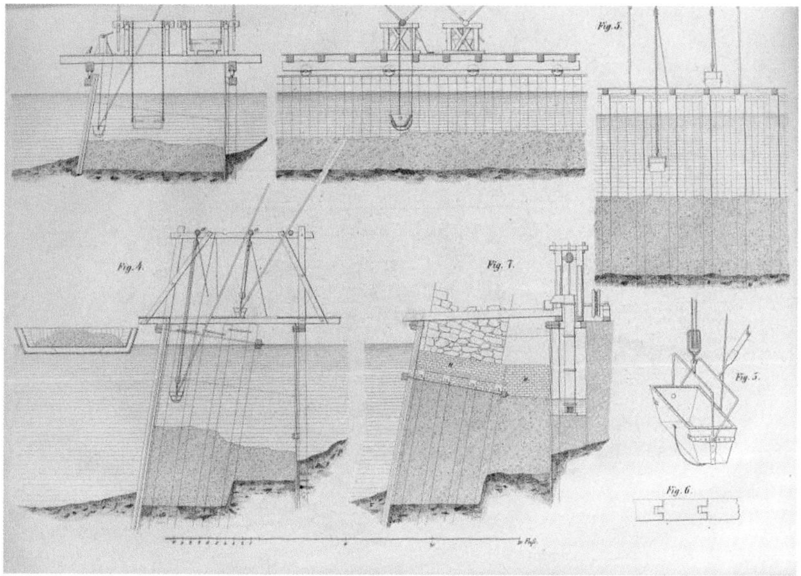

66. F. Neuhaus, L. Schwan, dock along the River Oder, Stettin, 1842–1844. The pouring process (L. Schwan, *Beschreibung des Baues einer Kaimauer zu Stettin an der Oder vorlängs des dortigen Eisenbahnhofes, insbesondere des dabei in Anwendung gebrachten Gründungsverfahrens mittelst Beton aus Wolliner Mergelkalk*, in, "Zeitschrift für Praktische Baukunst," VIII, 1848).

sandy and unsteady to rocky, and the depth of the foundation consequently changes as well. The building site is divided into six sections. A foundation consisting of a 10 feet high concrete basement standing straight on rocks is planned for the first two sections upstream, whereas the foundations along all other sections are made of timber piles and concrete poured among the upper parts of the piles, forming a basement between 13 and 40 feet high, which is finally covered with timber beams and boards.

Since transportion costs make trass too costly in Stettin, the use of hydraulic lime produced by burning marls from the nearby island of Wollin is considered.[29] The chemical composition of Wollin marl is investigated, proving that it mainly consists of lime (about 80%) and silica (about 10%), besides minor quantities of clay and iron oxides. The underwater hardening of mortar made with Wollin hydraulic lime is furthermore verified, checking the consistency of samples kept covered with water in glasses for both eight days and four weeks long. The concrete strength is tested as well by charging, using rails, a 3 square feet wide and 1 feet high specimen that has hardened four weeks long underwater in the Oder.[30]

The production of mortar and concrete is fully mechanized. Necessary installations are placed close to the riverside: a mortar mixer, eight drums to mix concrete and a steam engine to power all machineries. The mortar mixer is essentially composed of a circular pit provided with five fixed toothed spokes on the bottom and a turning wheel provided with six toothed spokes at the top. When the wheel turns, the teeth mix together sand and lime, while mortar flows through a side opening into a lower pit, pushed by the centrifugal force that the turning speed produces. Since the mortar mixer has a vertical rotating shaft, the connection with the flywheel is ensured by a chain

[29] See Schwan, *Beschreibung des Baues einer Kaimauer*...cit., coll. 160–161.
[30] *Ibid.*, col. 162.
[31] The building site of the *Neues Museum* in Berlin is considered one of the most advanced of that time. See Werner Lorenz, *Classicism and high technology - the Berlin Neues Museum*, in, "Construction History,"

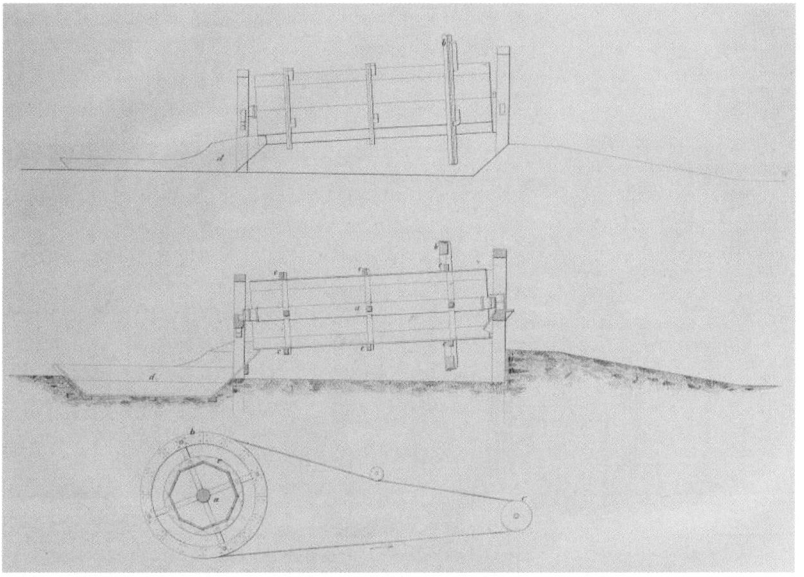

67. L. Schwan, design of a sloping concrete drum (L. Schwan, *Beschreibung des Baues einer Kaimauer zu Stettin an der Oder vorlängs des dortigen Eisenbahnhofes, insbesondere des dabei in Anwendung gebrachten Gründungsverfahrens mittelst Beton aus Wolliner Mergelkalk*, in, "Zeitschrift für Praktische Baukunst," VIII, 1848).

68. C. W. Hoffmann, drawing of a mortar mixer used on the building site of the *Neues Museum* in Berlin (Anon., *Kalkmaschine. Entworfen und ausgeführt durch den Baumeister Hoffmann in Berlin*, in, "Zeitschrift für Praktische Baukunst," II, 1842).

and a pulley, unlike the mortar mixers used for building the Ruhr locks. It is probably worth noting that, approximately at the same time that the Stettin quay is under construction, the engineer Carl Wilhelm Hoffmann uses a similar mortar mixer on the building site of the Berlin *Neues Museum* by Friedrich August Stüler (1800–1865).[31] This kind of machine actually originated in France, in the mortar mixer invented by German Boffrand, toward the middle of the 18th century, and made particularly famous by Perronet, who used it on the building site of the *Pont de Neuilly*.[32]

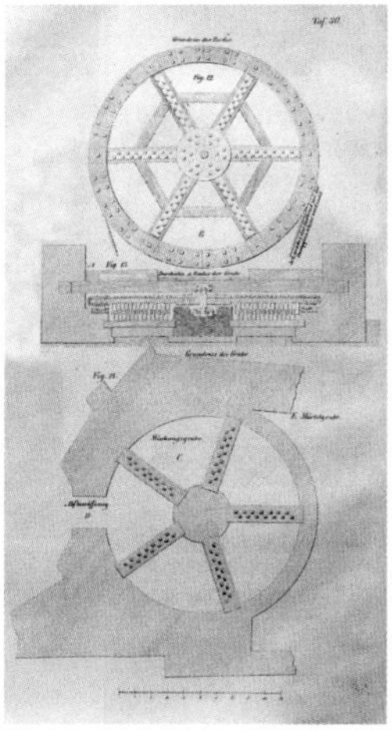

Once the mortar is ready, it is mixed in the concrete drums with crushed stones, bricks and rubble taken from the demolition of nearby fortification walls. Concrete then falls from the drums straight into wagons on rails, to be transported to the riverside and put into barges that deliver it all along the building site. Schwan asserts that he later improved the kind of concrete mixer used to build the foundation of the quay along the Oder, by installing the drum on a sloping structure, so that concrete components can be loaded from the upper extremity and dispensed from the lower one, once mixed.[33] This kind of sloping concrete drum, which Schwan claims to have invented, is actually reminiscent of the frustum-shaped concrete drums that are used in France for the building of the Paris fortification around 1840, and that the "Allgemeine Bauzeitung" publishes in 1843.[34]

The first construction phases of each stretch of the Stettin quay consist in ramming two rows of sheet piles into the riverbed; the one towards the river stream is stronger and slopes towards the landside, whereas the opposite one is lighter and stands vertically. Once this is achieved, concrete is poured by means of boxes into the foundation pits along the stretches where no timber piling is planned; otherwise, first piles are rammed and then concrete

vol. XV, 1999, (pp. 39–55), pp. 45–48.

[32] See Gargiani, *Concrete from archeology*...cit., pp. 150–151.

[33] Schwan, *Beschreibung des Baues einer Kaimauer*...cit., col. 155.

[34] See Alfred Lecointe, *Bemerkungen über einige mechanische Verfahrungsarten zur Bereitung des Mörtels und Betons*, in, "Allgemeine Bauzeitung," VIII, 1843, pp. 399–400.

is poured. Piles also slope, according to the angle of the sheet piles, and a special kind of box is created to pour concrete among them. It is narrow and connected to a long handle by means of a hinge, so that workers can push the box slantwise between the sloping poles.

The increasing need for underwater concrete foundations between the second half the 1830s and the first half of the 1840s emphasizes how crucial the availability of effective and affordable hydraulic binders is, especially in places where neither trass nor natural hydraulic lime are available. This gradually opens the way for the manufacturing of artificial cements. Such awareness is evident in a 1843 writing by Bavarian engineer Carl Maximilian von Bauernfeind (1818–1894). Charged by the Nurnberg Railway-Committee with trials that aim to produce artificial hydraulic lime from materials located near the towns of Hof and Münchberg, Von Bauernfeind states: "the need for hydraulic lime becomes particularly crucial […] when the conditions of the soil, on which bridges, culverts, locks and other masonry works need to be built, impose the use of concrete foundations." "In case no natural hydraulic lime is available," goes on Von Bauernfeind, "there are materials to produce it artificially, reaching the same quality at lower costs."[35] A few years later, the master builder Schwan reports on the construction of the Stettin quay and complains that, although "concrete is recognized as one of the most valuable materials to build underwater, the widespread use of it is limited by the fundamental need of costly trass, in places where natural hydraulic lime is not available."[36]

Lean conglomerates for rural buildings and the development of *Kalksand*

While concrete is being used in the area of underwater construction, some publications in technical literature report on above-ground concrete construction that has been realized abroad. From 1832 onward, the "Polytechnisches Journal" reports about the buildings entirely made of concrete that the architect François-Martin Lebrun has been constructing in southern France since 1828.[37] The book about concrete constructions that Lebrun publishes in 1835 in Paris is quickly translated into German and published in 1837 in Ulm (Wurttemberg).[38] While Lebrun describes concrete as being appropriate to all kinds of construction, the German editor highlights how crucial the availability of local hydraulic lime is to producing affordable concrete,

[35] "Der Mangel an hydraulischem Kalk wird bei Wasserbauten um so mehr fühlbarer, […], wenn der Baugrund für Brücke, Dürchlässe, Schleusen und anderes Mauerwerk, Betonfundationen bedingt." "Denn, wenn wir auch keinen natürlichen hydraulischen Kalk besitzen, so fehlt es uns doch nicht an Materialien, künstlichen zu bereiten, der, bei gleicher Qualität, viel billiger ist als jener" (Carl Maximilian von Bauernfeind, *Ueber hydraulischen Kalk und Mörtel aus Natur- und Kunstprodukten des Fichtelgebirgs*, in, "Kunst- und Gewerbe-Blatt des polytechnischen Vereins für das Königreich Bayern," XXVIII, 1843, (coll. 624–647), coll. 624–625).

[36] "Eines der schätzenswerthesten Materialien zum Wasserbau ist anerkannt der Beton, dessen allgemeiner Anwendbarkeit nur der Umstand entgegen steht, daß die Herstellung der Masse in größeren Quantitäten für alle diejenigen Gegenden zu theuer wird, wo sich ein natürlichen hydraulischen Kalk nicht vorfindet und zur Bildung des Mörtels die sehr kostspieligen Zuschläge von Trass angewendet werden müssen" (Schwan, *Beschreibung des Baues einer Kaimauer*...cit., coll. 160–161).

[37] See Charles-Pierre Gourlier, *Bericht des Hrn. Gourlier über eine Abhandlung des Hrn. Lebrun d. jüng., Maumeister zu Alby, Dept. du Tarn, die Anwendung des Grund oder Steinmörtel (béton) zum Baue ganzer Wohnhäuser betreffend*, in, "Polytechnisches Journal," XIII, vol. XLIV, 1832, pp. 114–119 (Id., *Rapport fait par M. Gourlier, au nom du Comité des arts économiques, sur un mémoire de M. Lebrun jeune, architect à Alby, département du Tarn, relatif à l'emploi du béton pour la construction entière d'une maison d'habitation*, in, "Bulletin de la société d'Encouragement pour l'Industrie Nationale," XXXI, 1832, pp. 99–103); Anon., *Ueber die Benutzung des Steinmörtels zu verschiedenen Bauten*, in, "Polytechnisches Journal," XVI, vol. LXII, 1836, p. 155; Charles-Pierre Gourlier, *Ueber das Verfahren des Hrn. Lebrun, Bauten aus Wassermörtel; ein der Societé d'Encouragement von Hrn. Gourluer erstatteter Bericht*, in, "Polytechnisches Journal," XXII, vol. LXXXV, 1842, pp. 431–433 (Id., *Rapport fait par M. Gourlier, au nom du Comité des arts économiques, sur les constructions en béton de M. Lebrun, architecte à Montauban (Tarn-et-Garonne)*, in, "Bulletin de la société d'Encouragement pour l'Industrie Nationale," XLI, 1842, pp. 283–286).

[38] François-Martin Lebrun, *Méthode pratique pour l'emploi du béton en remplacement de*

and stresses the prevalent suitability of concrete for hydraulic works and for vaults. These are essential aspects of the spread of concrete construction in Germany. We already stressed the importance given to the exploitation of local raw materials to producing hydraulic binders and a rising awareness of the importance of concrete in building underwater. Meanwhile, the ambition of building conglomerate vaults gradually develops during the second half of the 1840s with some references to the conglomerate vaults of Roman antiquity. To effectively grasp the German approach to Lebrun's way of building in concrete, it is worth quoting a passage from the preface to the aforementioned translation. "The concrete buildings by Mr. Lebrun," states the text, "are durable, as it has been certified by committees, and affordable, in comparison to all other kinds of masonry; they might therefore produce an essential change in the current way of building, especially in hydraulic works and in the construction of vaults. Wurttemberg and other German States have even an edge on France, since they have plenty of natural hydraulic lime, which is an essential material to produce concrete, and they do not need to produce it artificially."[39]

It is still worth noting that the term *béton* is translated with the old, German, compound word *Steinmörtel*, which is already found in some dictionaries from the 18th century.[40] As already observed in Wiebeking's *Wasserbaukunst*, the term *Béton* (later *Beton*) gradually enters the German language under the influence of French building culture, first in the specific lexicon of hydraulic engineering, and, later, in the general lexicon of construction. As a matter of fact, just a few years later, the second edition of Lebrun's book about concrete, which dates back to 1843, is translated into German using the term *Béton*.[41]

Also in 1837, Heinrich Friedrich Franz Körte (1782–1845), who teaches natural science at the *Landwirthschaftliche Akademie of Möglin* in Pomerania, translates and publishes in the periodical "Möglin'sche Jahrbücher der Landwirthschaft" a short Swedish book about a construction technique based on the use of rough-cut stones, timber and a sort of raw conglomerate, which the entrepreneur Carl August Rydin (1798–1877) has created to rapidly build affordable and fireproof houses after a fire had destroyed his town Borås in 1827.[42] Rydin recommends using rough-cut stones to build foundations,

toute autre espèce de maçonnerie dans le construction en général, Paris, Chez Carilian Goeury, 1835; Id., *Der Steinmörtel oder praktische Anweisung den Steinmörtel bei Gebäuden im Allgemeinen, besonders aber bei Bauten an und unter dem Wasser, bei Gewölben etc. statt jeder andern Art von Mauerarbeit Vortheil zu benutzen*, Ulm, Nübling, 1837.

[39] "Die Bauten, welche Herr Lebrün aus Steinmörtel aufführte, die Dauerhaftigkeit derselben, welche durch Commissionen beurkundet wurde, die entschiedene Wohlfeilheit derselben im Vergleich mit den andern Maurerarbeiten, dürften, besonders beim Wasser- und Gewölbebau, eine gänzliche Reform in der bisherigen Art zu bauen herbeiführen. Württemberg und andere deutsche Staaten haben überdieß noch einen bedeutenden Vorzug vor Frankreich, indem sie den hydraulischen Kalk, ein Hauptmaterial bei Verfertigung des Steinmörtels, von Natur aus in großer Menge besitzen, und nicht zu Verfertigung des künstlichen hydraulischen Kalkes ihre Zuflucht nehmen dürfen" (Lebrun, *Der Steinmörtel oder praktische Anweisung*...cit., n.p.)

[40] About the definition of *Steinmörtel* in 18th century, see Lukas Voch, *Baulexikon, oder Erklärung der teutschen und französischen Kunstwörter in der bürgerlichen Kriegs- und Schiffbaukunst, wie auch der Hydrotechnik und Hydraulik*, Augsburg, Leipzig, Matthäus Riegers sel. Söhnen, 1781, p. 285; Johann Karl Gottfried Jacobsson, Otto Ludwig Hartwig, Gottfried Erich Rosenthal, *Johann Karl Gottfried Jacobssons technologisches Wörterbuch oder alphabetische Erklärung aller nützlichen mechanischen Künste, Manufakturen, Fabriken und Handwerker, wie auch aller dabey vorkommenden Arbeiten, Instrumente, Werkzeuge und Kunstwörter, nach ihrer Beschaffenheit und wahrem Gebrauche*, Berlin, Stettin, Friedrich Nicolai, vol. IV, 1784, p. 279. Both dictionaries define *Steinmörtel* as a lime-based concrete, but Voch also introduces the entry "Betonmörtel," which is described as a sort of *Steinmörtel* containing pozzolana, trass or Dornick ash (see Voch, *Baulexikon*...cit., pp. 40–41).

[41] See François-Martin Lebrun, *Traité pratique de l'art de bâtir en béton*, Paris, Carilian et Goeury et Victor Dalmont, 1843, Id., *Practische Abhandlung über die Kunst mit Béton zu bauen*, in, "Journal für die Baukunst," vol. XIX, 1843, pp. 135–186, 195–253, 287–346.

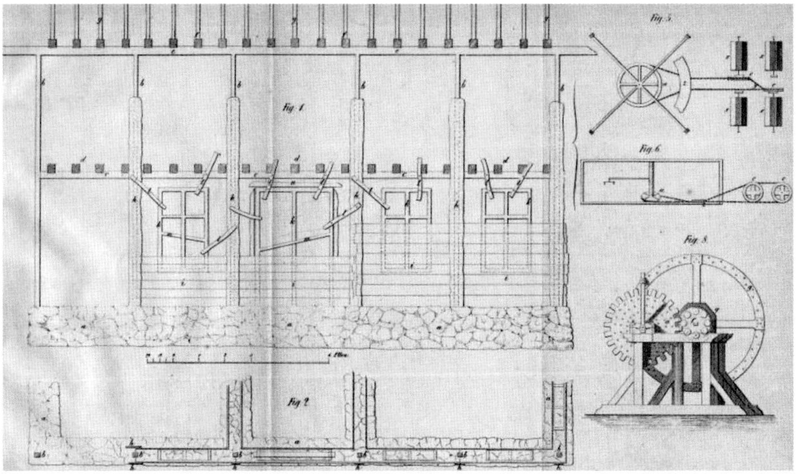

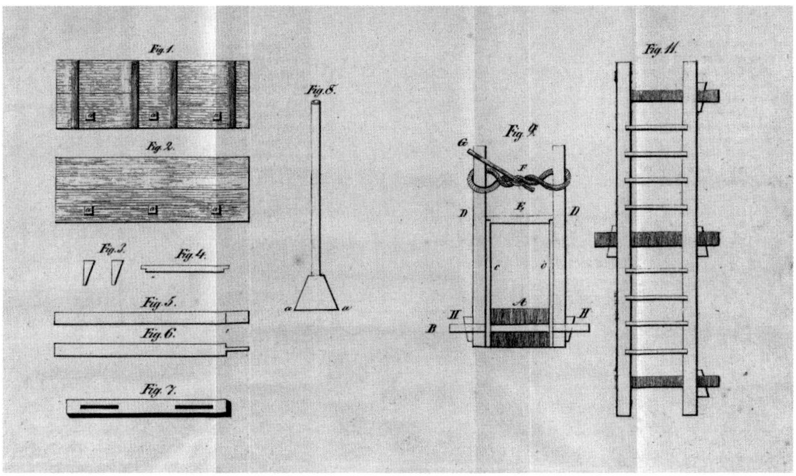

69. C. G. Rydin, construction drawings of a house made of stones, timber and lean conglomerate; J. D. F. Engel, mixer for lean conglomerate used in Cöslitz (J. D. F. Engel, *Der Kalk-Sand-Pisébau und die Kalksand-Ziegelfabrikation*, Leipzig, C. H. Seemann, Leipzig, 1865).

70. J. G. Prochnow, tools for building with lean conglomerate. Formwork for walls, pestle, mould for lean conglomerate-based artificial stones (J. G. Prochnow, *Anleitung zur Kunst, Wohnungen und Wirthschafts - Gebäude in sehr kurzer Zeit wohlfeil, Feuer- und Wetterfest zu erbauen aus reinem Sande und sehr wenigem Kalk*, Schwedt, J. C. W. Jantzen, 1842).

timber to build a framework structure from the foundation to the roof and conglomerate to build walls by filling the framework fields and, at the same time, covering the timber elements in order to protect them in case of fire. Conglomerate is made of stones, pebbles and a kind of quite lean mortar made of four parts sand and one part lime. Rydin first pours mortar in layers between provisional timber boars that have been fixed at the framework poles, and, then, he arranges stones in the mortar with the goal of saving mortar.

Johann Gottlieb Prochnow, a farmer from the village of Bahn in Pomerania, is aware of the aforementioned publications, in addition to the *Traité sur l'art de faire des bons mortiers* by Raucourt de Charleville and the *Cahiers de l'Ecole d'Architecture Rurale* by François Cointeraux, which deals with the building technique of *pisé* or rammed earth.[43] Searching for an affordable technique to build fire- and weather-proof walls, Prochnow adjusts the techniques he has read about according to the local availability of building materials, and conceives walls entirely made of a very lean conglomerate produced with sand and ordinary lime. Around 1840, he applies this technique to build a stable and a shed at his estate. Shortly thereafter, the ruler of the Gryfino province, Carl Freiherr von Steinaecker, asks Prochnow to write some instructions about the technique he conceived. The instructions are published in 1842 and represent the main source of knowledge about Prochnow's buildings.[44] They have foundations made of rough-cut stones, timber roofs and walls made just of lean conglomerate poured and tamped in quite rudimentary timber formworks, which Prochnow actually adopts from Cointeraux's illustrations of rammed earth constructions. They consist of two longitudinal faces made of boards kept at an appropriate distance by means of three crosspieces, at the bottom, and six stakes, which are connected into the crosspieces and bound two by two at the top by means of ropes. The conglomerate is made of ten measures of gravel, two and a half measures of sand and one measure of lime. Unlike Lebrun, Prochnow abandons hydraulic lime, since it is not easily available in his region.[45] Nevertheless, in case rough-cut stones for foundations were unavailable, Prochnow suggests using a kind of lean conglomerate containing some pebbles and brick rubble, which are supposed to endow "mortar with the property of rapidly hardening in presence of humidity."[46] Prochnow's conglomerate can be considered a sort of non-hydraulic fine-grained concrete, which, in the case of foundations, becomes a kind of real hydraulic concrete. In spite of that, Prochnow uses neither the term *Béton*, nor *Steinmörtel*, and resorts to the compound word *Formmasse*, which essentially means moulding compound. Prochnow also proposes pouring lean conglomerate in appropriate moulds to produce artificial stones. This idea very probably derives from Cointeraux's book, but it also makes sense in light of the German ambition to produce mortar-based artificial stones, and we cannot rule out the idea of a

[42] Carl Gustaf Rydin, *Gjutna kalkbrukshus. Ett nytt byggnadssätt*, Stockholm, Georg Scheutz, 1834; Id., *Das gegossene Kalkmörtelhaus. Eine neue Bauart von C. H. Rydin* [sic!], ed. Heinrich Friedrich Franz Körte, in, "Möglin'sche Jahrbücher der Landwirthschaft," vol. II, 1837, pp. 151–181. See also Carl Gustaf Rydin, *Gegossenes Kalkmörtel-Haus. Eine neue Bauart von Rüdin* [sic], ed. Ström, in, "Kunst- und Gewerbeblatt des Polytechnischen Vereins für das Königreich Bayern," XXVIII, 1842, coll. 502–513, and in, "Polytechnisches Journal," 1842, vol. LXXXV, pp. 423–431.

[43] Antoine Raucourt de Charleville, *Traité sur l'art de faire de bons mortiers et notions pratiques pour bien en diriger l'emploi: précédé d'expériences recentes faites sur les chaux de France et de Russie*, St. Pétersbourg, Voies de communication, 1822; Id. *Traité sur l'art de faire de bons mortiers et d'en diriger bien l'emploi, ou Méthode générale pratique pour fabriquer en tous pays la chaux, les ciments et les mortiers les meilleurs et les plus économiques*, Paris Malher et Cie, 1828^2. Cointeraux, *Ecole d'Architecture rurale*, cit.

[44] Johann Gottlieb Prochnow, *Anleitung zur Kunst, Wohnungen und Wirthschafts - Gebäude in sehr kurzer Zeit wohlfeil, Feuer- und Wetterfest zu erbauen aus reinem Sande und sehr wenigem Kalk*, Schwedt, J. C. W. Jantzen, 1842.

[45] *Ibid.*, p. 7.

[46] *Ibid.*, p. 16.

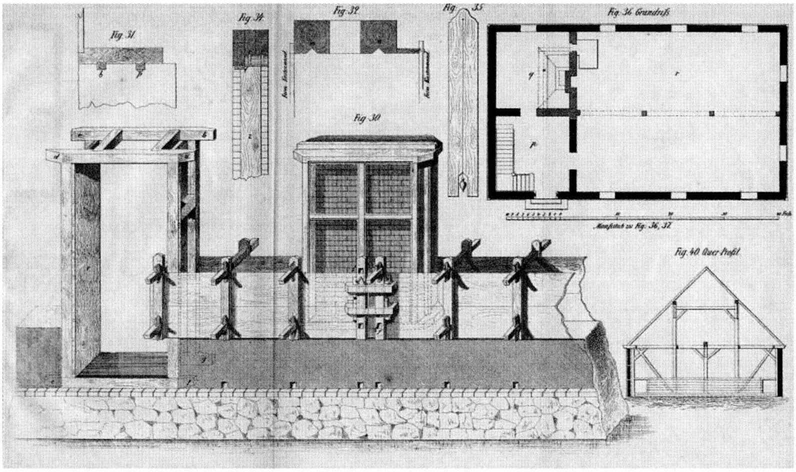

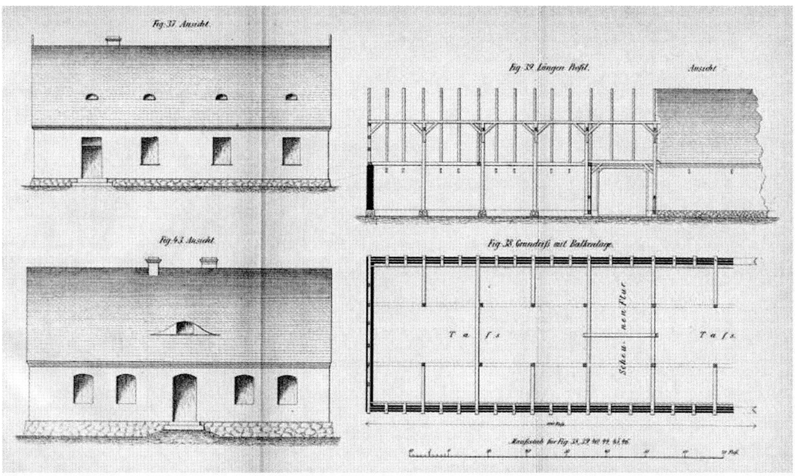

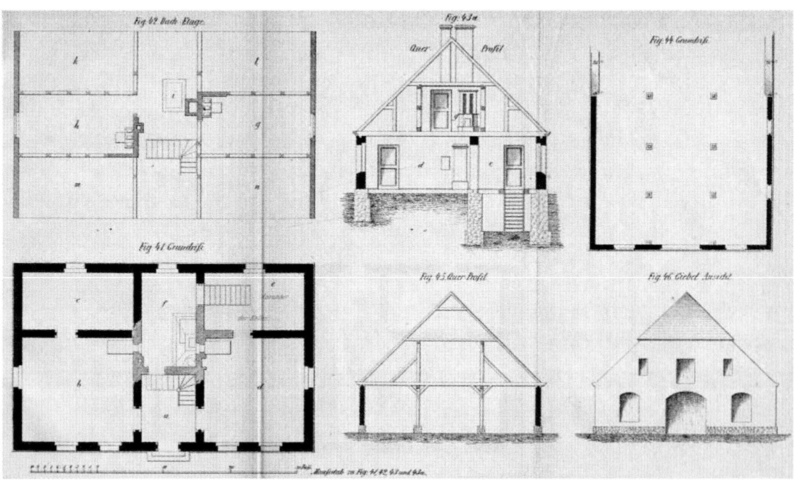

71-73. J. D. F. Engel, construction of rural houses and sheds made of stones, timber and lean conglomerate (J. D. F. Engel, *Der Kalk-Sand-Pisébau und die Kalksand-Ziegelfabrikation*, Leipzig, C. H. Seemann, Leipzig, 1865).

possible influence of Sachs's artificial stones. Even if it seems to have limited consequences in the short term, the production of artificial stones made of lean conglomerate will be implemented about fifteen years later, opening the way to the manufacturing of sand-lime bricks.[47]

Since about the mid 1840s, the use of lean conglomerate spreads from Bahn to some other villages of Pomerania, Brandenburg, Silesia and even Bavaria, producing a number of occasions to experiment with variations and even, in some cases, to improve the mixture and the formworks. Around 1845, the entrepreneur Emil Berndt builds two factories and fourteen houses in poured lean conglomerate in "Deuben near Dresden."[48] Berndt uses a kind of conglomerate made of lime, coal slags and mud. Coal slags are reminiscent of the Dornick ash and seem to anticipate the production of slag-based concrete and cement that develops about twenty-five years later. The master builder Johann Daniel Friedrich Engel (1821–1890), who is a major promoter of lean conglomerate, builds a rural house in Brandenburg in 1847, and Albrecht Daniel Thaer (1752–1828), the agronomist who founded the aforementioned *Landwirthschaftliche Akademie of Möglin*, builds a house and a barn in Möglin.[49] In some cases, however, traditional timber framework supports the roof and stabilizes the whole building. Between 1845 and 1847, the entrepreneur Johann Carls Leuchs erects several buildings in lean conglomerate in "weißen Au, près de Nuremberg."[50] The main one is an imposing building that covers a surface of about 350 feet by 36 feet and houses a factory and residences, rather than the far more common rural activities. Foundations are made of hydraulic lean conglomerate, which is produced using black lime from the nearby town of Bamberg, whereas the walls are built using a mixture of fat lime, sand and pebbles. The use of lancet windows appears to be original for lean conglomerate buildings. It enables the removal of load-bearing wooden frames once the conglomerate has hardened, whereas ordinary windows

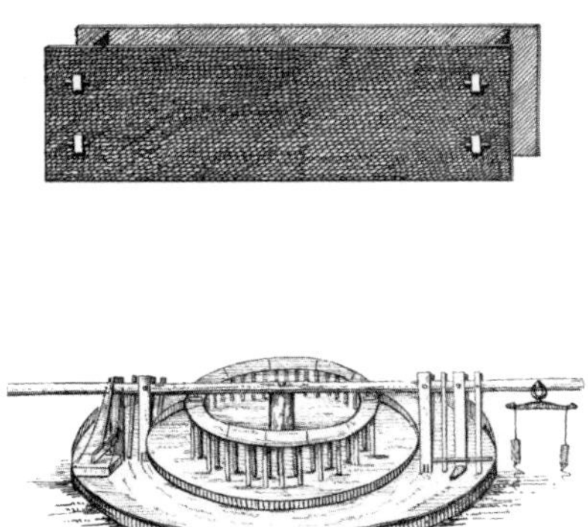

74. J. C. Leuchs, drawings of a formwork and a mortar mixer used to build a factory in lean conglomerate in Weissen Au, approx. 1845–1847 (J. C. Leuchs, *Verbesserte Darstellung der Kitte, Mörtel, künstlichen Steine, künstlichen Massen, des Siegellaks, der Oblaten, der Löthe, und Anwendung der ersteren zum wohlfeilsten Bau von Häusern, Mauern, Kellern, Fußboden, Dächern, Rauchfängen, Wasserbecken, Wasserröhren, Brücken und Strassen*, Nürnberg, C. Leuchs & Co., 1848).

75. M. K. E. von Von Prittwitz, mortar mixer used to build fortresses in Silesia (M. K. E. von Prittwitz, *Beschreibung einer mechanischen Vorrichtung zur Bereitung des Mörtels*, in, "Verhandlungen des Vereins zur Beförderung des Gewerbfleisses in Preußen," XI, 1832).

[47] *Ibid.*, pp. 28-30. In 1854, the physician Anton Bernhardi conceives a press for the mass production of lean conglomerate bricks that he considers suitable to build affordable and salubrious houses for workers; see Johann Daniel Friedrich Engel, *Ueber Kalkziegelfabrikation und Kalkziegelbau*, in, "Polytechnisches Journal," vol. CLIII, 1859, pp. 100–09; Id., *Der Kalk-Sand-Pisébau und die Kalksand-Ziegelfabrikation…* cit., pp. 99–118; Bernhard Liebold, *Zement*, Halle, Hendel, 1874, p. 53.

[48] See C. Gebhardt, *Der Aschenstampfbau*, in, "Zeitschrift für Bauhandwerker," XVII, 1873, pp. 159–160, 174–175.

[49] Engel, *Der Kalk-Sand-Pisébau und die Kalksand-Ziegelfabrikation…*cit., pp. 94–95.

[50] Johann Carl Leuchs, *Verbesserte Darstellung der Kitte, Mörtel, künstlichen Steine, künstlichen Massen, des Siegellaks, der Oblaten, der Löthe, und Anwendung der ersteren zum wohlfeilsten Bau von Häusern, Mauern, Kellern, Fußboden, Dächern, Rauchfängen, Wasserbecken, Wasserröhren, Brücken und Strassen*, Nürnberg, C. Leuchs & Co., 1848, pp. 114–129. See also Anon., *Ueber Häuserbau mit Mörtel*, in, "Kunst- und Gewerbeblatt des Polytechnischen Vereins für das Königreich Bayern," XLII, 1856, p. 576.

76. J. C. Leuchs, factory in lean conglomerate, Weissen Au, approx. 1845–1847. Elevation, drawing of a stretch of wall in process of building (J. C. Leuchs, *Verbesserte Darstellung der Kitte, Mörtel, künstlichen Steine, künstlichen Massen, des Siegellaks, der Oblaten, der Löthe, und Anwendung der ersteren zum wohlfeilsten Bau von Häusern, Mauern, Kellern, Fußboden, Dächern, Rauchfängen, Wasserbecken, Wasserröhren, Brücken und Strassen*, Nürnberg, C. Leuchs & Co., 1848).

77. J. D. F. Engel, drawings of mixers for lean conglomerate. Rydin's mixer, screw-mixer used in Fiddichow, Pomerania. In addition, a rake and the face of a formwork (J. D. F. Engel, *Der Kalk-Sand-Pisébau und die Kalksand-Ziegelfabrikation*, Leipzig, C. H. Seemann, Leipzig, 1865).

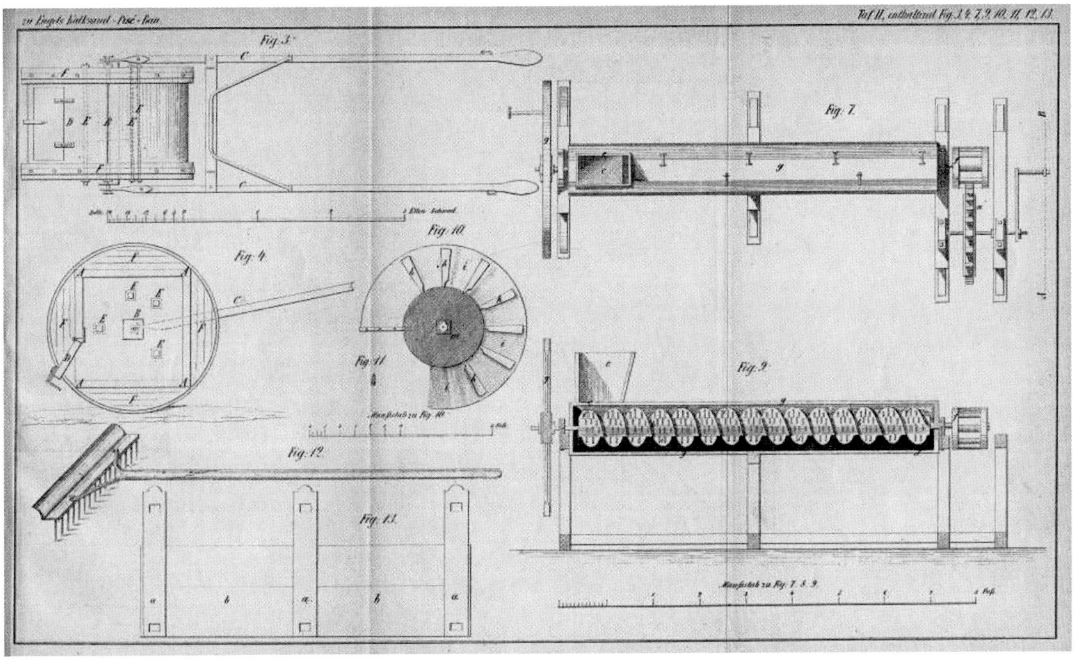

78-79. J. D. F. Engel, drawings of formworks and tools to build with lean conglomerate (J. D. F. Engel, *Der Kalk-Sand-Pisébau und die Kalksand-Ziegelfabrikation*, Leipzig, C. H. Seemann, Leipzig, 1865).

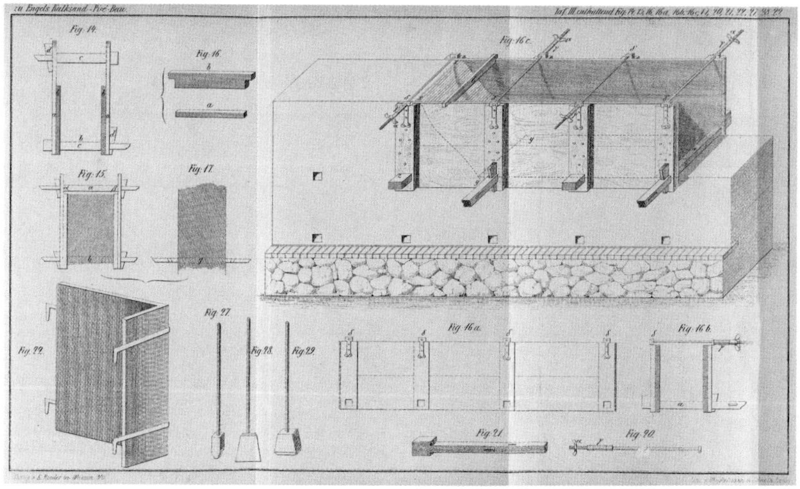

in lean conglomerate buildings are generally provided with permanent-bearing frames, similar to those in rammed earth buildings. A certain Mr. Badeke from the village of Fiddikow also develops an original way of building windows. Badeke first builds full walls and then cuts the openings of doors and windows once the conglomerate has hardened.[51]

According to Engel, early attempts at building lean conglomerate vaults date back to the mid 1840s. Badeke is said to have built 11-feet-span and 15-inches-thick vaults of lean conglomerate between brick masonry arches, which are similar to certain ancient Roman barrel vaults; whereas a certain Mr. Blut from Cöslitz, in Pomerania, is said to have tried the contrary, that is, masonry vaults between lean conglomerate arches.[52]

Similarly to concrete, lean conglomerate is sometimes produced mechanically. Prochnow suggests using a singular mixer that he adopts from Rydin, in a text that he publishes in 1845 as an update to the 1842 instructions. It essentially consists of a box outfitted with some internal crosspieces and an opening to charge loose material and take out the conglomerate. This box is fixed on two wheels of a horse-cart, so that when the cart is drawn, the wheels turn and the conglomerate components mix together. In the village of Cöslitz in Silesia, lean conglomerate is prepared in turning drums, which are similar to the ones used for concrete, whereas a singular kind of mixer is used in the village of Fiddichow in Pomerania. It is a sort of Archimedes screw placed in a long, horizontal box, provided with an upper opening at an extremity, to charge lime and aggregates, and a bottom opening at the opposite extremity, to collect the conglomerate that the turning screw has mixed. Leuchs uses a kind of mixer similar to the ones used on the building sites of the Pomeranian fortresses.[53] This animal-powered machine consists of a circular basin, a central pole and a beam, which is hinged on the central pole and holds poles and brushes to mix lime and aggregates in the basin.

Improvements of formworks mainly aim at reinforcing the side faces,

[51] See Johann Daniel Friedrich Engel, *Der Bau in Kalk-Sand-Pisée*, in, "Zeitschrift für praktische Baukunst," VII, 1847, (coll. 9–14), col. 12.
[52] See Engel, *Der Bau in Kalk-Sand-Pisée*...cit., col. 14.

[53] See Moritz Karl Ernst von Prittwitz, *Beschreibung einer mechanische Vorrichtung zur Bereitung des Mörtels*, in, "Verhandlungen des Vereins zur Beförderung des Gewerbfleisses in Preußen," XI, 1832, pp. 283–284. A similar kind of mortar mixer will be also used during the 1850s, on the construction site for the lock and the fortifications of Spandau near Berlin (see Cremer, *Mörtelmachine, welche bei idem Bau des Schifffahrtsschleuse des Berlin-Spandauer Kanals bei Plötzensee in Anwendung gekommen ist*, in, "Zeitschrift für Bauwesen," VI, 1856, coll. 192–194).

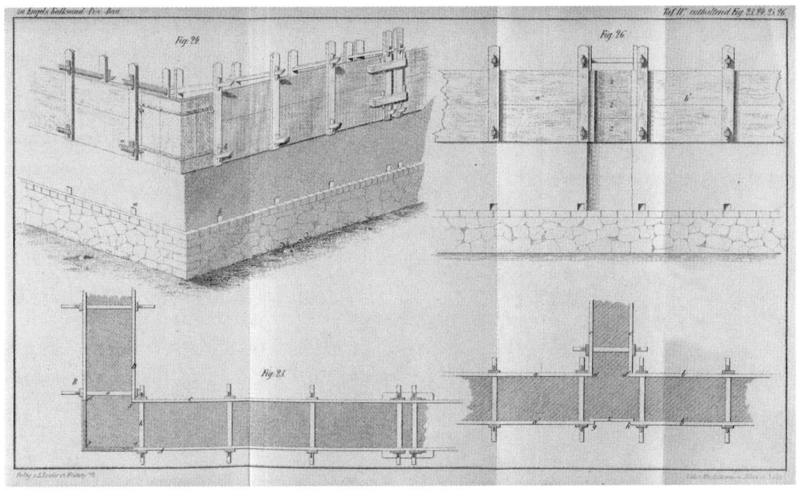

which otherwise bend easily. For this purpose, Prochnow imagines fixing the vertical poles to the horizontal boards, and connecting the two sides of the formwork by means of crosspieces at the bottom and at the top, instead of using ropes. Further improvements concern the replacement of timber crosspieces with metal adjustable screws, and the design of different formwork pieces expressly conceived to build wall corners.

Thanks to the previously described developments, interest in lean conglomerate construction increases during the second half of the 1840s and, between about 1847 and the very early 1850s, a number of publications, including monographs, describe the practical rules to apply this way of building.[54] In some of these publications, lean conglomerate is named *gestampfter Mörtel* (rammed mortar), in others *Kalksand*, and the building technique is defined as *Kalksandbau*, *Kalksandpisé*, or *Kalksandpisébau*. Kalksand is the name under which lean conglomerate acquires its place in the panorama of building techniques used in Germany and, simultaneously, a certain awareness about the similarity between *Kalksand* and concrete begins to develop, opening the way for the spread of concrete use in above-ground construction during the following two decades.

Concrete and cement in Hamburg, between Dutch, French and English influences

Significant development concerning cements, hydraulic mortar and concrete occur in Hamburg between 1843 and 1846, against the backdrop of the reconstruction and modernization of the city after the great fire of 1842. Around 1843, the merchant Emil Müller is licensed to sell Portland cement from the factory J. M. Maude, Son and Co. in London.[55] In 1846, Müller himself writes about Portland cement and its properties in a report that he sends to the "Journal für die Baukunst," mentioning, inter alias, about the flexural strength tests on beams and cantilevers that the engineer Eduard J'Anson junior executes in 1845 in Maude's factory.[56] According to Müller, Portland

[54] Engel, *Der Bau in Kalk-Sand-Pisée*...cit.; Leuchs, *Verbesserte Darstellung*...cit.; Johann Christian Wedeke, *Der Bau mit gestampftem Mörtel*, Quedlinburg, Leipzig, Gottfried Basse, 1850; Johann Daniel Friedrich Engel, *Der Kalk-Sand-Pisébau*, Wriezen, Roeder, 1851; [Friedrich Conrad Theodor] Krause, *Anleitung zur Kalk-Sand-Bau-Kunst*, Glogau, Flemming, 1851.

[55] In an early book about Portland cement, dating back to 1853, the Berlin master builder [Wilhelm Andreas] Becker asserts that Portland cement does not appear in Germany before 1843; see [Wilhelm Andreas] Becker, *Erfahrungen über den Portland-Cement*, Berlin, Gebauersche Buchhandlung (I. Petsch), 1853, esp. pp. 3–4, 46. About the factory J. M. Maude, Son & Co., see Francis, *The cement industry*...cit., pp. 110–113.

80-81. T. Bülau, *Haus der Patriotischen Gesellschaft*, Hamburg, 1844–1847. Plans and sections of the building site (T. Bülau, *Das Haus der patriotischen Gesellschaft in Hamburg oder: Einiges über einen Neubau*, Hamburg, I. I. Nobiling, 1849).

cement has been used in Hamburg to plaster parts of the façades of some important buildings such as the *Alster-Arkaden* (1843), the *Börsen-Arkaden* (1845) and the *Wachtgebäude* at the St. Paul harbour (1845).

Crucial to the reconstruction of Hamburg is the improvement of the connection between the Rivers Alster and Elbe. It requires the building of a lock in the *kleiner Alster*, under the directorship of the *Bauinspector* Johann Hermann Maack (1809–1868), from 1843 to 1846.[57] Mack has travelled between 1835 and 1838 to Holland, northern France, England and Scotland to study the most advanced hydraulic works of that time, and it is probably in light of what he learned during these research journeys that he plans to build the lock in the *kleiner Alster* on a concrete foundation.[58] He undertakes trials to test the effectiveness of concrete underwater and experiments with specimens prepared according to what he considers to be the typical English way to produce concrete, namely by mixing the different components kept dry, then adding water just at the final step. The mixture is composed of hydraulic lime burnt from Blue Lias stones from the south of England, which have a particularly good reputation, together with egg-sized pebbles, sand and iron powder – this last one also coming from England.[59] After having been kept underwater for several months, the specimens do not give encouraging outcomes. Maack therefore gives up the English way of producing concrete and resorts to trass-based concrete, which he has studied from Jean-Baptiste Rondelet and observed in Holland.[60] To prepare it, he uses five measures of slaked lime from Prussia, six of trass, eight of sand and an appropriate amount of water to get a kind of hydraulic mortar that he then mixes with twice the amount of rubble.

In 1844, while the lock in the *kleiner Alster* is still under construction, the *Gesellschaft zur Beförderung der Künste und Nützlichen Gewerbe* (Society for the Encouragement of Arts and Useful Crafts), better known as *Patriotische Gesellschaft*, holds a competition for the design of the new seat of the Society, which becomes known as the *Haus der Patriotischen Gesellschaft*. The new building is planned to be erected on the site of the old town hall, which has been destroyed by the fire of 1842. Maack and the architect Heinrich Wilhelm Burmster (1802–1849), who were charged with examining the soil of the plot, suggest pouring a concrete bed foundation over the entire site and exploiting the timber piles of the old town hall, which are still in the foundation's soil.[61] Since the design of the building, which has been entrusted to the architect Theodor Bülau (1800–1861), is expected to still be in progress at the beginning of construction, the concrete bed laid all over the site is intended to be a sort of raft slab that enables a free placement of the walls and columns of the basement.[62]

[56] See Emil Müller, *Vom Portland-Zement*, in, "Journal für die Baukunst," vol. XXIII, 1846, pp. 368–72; the same article is also published in the Austrian "Allgemeine Bauzeitung" see Anon., *Portland-Zement*, in, "Allgemeine Bauzeitung," XI, 1846, pp. 111–113.

[57] It is the *Rathausschleuse* of today; in this regard, see Jörn Gutbrod, *Wiederherstellung der biologischen Durchgängigkeit der Alster an der Rathausschleuse in Hamburg*, master degree dissertation, Hamburg, Technische Universität Hamburg-Harburg, 2003, esp. pp. 14–16. About Johann Hermann Maack and his hydraulic works in Hamburg, see Stephan Rolph, *Johann Hermann Maack: 1809 bis 1868: Bauinspektor der Brücken und Schleusen in Hamburg von 1841 bis 1868: sein Lebensweg und seine Bauwerke*, Hamburg, C. W. Dingwort, 1987.

[58] About Maack's education as hydraulic engineer, see Ralph, *Brücken für Hamburg*...cit., pp. 13–16.

[59] See Johann Hermann Maack, *Einiges über die Bereitung und Anwendung des Concrets oder Betons, mit besonderes Rücksicht auf die Fundirung des Gebäudes der Patriotischen Gesellschaft*, in, *Verhandlungen und Schriften der*

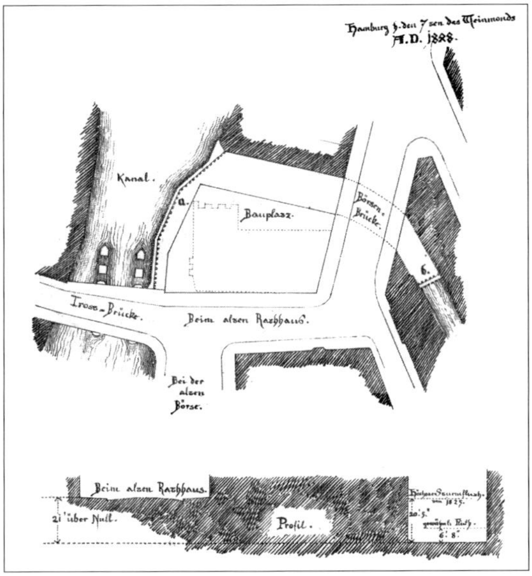

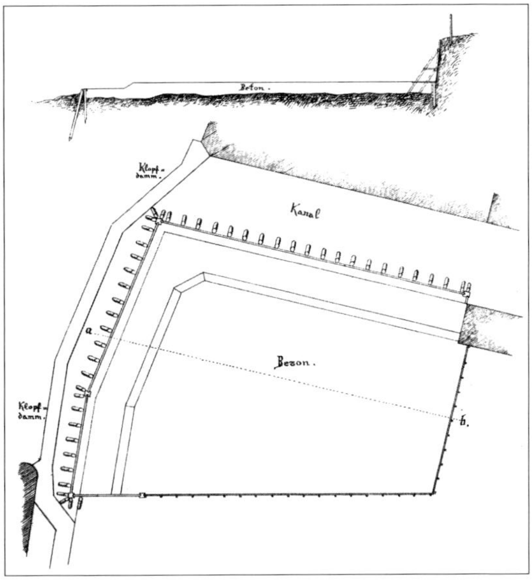

The site is surrounded by two streets, the canal named *Nikolaifleet* and a secondary smaller canal. The side along the *Nikolaifleet* is the most exposed to tide and is therefore provided with a provisional cofferdam in the spring of 1844. It is a kind of cofferdam covered with boards serving as a gangplank during periods of low tide, and it is to be removed once the walls of the basement are completed.[63] Excavation takes place during the summer of 1844; sheet piles reinforced by slanting poles are rammed along the two watersides, while the two street-sides are shored. Due to the lack of free space around the foundation pit, the small canal is covered with boards and provided with sheds for storing materials and preparing mortar and concrete. Seven measures of slaked lime, five of trass and thirteen of sand compose the mortar. The concrete is initially produced adding pebbles to mortar based on a 2:1 ratio; however, given the lack of pebbles at an affordable price, these are later replaced with gravel and brick rubble.[64] Maack claims to have used a lower amount of trass in comparison to the foundation of the lock in the *kleiner Alster*. He believes that the foundations of buildings are less demanding than those of locks and considers it advantageous to use a larger amount of lime in the presence of brick rubble, seeing that lime and bricks bind well together.[65] Concrete is transported by means of wheelbarrows from the sheds over the small canal onto a

Hamburgischen Gesellschaft zur Beförderung der Künste und Nützlichen Gewerbe. Verhandlungen vom Jahre 1845, vol. I, 1846, (pp. 263–270), pp. 264–266. About the reputation of hydraulic lime from Blue Lias and the state of art of knowledge and practice concerning cement and concrete in England, see Charles William Pasley, *Observations on limes, calcareous cements, mortars, stuccos and concrete, and on puzzolanas, natural and artificial*, London, John Weale, 1838.

[60] For Maack's opinion about hydraulic lime, trass-based mortar and its use in Holland, as well as for the reference to Rondelet, see Franz Gustav Joachim Forsmann, Johann Hermann Maack, "Bericht den Vorschlag des Ingenieurs Giles über die Anwendung des hydraulischen Kalkes beim Bau der Hammerbrook-Schleuse u. s. w. betreffend," April 14, 1845 (SAH, 321–2 Baudeputation, B451, n. fol.).

[61] See Johann Friedrich Voigt, *Bericht der am 2. Mai 1844 niedergesetzten Commission, zur Beratung über den Bauplan für das Gebäude der Patriotischen Gesellschaft*, in, *Verhandlungen und Schriften der Hamburgischen Gesellschaft zur Beförderung der Künste und Nützlichen Gewerbe. Verhandlungen vom Jahre 1844*, vol. I, 1845, (pp. 120–137), p. 136; Theodor Bülau, *Das Haus der patriotischen Gesellschaft in Hamburg oder: Einiges über einen Neubau*, Hamburg, I. I. Nobiling, 1849, p. 1.

[62] *Ibid.*

[63] *Ibid.*, p. 2.

[64] *Ibid.*, p. 3.

[65] See Maack, *Einiges über die Bereitung und Anwendung des Concrets oder*...cit., p. 266.

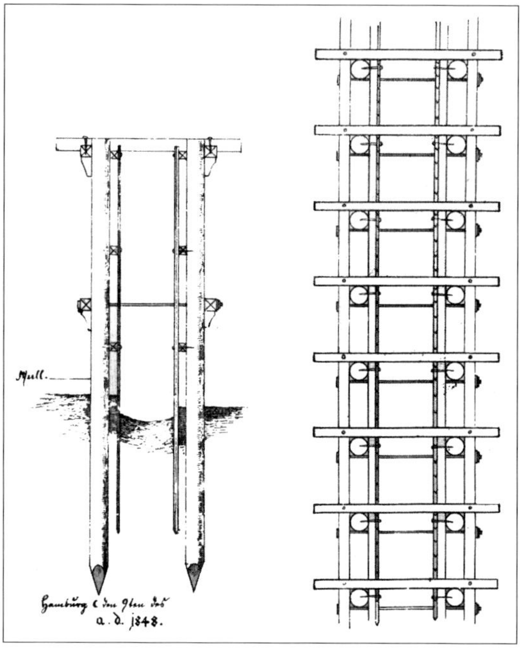

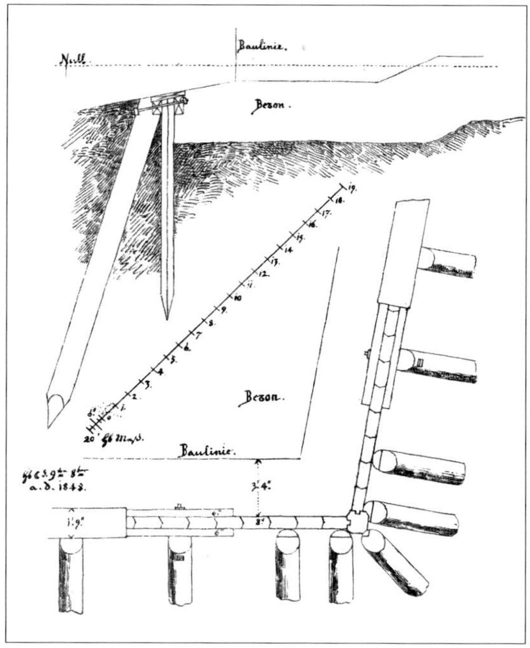

scaffolding-bridge placed over the foundation pit and poured down by means of shovels from a height of about 20 feet, in parallel strips, each one measuring 4 feet in height. The scaffolding-bridge consists of two 8 feet large gangplanks fixed on a framework of bars and crossbars, which can be shifted each time a strip is completed. Part of the concrete foundation is finished in December 1844, when works must be stopped because of the onset of frost. The concrete is kept underwater until April 1845 when work resumes, and the foundation is completed in the summer of the same year.[66]

While the building of the *Haus der Patriotischen Gesellschaft* advanced, in April 1845, Maack begins to plan the lock at Hammerbrook, east of Hamburg. Supported by the master builder Franz Gustav Joachim Forsmann (1795–1878), he engages in a dispute with the British engineer Francis John William Thomas Giles about whether hydraulic lime or cement made of fat lime and trass is the most appropriate binder for producing hydraulic mortar. This argument is similar to Zimmermann's on the same matter, which dates back to the end of the 1820s and uses Treussart and Vicat as its main references, doubly so, considering that the reference to Vicat is still current in the controversy between Maack and Giles. In a report to the Hamburg *Baudeputation*, Giles asserts that both hydraulic lime and cement made of fat lime and pozzolana or trass are suitable for underwater construction, whereas only hydraulic lime is appropriate for construction exposed to the

[66] See Bülau, *Das Haus der patriotischen Gesellschaft...* cit., pp. 2–4; Voigt, *Bericht der am 2. Mai 1844...*cit., p. 136; W. A. Kramer, *Berich über die Deliberation-Versammlung vom 31 October, 1844*, in, *Verhandlungen und Schriften der Hamburgischen Gesellschaft zur Beförderung der Künste und Nützlichen Gewerbe. Verhandlungen vom Jahre 1844*, vol. I, 1845, (pp. 167–171), p. 169; Heinrich Hübbe, *Bericht über die 362ste Deliberations-Versammlung vom 3. April 1845*, in, *Verhandlungen und Schriften der Hamburgischen Gesellschaft zur Beförderung der Künste und Nützlichen Gewerbe. Verhandlungen vom Jahre 1845*, vol. I, 1846, (pp. 358–362), p. 360.

air. Thus, to build the lock at Hammerbrook, he suggests using hydraulic lime burnt from Blue Lias, which he considers suitable for both the foundation and the masonry of the lock chamber.[67] Maack and Frosmann, on the other side, believe in the suitability of trass-based mortar for masonry works exposed to air, and mention, as support, the Dutch houses, which they say were typically built with trass-based mortar, and they reference as well the Italian pozzolana-based mortar, which is believed to work both underwater and above ground.[68] Giles diplomatically answers that, although he is unfamiliar with Dutch buildings, and although he does not consider Italian buildings as appropriate examples because of the differences in weather and temperature between Italy and Germany, he considers both trass-based and hydraulic lime-based mortar as generally appropriate in humid places and, consequently, to the lock at Hammerbrook.[69]

Meanwhile, the construction of the walls and columns in the basement of the *Haus der Patriotischen Gesellschaft* proceeds, but, during the summer of 1845, a crack appears in the concrete foundation, causing water to seep into the basement and further visible cracks in the masonry under construction. The leak is covered with some layers of bricks laid in cement-based mortar, and the water seeping stops. The problem is resolved, but it is further discussed at the assembly of the *Patriotische Gesellschaft* in December, following a request for information from a participant.[70] Three eminent members of the *Patriotische Gesellschaft* answer questions and comment on the matter: Bülau, the jurist Johann Friedrich Voigt (1806–1886) and the hydraulic engineer Heinrich Hübbe (1803–1871). Bülau and Hübbe reassure the public that the accident has caused no further consequences, and that the construction is solid.[71] According to Bülau, the crack was caused by different hardening rates between the mass of concrete that has been poured in 1844 and the one poured in 1845, whereas Hübbe believes that the difference of load between

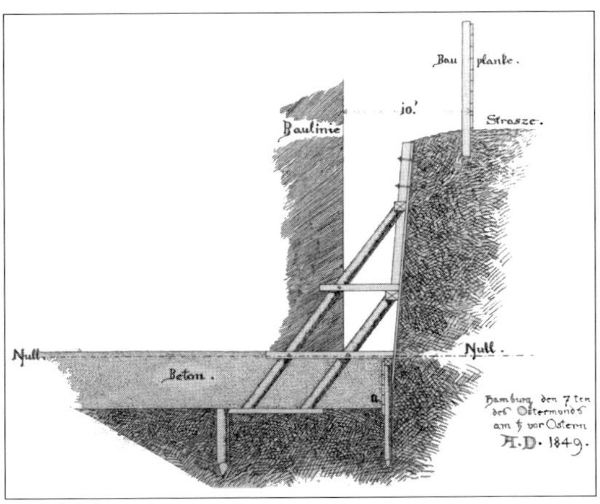

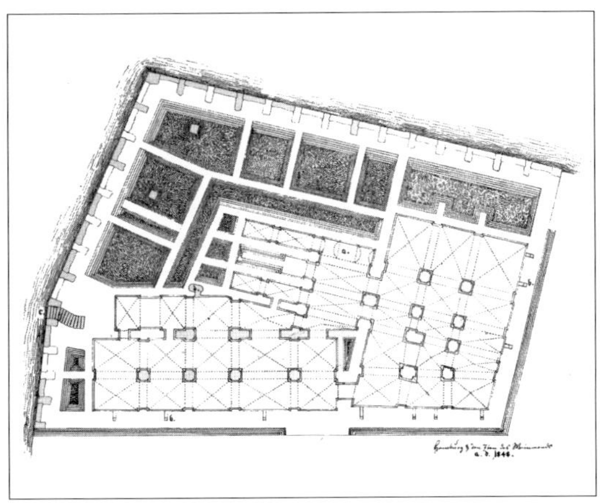

82-85. T. Bülau, *Haus der Patriotischen Gesellschaft*, Hamburg, 1844–1847.

82. The provisional cofferdam, section and view from above (T. Bülau, *Das Haus der patriotischen Gesellschaft in Hamburg oder: Einiges über einen Neubau*, Hamburg, I. I. Nobiling, 1849).

83. Sheet piling, section and view from above (T. Bülau, *Das Haus der patriotischen Gesellschaft in Hamburg oder: Einiges über einen Neubau*, Hamburg, I. I. Nobiling, 1849).

84. Shoring (T. Bülau, *Das Haus der patriotischen Gesellschaft in Hamburg oder: Einiges über einen Neubau*, Hamburg, I. I. Nobiling, 1849).

85. Plan of the basement (T. Bülau, *Das Haus der patriotischen Gesellschaft in Hamburg oder: Einiges über einen Neubau*, Hamburg, I. I. Nobiling, 1849).

[67] See John William Thomas Giles, "Bericht an eine hochlöbliche Bau-Deputation," April 8, 1845 (SAH, 321–322 Baudeputation, B451, n. fol.).
[68] See Forsmann, Maack, "Bericht den Vorschlag des Ingenieurs Giles..." cit.
[69] See John William Thomas Giles, "Sr. Wohlweisheit Herrn Senator Ienisch Präfes hochlöblicher Bau-Deputation," April 22, 1845 (SAH, 321–2 Baudeputation, B 451, n. fol.).
[70] See Heinrich Hübbe, *Bericht über die 368ste Deliberations-Versammlung vom 18. Dezember 1845*, in, *Verhandlungen und Schriften der Hamburgischen Gesellschaft zur Beförderung der Künste und Nützlichen Gewerbe. Verhandlungen vom Jahre 1845*, vol. I, 1846, (pp. 593–600), pp. 595–600.
[71] Nevertheless, Bülau express hard criticism against the effectiveness of concrete foundation four years later, see Bülau, *Das Haus der patriotischen Gesellschaft*...cit., pp. 1–2.

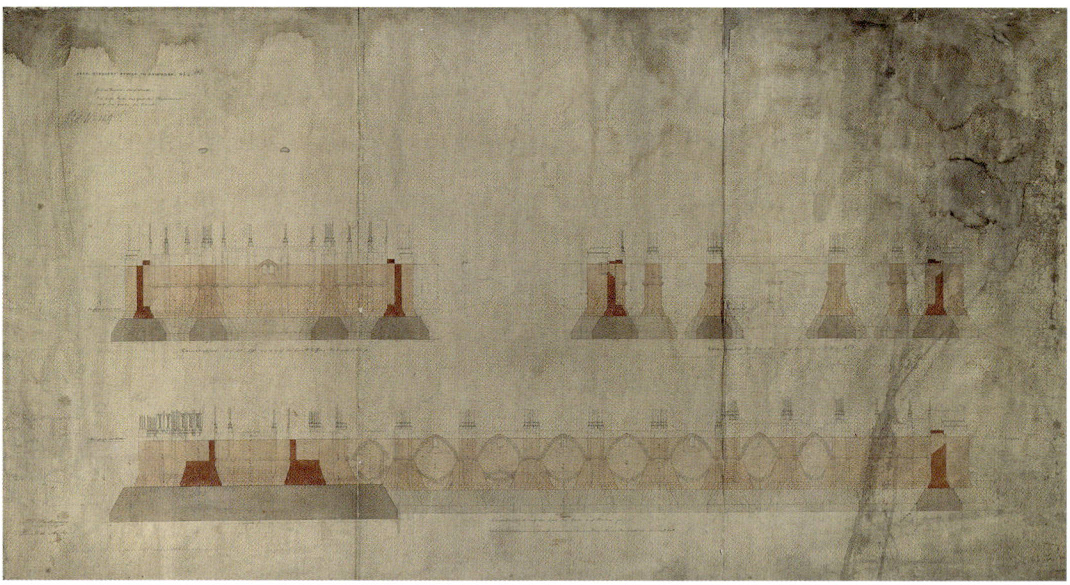

86-89. G. G. Scott, Church of St. Nicholas, Hamburg, 1845–1874.

86. Concrete foundation, plan (SAH-Kartensammlung-Nikolaikirche 720-1 361-2 XII 35).

87. Concrete foundation and walls of the basement, sections (SAH-Kartensammlung-Nikolaikirche-720-1 361-2 XII 49).

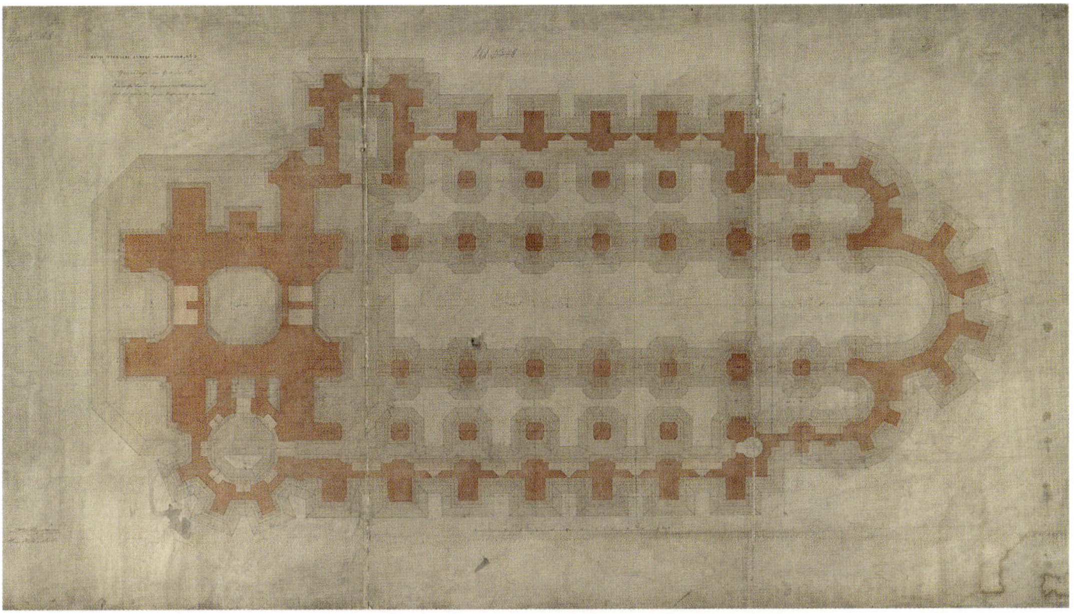

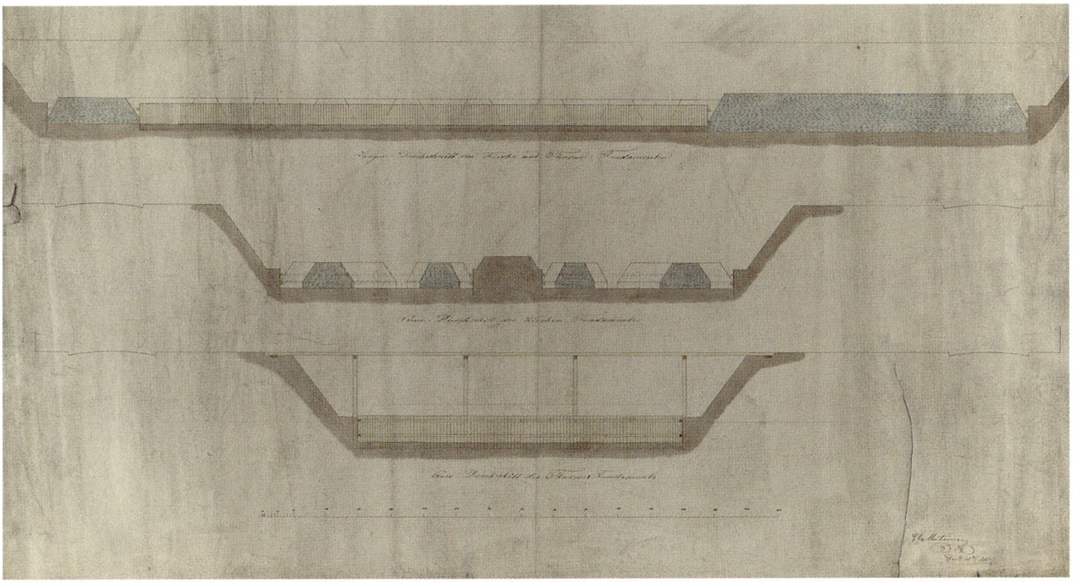

88. Plan of the basement on the concrete foundation (SAH-Kartensammlung-Nikolaikirche-720-1 361-2 XII 48).

89. Shoring and formworks for the concrete foundation, longitudinal and cross sections (SAH-Kartensammlung-Nikolaikirche-720-1 361-2 III 35a).

the perimeter walls and the columns supporting the central hall is at the origin of the crack. Voigt, differing again, expresses worry about the solidity of all concrete foundations, even more so since a huge building in Hamburg is going to be built on a concrete foundation, namely the new *Nikolaikirche* (Church of Saint Nicholas), which is intended to replace its predecessor, destroyed during the fire of 1842.[72]

Discussion about the reconstruction of the Church of Saint Nicholas begins just a few days after the fire, and a technical committee gathering eminent Hamburg engineers, architects and master builders is soon established to study the matter.[73] After some urban planning studies, which take about two years, an architectural competition is launched in May of 1844. Forty-four projects are submitted within the established time limit of six months, and a public exhibition opens on December 19. A few days after the opening, a further project arrives from London on a ship that had been delayed by ice for several days in Cuxhaven. It is by the English architect George Gilbert Scott (1811–1878), and Müller introduces and supports it in Hamburg. The projects is admitted and, in May 1845, Scott is officially appointed architect for the building of the new Church of Saint Nicholas, while discussions about some changes and improvements of the plans are still ongoing.[74] Scott appoints Henry Green Mortimer as clerk of works in Hamburg and develops further plans in London.[75] Henceforth, Saint Nicholas Church becomes a sort of melting pot of influences from British, Dutch and France building culture, with significant consequences for the construction of the concrete foundation.

In October 1845, excavation begins on the front side of the church, where the building of a high tower is planned.[76] Scott plans a square bed of concrete with each side measuring 30 meters under the tower; for all the other walls and pillars of the church, he plans concrete footings, which are two-and-a-half times wider than the walls and pillars standing above.[77] A complex design of shored earth and timber carpentry is expressly conceived to build up the formwork to pour concrete, especially for the footings, whose transversal sections taper from the bottom to the top. According to the Hamburg historian Ferdinand Stöter (1811–?), Scott travels to Holland to visit some hydraulic works under construction and to learn about how to produce concrete, before performing some trials and developing a concrete mixture

[72] See Hübbe, *Bericht über die 368ste Deliberations-Versammlung*...cit., p. 597. About the concrete foundation of the Church of Saint Nicholas see also: Joseph Egle, *Notizen über die Kunkurrenz zur Lieferung eines Bauplanes für die St. Nikolai-Kirche in Hamburg*, in, "Allgemeine Bauzeitung," XIII, 1848, (pp. 123–127), p. 127; Ferdinand Stöter, *Geschichte und Beschreibung des St. Nikolai Kirchenbaues in Hamburg*, Hamburg, Boysen, 1883, pp. 37–38; Julius Faulwasser, *Die St. Nikolai-Kirche in Hamburg*, Hamburg, Boysen & Maasch, 1926, pp. 82–83; Stefan M. Holzer, *Statische Beurteilung historischer Tragwerke*, Berlin, Ernst & Sohn, 2013, pp. 267–268.

[73] See Faulwasser, *Die St. Nikolai-Kirche*...cit., p. 48.

[74] Before designing the plans for the competition, Scott visits the building site of the Cologne Cathedral and makes direct contact with the building master Ernst Friedrich Zwirner (1802–1861) who is later appointed consultant of the committee that judges the competition projects in Hamburg (Faulwasser, *Die St. Nikolai-Kirche*...cit., pp. 60, 71).

[75] Mortimer is killed in a fall on the building site of the church in 1849. Scott erects a memorial window to him in 1850 in the church of Saint Nicholas in Witham; see "Monumental inscriptions at St. Nicolas, Witham," Essex Record Office, T/Z 151/29. Aside from Mortimer, Scott's representative in Hamburg is William Burlison, the building contractors are Johann Philipp Friedrich Heydtmann and Heinrich Wilhelm Müller (Faulwasser, *Die St. Nikolai-Kirche*...cit., p. 82).

[76] Excavation involves the removing of "more the one million and half of cubic feet of earth." This provides an opportunity to discover details about the history of the area, which seems to have once been the landfill of an abattoir, and to discover, furthermore, a rare mineral named Struvite, after the diplomat Heinrich Christian Gottfried von Struve (1772–1851). See, in this regard, Stöter, *Geschichte und Beschreibung des St. Nikolai Kirchenbaues*...cit., p. 37).

[77] See Faulwasser, *Die St. Nikolai-Kirche*...cit., p. 82.

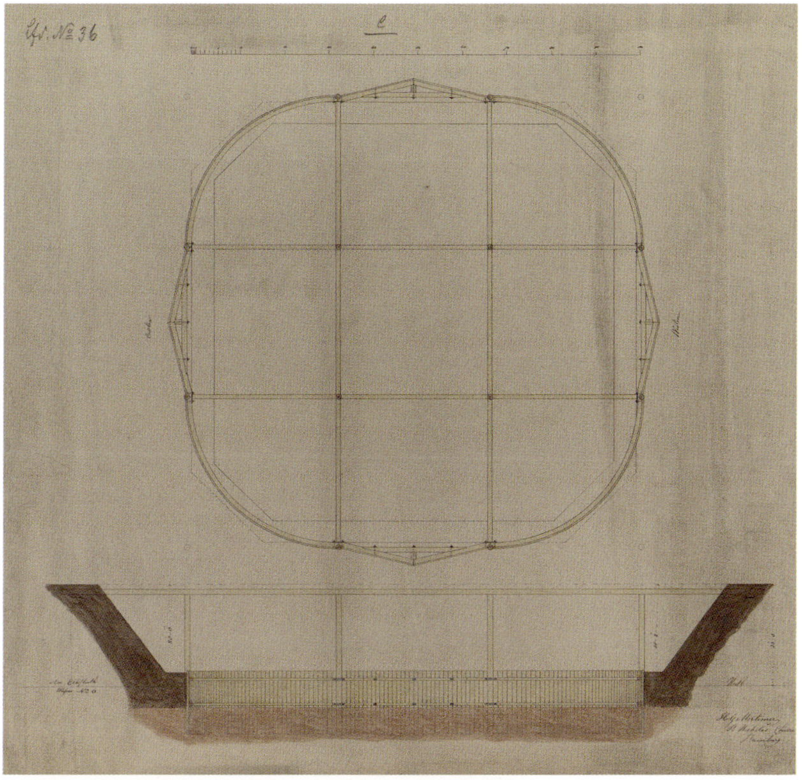

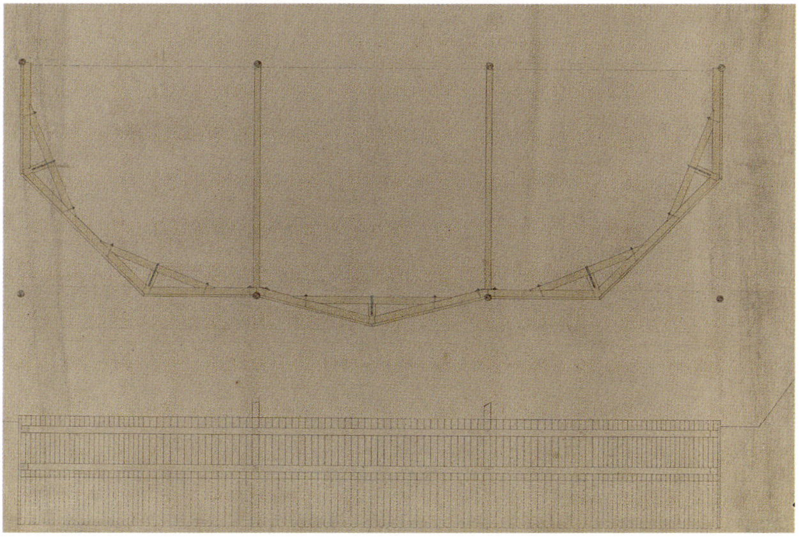

90-91. G. G. Scott, Church of St. Nicholas, Hamburg, 1845–1874. Concrete foundation of the tower, shoring, cross sections (SAH–Kartensammlung–Nikolaikirche–720–1 361–2 XII 36b; 2 XII36c).

made of lime, trass, sand, gravel and crushed bricks.[78] Carl Julius Faulwasser (1855–1944), a later Hamburg historian and architect, confirms Stöter's account: "The concrete mixture is based on Dutch examples, being made of crushed bricks, gravel, slaked lime powder and trass."[79] In a short note on a plan illustrating the formwork of the external side of the tower foundation, trass-based concrete is instead related to French models and is described as a kind of "French Beton" made of "12 measures of pozzolana or Dutch Trass, 6 of good sand, 9 of unslaked lime, 13 of stone [...] not-exceeding the size of an egg, 3 of iron scales from the forge."[80] Concrete is poured and tamped in layers from scaffolding, while iron bars with bent extremities are laid between the concrete layers, forming a kind of rudimentary reinforced concrete. The use of iron bars is reminiscent of the experiments that Brunel and Charles Francis from the Francis & White carried out in London, but we cannot rule out the idea that the accident that occurred during the building of the foundation of *Haus der Patriotischen Gesellschaft* also played a role. Once the concrete bed and footings are achieved, the foundation is completed with a basement made of bricks and a kind of mortar that, according to what Müller later maintains in a commercial leaflet, is Portland cement-based.[81] In the same leaflet, Müller also claims that some Portland cement is used to produce the concrete for the foundation of the tower, but this assertion is not confirmed in the previously mentioned historical sources by Stöter and Faulwasser. The foundation is finally achieved in summer 1846, and on September 24, the cornerstone ceremony takes place on the site of the chancel.

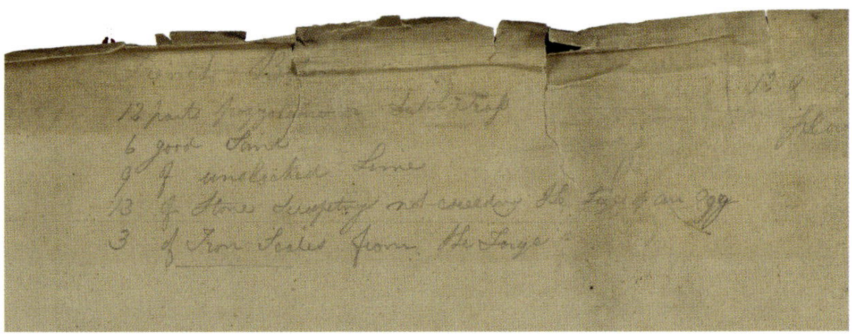

92. G. G. Scott, Church of St. Nicholas, Hamburg, 1845–1874. Note about the concrete mixture for the foundation of the tower (SAH-Kartensammlung-Nikolaikirche – 720-1 361-2 XII 36b).

[78] See Stöter, *Geschichte und Beschreibung des St. Nikolai Kirchenbaues*...cit.

[79] "Für die Betonmischung wurden nach holländischem Vorbild geschlagene Ziegel und Kies mit zerstäubtem Kalk und Traß verwandt" (Faulwasser, *Die St. Nikolai-Kirche*...cit., p. 82).

[80] SAH, pl. 720-1, 361-2, XII 36b, s.d.

[81] See Emil Müller, commercial leaflet about Portland cement, in, "Angebote von Zement durch in- und ausländische Firmen, 1834–1869," (SAH, 326-2 I Strohm- und Hafenbau – Geräte und Baumaterialein 160).

[82] A short notice about William Ranger's artificial stones appears in the "Polytechnisches Journal" in 1832; the same periodical also reports Ranger's patent for the production of cement-based artificial stone. See, on this matter, Anon., *Ein neues Baumaterial*, in, "Polytechnisches Journal," XLVII, 1832, pp. 73–74; Anon., *Verzeichniß der vom 24. Nov. 1832 bis 22. Jan. 1833 in England ertheilten Patente*, in, "Polytechnisches Journal," XLVII, 1832, (pp. 308–310), p. 308. In German technical literature, information about the artificial stones produced by Poirel for the harbour of Algiers are given in the translation of Lebrun's second book about concrete constructions and in a book by the General Major Carl von Decker about warfare in Algeria. See, on this topic, Lebrun, *Practische Abhandlung über die Kunst mit Béton zu bauen*...cit. p. 173, Carl von Decker, *Algerien und die dortige Kriegführung*, Berlin, Friedrich August Herbig, 1844, pp. 384–389).

[83] About Leube's life, see Edgar Breuss, Wilhelm Günther, Kurt Kaindl, Rudolf Zrost, *175 Jahre Leube. Chronik eines Familienunternehmens 1838 bis 2013*, Leube, St. Leonhard, 2013, especially pp. 8–16, 106–122. See also Robert Oedl, Werner Leube, Gustav Schefold, *125 Jahre Gebrüder Leube in Ulm und 100 Jahre Gartenau*, Gartenau, Zementwerk Gebrüder Leube, 1964. While Leube is developing the production of cement, foreman Johann Daniel Weil (1791–1862), from the village of Blaubeuren, carries out empirical experiments with lime and trass. Since 1838, Weil has his own marl quarry and produces hydraulic lime in a kiln. In the presence of *Kreisbaurat* Bühler, Leube and Weil test some marls shortly before each of them starts his own hydraulic lime manufacturing; however, it is not clear whether Weil asks Leube to scientifically analyse the efficacy of his marl, or

Moulded cement-based artificial stones and mortar tiles

The forefront of research for the development of mortar-based moulded artificial stones goes through a crucial period during the 1840s, driven by the availability of more efficient hydraulic lime and cement and by the examples of artificial stones developed abroad, such as the ones patented by William Ranger from Brighton in England and the ones developed by Victor Poirel for the Algiers harbour.[82]

Since the availability of hydraulic lime or cement is essential to producing this kind of artificial stone, important contributions come from some cement makers, especially Gustav Ernst Leube (1808–1881) from Ulm in Württemberg. Leube is actually a chemist who has a particular interest in mineralogy and geology. Aware of the works by John, Vicat and Fuchs, Leube discovers a kind of limestone that is suitable to produce hydraulic lime while investigating stones from the Blautal, west of Ulm. Around 1838, together with his brother Julius, Leube sets up a kiln, which he later gradually develops into a real cement and artificial stone factory, and he experiments with hydraulic lime to build a pavement floor made of a kind of concrete outside his apothecary.[83] "I let a pavement be built in front of my apothecary shop," Leube writes in his personal journal, "under the guidance of the city *Bauinspektor* Frühwirt. A one-and-a-half inch thick coat of mortar made of 1 Simri (20 litres) of hydraulic lime and 2 Simri of gravel was laid on a good base, and it quickly hardened. I have no doubts that this type of pavement is going to become usual in this town."[84] As a matter of fact, Leube is entrusted with the building of a concrete floor in the cathedral of Ulm in 1840, and with further ones in Stuttgart and Munich in the following years. The use of such floors gradually increases in the wake of the development of public spaces in nineteenth-century towns, serving as the prelude to the development of concrete floors for roads and to the production of ready-made concrete floor-tiles. In some cases, there is a deliberate intention to imitate natural stone for decorative purposes. The architect Johann Heinrich Friedrich Adler (1827–1908) reports that the floor in the octagonal northwest hall of the *Neues Museum* by Stüler is made of "marble and Portland-cement," that is, a base of Portland cement-based mortar finished with marble cement, which is a finer mortar made of double burnt gypsum treated with alum.[85]

Leube asks Weil for a burn test in the kiln, before he builds his own in Ehrenstein. Weil's business grows after 1838, and in 1845, Weil's son-in-law, who is a builder, joins the business and transforms the company into Weil & Sigloch. In this regard, see Albrecht, *Vom Caementum zu Zement*...cit., p. 129; Oedl, Leube, Schefold, *125 Jahre Gebrüder Leube*...cit., pp. 9–14.

[84] "Vor meiner Apotheke wurde unter Leitung von städt. Bauinspektor Frühwirt ein Trottoir gemacht. Nachdem ein guter Grund gemacht war, wurde eine 1 ½ Zoll dicke Schicht von 1 Simri (= 20 Liter) hydraulischem Kalk und 2 Simri Kies eingebaut, die nach kurzer Zeit erhärtete. Ohne Zweifel wird die Anwendung zu Trottoirs allgemein in der Stadt werden" (Ernst Gustav Leube's journal, November 8, 1839, published in Oedl, Leube, Schefold, *125 Jahre Gebrüder Leube*...cit., p. 38). About the construction of road floors made of asphalt on concrete bases in France, see Anon., *Ueber die Verwendung des natürlichen und nachgeahmten Erdharzes zu Fußpfaden, Fahrbahnen und architektonischen Zweken in Frankreich*, in, "Polytechnisches Journal," XIX, vol. LXIX, 1838, pp. 426–432), p. 427. About the use of concrete for street-paving in the United Kingdom, see Anon., *Ueber di Anwendung von Steinmörtel zum Straßenbaue*, in, "Polytechnisches Journal," XIX, vol. LXX, 1838, pp. 459–460.

[85] See Friedrich Adler, *Das Neues Museum in Berlin*, in, "Zeitschrift für Bauwesen," III, 1853, (coll. 23–34, 571–586), col. 576. In the *Neues Museum*, Stüler also makes wide use of *terrazzo* floors, taking inspiration from the technique of the Venetian *terrazzo*. However, as proven by analyses carried out during recent restoration, the terrazzo floors in the *Neues Museum* are made using gyps-based mortar instead of the cement-based mortar, which was, instead, usually used in Italy. Moreover, Stüler uses marble cement on columns. According to Adler, the ones in the treble hall next to the *Ägyptische Hof* are made of bricks bound with Portland cement-based mortar and covered with marble cement. See, on these topics, *ibid.*, col. 29; Staatliche Museen zu Berlin - Stiftung Preußischer Kulturbesitz, ed., *The Neues Museum Berlin*, Leipzig, Seemann, 2009, p. 124, pp. 176–182.

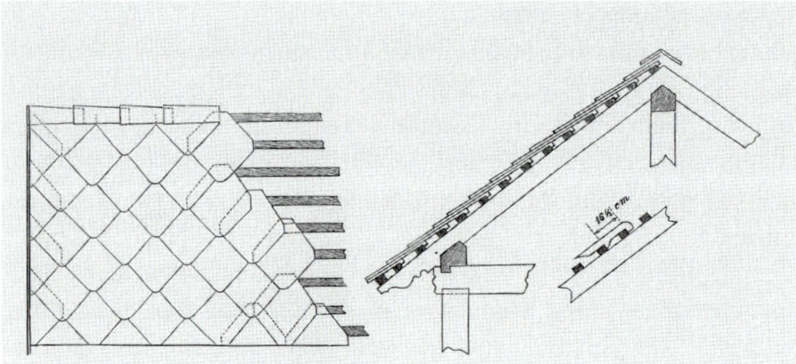

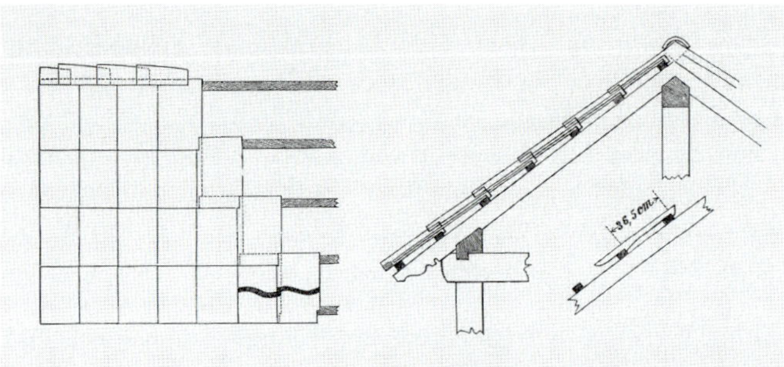

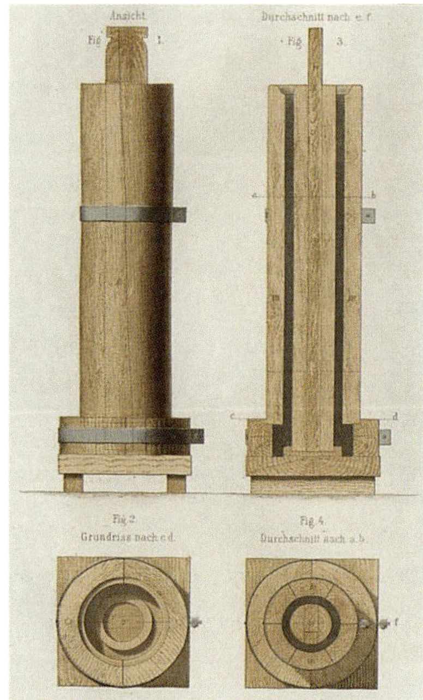

In addition to floors made of mortar coats or concrete, in 1840 Leube also produces the artificial ashlars that are used to restore a bridge along the road between Ulm and Gögglingen as well as the ones that are used to build a bridge along the road between Ulm and Wieblingen. Such stones consist of a mixture made of one measure of hydraulic lime and three measures of sand.[86] Around the mid 1840s, Leube delivers hydraulic lime and natural stones for the restoration of the Ulm Cathedral that the master mason Ferdinand Trän (1811–1870) is carrying out, and for some other works at the churches in the villages of Pfuhl and Blaubeuren.[87] On these occasions, besides natural stones, Leube also offers samples of artificial stone columns and capitals, which are first moulded and then chiselled finer. Some of these elements are even said to have been installed in the Ulm Cathedral.[88] During

[86] See Breuss, Günther, Kaindl, Zrost, *175 Jahre Leube*...cit., p. 112.

[87] See Oedl, Leube, Schefold, *125 Jahre Gebrüder Leube*...cit., p. 38.

[88] See Breuss, Günther, Kaindl, Zrost, *175 Jahre Leube*...cit., p. 114. Leube also delivers hydraulic lime to build part of the brick masonry of the fortress of Ulm (see Otmar Schäuffelen, *Die Bundesfestung Ulm und ihre Geschichte*, Ulm, Vaas, 1980, p. 43).

[89] See Oedl, Leube, Schefold, *125 Jahre Gebrüder Leube*...cit., pp. 38–39. Some years prior, the "Polytechnisches Journal" had reported about the use of concrete for the construction of bases for rails, in combination with other materials like timber and slates, and about a proposition for the construction of concrete-ways instead of railways (see Anon., *Ueber die Anwendung von Schieferplatten als Unterlagen für Eisenbahnen*, in, "Polytechnisches Journal," XV, vol. LIII, 1834, p. 457; Joseph Jopling, *Ueber Platformen für Eisenbahnen*, in, "Polytechnisches Journal," XVII, LX, 1836, pp. 2–5 (Id., *Jopling's improved railway platforms*, in, "Mechanics' Magazine," vol. XXIV, 1836, pp. 97–99); Charles Blacker Vignoles, *Ueber ein Eisenbahnsystem mit hölzernen Längebalken als Unterlage*, in, "Polytechnisches Journal," XVIII, vol. LXIV, 1837, (pp. 2–12), p. 9 (Id., *Mr. Vignoles' system of longitudinal timbers for the upper works of railways*, in, "Mechanics' Magazine," vol. XXVI, 1837, pp. 257–261); Anon., *Ueber eine neue Art von hohlen Schienen für Eisen- oder Schienenbahnen*, in, "Polytechnisches Journal," XVIII, vol. LXV, 1837, pp. 9–11 (J. R., *Hollow cylindrical rail for railways*, in, "Mechanics' Magazine," vol. XXVII, 1837, pp. 49–50); Charles-Chrétien-Constant

93-94. A. Kroher, instructions to install rhombic and S-shaped tiles (A. Kroher, *Cement und Cementwaaren-Fabrik in Staudach*, Augsburg, Rackl & Lochner, 1878).

95. [W. A.] Becker, drawings of a mould for poured, circular pipes, similar to the one that Karlinger from Miesbach patents in 1848 ([W. A.] Becker, *Praktische Anleitung zur Anwendung der Cemente zu baulichen, gewerblichen, landwirthschaftlichen und Kunst-Gegenständen*, Berlin, Nicolaische Verlagsbuchhandlung, 1860).

the building of the railway between Ulm and Friedrichshafen, Leube, who has been entrusted to provide cement for the ordinary masonry and for the very demanding construction of the Rosenstein Tunnel, proposes replacing the wooden railway sleepers with concrete artificial stones, and he even manages to install a number of these artificial stone sleepers in the station of Einsigen, south-west of Ulm.[89]

Also dating back to the early 1840s is an eminent but unsuccessful attempt at producing artificial rocks to build two waterfalls imitating Alpine scenery in the park of Glienicke in Potsdam, according to the Romantic taste of the time. Friedrich Carl Alexander von Preußen (1801–1883), the third born son of the king Friedrich Wilhelm III, plans these waterfalls together with the landscape painter August Wilhelm Ferdinand Schirmer (1802–1866). In this regard, the poet and landscape painter August Kopisch (1799–1853) reports: "They tried to reproduce alpine nature by building artificial rocks of every kind of material, but they finally had to give up this attempt and build a base made of stones from Rüdersdorf and bricks in Roman cement, over which they erratically stacked many small and large stones until the envisaged effect was achieved."[90]

In such a context, and after the trials of the mid-1830s, pioneering serial productions of ready-made small construction elements and decorations made of moulded cement-based mortar develop during the 1840s. These advances especially concern roofing tiles. At least since 1841, Heinrich Engelhardt, a builder from Nordhausen in Thüringen, produces his own cement and claims to be able to also produce "beaver tail tiles in different shapes, colours and sizes, even polished ones, to hang houses," as well as, "arabesques, rosettes and all kinds of decorations made of cement, for inside and outside, according to any sketch, rapidly and at affordable costs."[91] In 1842, Engelhardt submits a sample of his cement to the Prussian *Gewerbverein*, in order to take part in the competition promoted in 1833 concerning the reproduction of Roman cement with inland raw materials; the *Gewerbverein* considers the quantity of cement that Engelhardt has sent too small to undertake trials, and the submission bears no other consequence than a mention in the magazine of the *Gewerbverein*.[92] Approximately in the same years that Engelhardt promotes his cement and moulded objects, Adolph Kroher (1825–1892), a paper trader from Augsburg in Bavaria, undertakes the production of hydraulic lime in Staudach, after having heard of local limestone and marl deposits

Vauvilliers, *Bericht des Hrn. Vauvilliers über eine Abhandlung des Hrn. Thomassin, in welcher mit Steinmörtel gebaute Bahnen als vorzüglicher als die Eisenbahnen gepriesen werden*, in, "Polytechnisches Journal," XVI, vol. LXI, 1835, pp. 225–228 (Id., *Rapprt fait par M. Vauvilliers, au nom du Comité des arts mécaniques, sur un ouvrage de M: Thomassin, ayant pour titre: De la supériorité des chemins de béton sur les chemins de fer*, in, "Bulletin de la Société d'Encouragement pour l'Industrie National," XXXIV, 1835, pp. 227–230).

[90] "Man versuchte die Alpennatur hervorzubringen, künstliche Steine von allerlei Stoffen zu bilden, kam indessen zuletzt davon zurück und zog ein Unterbauen mit Ziegeln und Rüdersdorfer Steinen und Roman-Zement vor, worauf man größere und kleinere erratische Blöcke türmte und schichtete, bis der gewünschte Effekt wirklich bewunderungswürdig erreicht wurde" (August Kopisch, *Die Königlichen Schlösser und Gärten zu Potsdam. Von der Zeit ihrer Gründung bis zum Jahre MDCCCLII*, Berlin, Ernst & Korn, 1854, p. 217); in this regard see also Johann Georg Ludwig Sievers, *Bauten für den Prinzen Kalr von Preussen*, Berlin, Deutscher Kunstverlag, 1842, p. 72.

[91] "Biberschwanz-Ziegeln, zum behängen der Häuser, von verschiedener Form, Farbe und Größe, selbst poliert; [...] Arabesken, Rosetten und alle Arten Verzierungen, in und an die Wohnhäuser aus erwähnter Zementmasse nach jeder beliebigen Zeichnung schnell und sehr billig gefertigt" (Heinrich Engelhardt, announcement in, "Nordhäuser Nachrichtenblatt," March 4, 1841, GStA PK I.HA Rep.120 XXII Nr. 3, cit. in, Brunsch, *Die historische Verwendung...* cit., p. 66).

[92] See Anon., *Auszug aus den Protokollen der Versammlungen des Vereins in den Monaten September und Oktober d. l. J.*, in, "Verhandlungen des Vereins zur Beförderung des Gewerbfleisses in Preussen," XXI, 1842, p. 139; Anon., *Auszug aus den Protokollen der Versammlungen des Vereins in den Monaten Januar und Februar d. l. J.*, in, "Verhandlungen des Vereins zur Beförderung des Gewerbfleisses in Preussen," XXIII, 1844, p. 29.

from two local plaster makers. The business opens around 1844 and grows substantially in the following years. Aside from cement, Kroher also produces rhombic, moulded tiles and a kind of double-hollow, S-shaped roofing tile. Kroher uses a mixture made of two equal measures of hydraulic lime and sand, which is poured, spread and pressed in moulds. For rhombic-shaped tiles, he designs a special open mould, which hinged on a table called a *Handschlagtisch*.[93]

In addition to tiles, in 1848, the aforementioned Karlinger from Miesbach, who exhibited moulded figures at the Munich industrial exhibition in 1834, patents a system to produce pipes made of hydraulic lime-based mortar, very probably moving from Fleuret's concept of pipes made of *pierre factice*. The method is based on the use of a complex contraption made of two concentric cylinders, a pole and two spacers assembled together so that they form a round mould for pipes. A first cylinder made of brass or copper sheeting is wrapped in cloth, laid into the longitudinal half of a solider counter cylinder and covered with the other half of the same counter cylinder. The two halves are fixed together by means of ropes or clasps, and a pole is introduced into the metallic cylinder, kept exactly in the middle by means of wooden steering spacers placed at the two extremities. The whole contraption is then raised and set vertically before cement-based mortar is poured into the space between the pole and the metallic cylinder. Once the mortar has set a little, the contraption is laid down and the metallic cylinder is extracted first, then the pole, and finally the counter cylinder is opened.[94] Toward the mid-1850s, this method will be criticized for its complexity, when further production of cement mortar pipes develop, giving rise to the manufacture of moulded objects made of cement-based mortar.

Concrete fillings for bridge decks, and to restore masonry cracks

Besides underwater foundations, in the second half of the 1840s, concrete is also used as filling material to build the deck of some stone masonry viaducts. An example is seen in the railway viaduct built between 1844 and 1847 near Görlitz in Silesia.[95] At 1,494 feet long, it crosses the valley of the River Neisse, being partially built on the riverbed. The concrete is made of cement, sand and sandstone rubble. It is first poured among the pile heads of the timber piling foundations of the three piers standing in the riverbed, and then as filling material for the bridge deck. This is composed of groups of three lengthwise and frustum-shaped vaults standing on the half of each 30 feet wide bridge-vault and covered first with concrete, then with sandstones forming the road floor. Being frustum-shaped, the vaults slope from the piers to the key of the vaults in order to allow rainwater drainage from the surface of the deck through a system of internal drains. A similar structure is adopted for the deck of the bridge built between 1844 and 1848 in Frankfurt, along the Main-Neckar railway. That deck is composed of groups of five lengthwise vaults covered with a kind of concrete that is made of local hydraulic lime, taken from the nearby village of Somborn, trass and crushed bricks.[96]

[93] See Adolph Kroher, *Cement und Cementwaaren-Fabrik in Staudach*, Augsburg, Rackl & Lochner, 1878; Kroher, *Staudacher Zementdachsteine. Ein Jubiläum*, in, "Tonindustrie Zeitung," XLIII, 1919, pp. 443–444; Charles George Dobson, *The history of the concrete roofing tile. Its origin and development in Germany*, London, Batsford, 1959, (Id., *Geschichte des Betondachsteins - Seine Herkunft und Entwicklung in Deutschland*, Hildesheim, Hagemann, 1959).

[94] See J. Karlinger, *Verfahren zur Verfertigung von Wasserleitungröhren aus hydraulischem Kalke*, in, "Kunst- und Gewerbe-Blatt des polytechnischen Vereins für das Königreich Bayern," XL, 1854, coll. 99–102, and in, "Polytechnisches Journal," XXXV, vol. CXXXII, 1854, pp. 202–204.

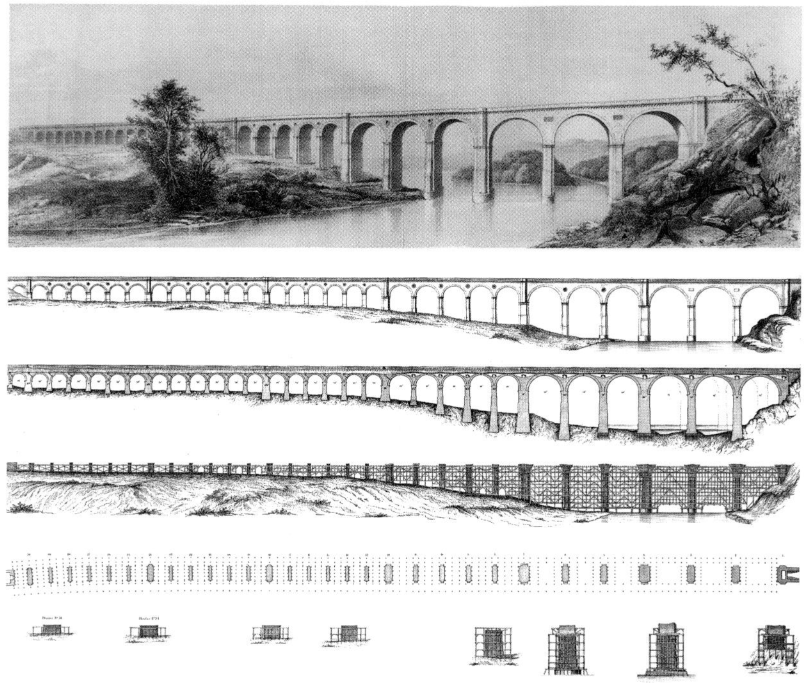

96-97. Weishaupt, Fischer, Viaduct of Görlitz, Silesia, 1844–1847. Perspective view, elevation, section, centrings, plan of the central part, views of some piers under construction (L. B. Henz, *Der Bau des Neisse-Viaducts bei Görlitz in der Niederschlesisch-Märkischen Eisenbahn*, in, " Zeitschrift für den Bauwesen," V, 1855).

Hydraulic mortar and concrete fillings are also used during the restoration work of the bridge over the valley of the River Diemel (1847–1851), close to the town of Warburg in Rhineland, undertaken around 1849, since the dismantling of the vault centrings produces serious damage to the piers. Cement-based mortar is used to set in replacing ashlars and to fill cracks, in order to restore the structural continuity of the work. All piers stand on stone masonry foundations, except the one on the left bank, which stands on a stone-faced wall filled with concrete. The fact that this pier still stands although several ashlars of the wall faces have fallen apart is considered evidence of the solidity of concrete.[97]

The concrete foundations of the piers for the railway bridges in Dirschau and Marienburg, the development of local artificial cement by Lentze

The engineer Johann Carl Wilhelm Lentze (1801–1883)[98] contributes on an essential level to the manufacturing development of artificial hydraulic binders in the mid 1840s, as the construction of two railway bridges along the so-called *Ostbahn* (the Eastern Railway connecting Berlin and Königsberg) is planned to cross the River Vistula near the town of Dirschau, and the River Nogat near to town of Marienburg.[99] Lentze, who has worked as a hydraulic engineering director in Münster and Düsseldorf, and Wiebe, who has just completed the building of the aforementioned bridge in Sonnborn, sit on

[95] See Ludwig Benjamin Henz, *Der Bau des Neisse-Viaducts bei Görlitz in der Niederschlesisch-Märkischen Eisenbahn*, in, "Zeitschrift für den Bauwesen," V, 1855, coll. 281–344.

[96] See Ad. Braubach, *Main-Brücke bei Frankfurt in der Main-Neckar-Eisenbahn*, in, "Zeitschrift für den Bauwesen," VI, 1856, coll. 479–488.

[97] See Ludwig Benjamin Henz, *Die Restauration des Diemel-Viadukt*, in, "Zeitschrift für den Bauwesen," II, 1852, coll. 15–34.

[98] See A. W., *Geheimer Oberbaurath a. D. Karl Lentze †*, in, "Centralblatt der Bauverwaltung," III, 1883, n° 26, pp. 233–234.

[99] Today: Tczew and Malbork, in Poland.

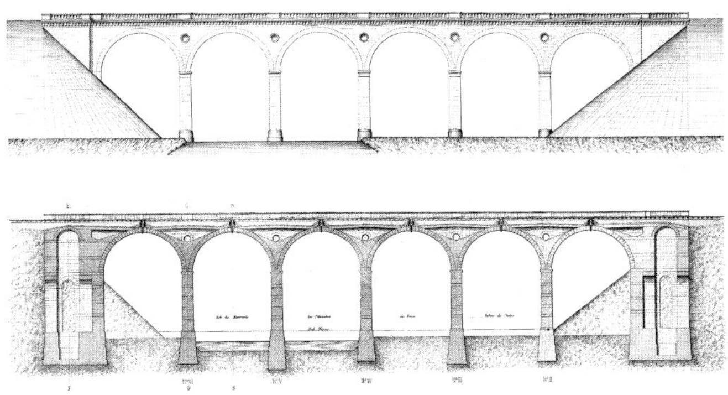

98. Weishaupt, Fischer, Viaduct of Görlitz, Silesia, 1844–1847. The bridge deck, strecth of the longitudinal section, cross sections (L. B. Henz, *Der Bau des Neisse-Viaducts bei Görlitz in der Niederschlesisch-Märkischen Eisenbahn*, in, "Zeitschrift für den Bauwesen," V, 1855).

99-100. Friedrich–Wilhelms–Nordbahn Actien–Gesellschaft, Bridge over the River Diemel, close to Warburg, Rhineland, 1847–1851.

99. Elevation and longitudinal section (L. B. Henz, *Die Restauration des Diemel-Viadukt*, in, "Zeitschrift für den Bauwesen," II, 1852).

100. Pier n. IV, longitudinal and frontal view (L. B. Henz, *Die Restauration des Diemel-Viadukt*, in, "Zeitschrift für den Bauwesen," II, 1852).

a committee entrusted with performing the preliminary studies in 1844.[100] The committee drafts a first plan for the bridges and calculates that the one in Dirschau should be divided into 10 spans, whereas the bridge in Marienburg should be divided into 5 spans.[101] Once preliminary studies are accomplished, Lentze is tasked with the further design of the bridges, and he is asked by the finance minister Eduard Heinrich von Flottwell to go on a journey to France and to the United Kingdom to study recent "iron bridges." Lentze spends the winter abroad between 1844 and 1845. Some existing documents prove that, beyond iron bridges, he also studies how to produce hydraulic mortar and concrete, especially in the United Kingdom. From the "Papers of the corps of royal engineers," Lentze takes some notes about a kind of concrete conceived by George Godwin (1789–1863) as a mixture of sand, ballast and lime taken from the villages of Dorking and Halling.[102] Visiting the building site of the Albert Docks in Liverpool, Lentze also takes note of two different hydraulic mortar compositions, a stronger one made of three measures of lime, one of sand, half of pozzolana and half of slags, and a weaker one made of one and three quarters measures of lime, three of sand, and one quarter of pozzolana.[103] On the same building site, Lentze observes a mortar mixer composed by a circular container and two millstones turning in the container around a vertical shaft.[104] Such a machine is similar to the so-called *machine française de M. Saint-Léger*, which is used on the building site of the *Canal de Saint-Martin* in Paris after the *machine à mortier anglaise* is set aside. An existing sketch proves that Lentze studies the *Canal de Saint-*

101. J. C. W. Lentze, notes about the production of concrete in England, taken from the Papers of the Royal Corps of Enginners, approx. 1846–1847 (SAS, P.1.436, fol. 25).

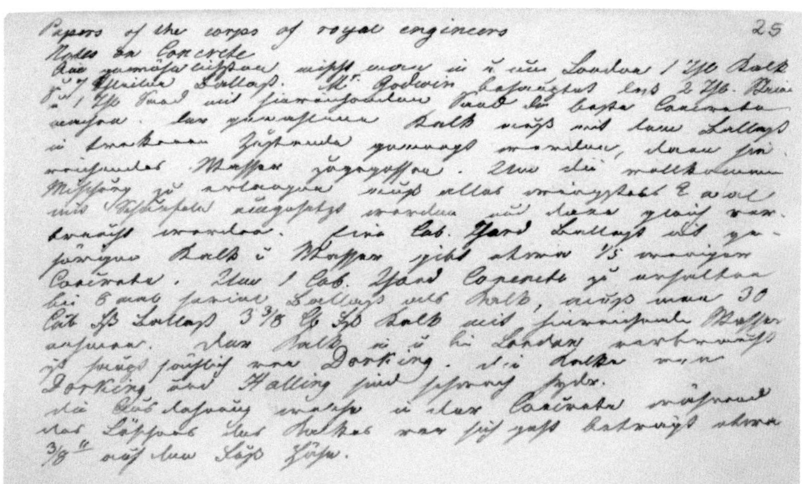
Papers of the corps of royal engineers 25
Notes on Concrete

[100] See Friedrich Wilhelm Freiherr von Reden, *Die Eisenbahnen Deutschlands. Statistisch-geschichtliche Darstellung ihrer Entstehung, ihres Verhältnisses zu der Staatsgewalt, so wie ihrer Verwaltungs- und Betriebs-Einrichtungen*, Berlin, Posen, Bromberg, Ernst Siegfried Mittler, 1846, pp. 315–321; Anon., *Stenographische Berichte über die Verhandlungen der durch die Allerhöchste Verordnung vom 4. November 1851 einberufenen Kammern*, Zweite Kammer, vol. III, Berlin, Deckersche Geheime Ober-Hofbuchdruckerei, 1852, n° 162, pp. 1195–1197; Georg Christoph Mehrtens, *Zur Baugeschichte der alten Eisenbahnbrücken bei Dirschau und Marienburg*, in, "Zeitschrift für Bauwesen," XLIII, 1893, (coll. 97–122), coll. 99–100.

[101] See *ibid.*, col. 100.

[102] See [ascribed to] Johann Carl Wilhelm Lentze, notebook (SAS, P.1.436, fol. 25).

[103] See Johann Carl Wilhelm Lentze, *Bemerkungen über die grösseren Brücken – Bauwerke in Frankreich, England und Irland, aud einer Reise im Winter 1844/45 gesammelt*, Berlin, I. Petsch, 1846, p. 26, (SAS, P 1. 813).

[104] *Ibid.*, pl. XVI.

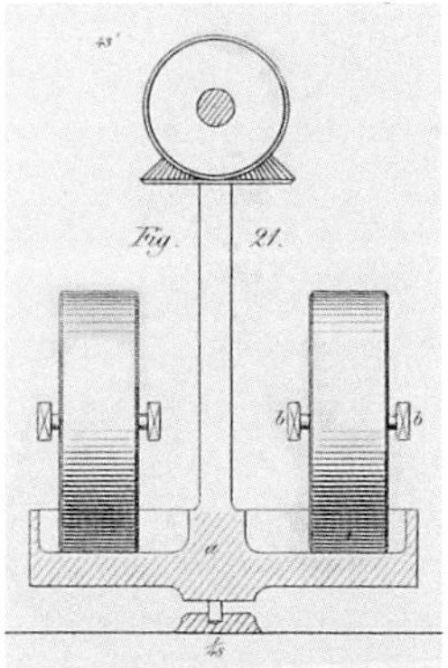

102. J. C. W. Lentze, drawing of the mortar mixer observed on the building site of the Liverpool docks (J. C. W Lentze, *Bemerkungen über die grösseren Brücken-Bauwerke in Frankreich, England und Irland, aud einer Reise im Winter 1844/45 gesammelt*, Berlin, I. Petsch, 1846).

Martin.[105] It comes therefore as no surprise that Lentze later uses similar mixers in Dirschau and in Marienburg.

Once back in Prussia, Lentze, together with the building councillor Spittel from Danzig, designs the plans of the two bridges.[106] The one in Dirschau is conceived as a 5-span suspension bridge, whereas the bridge in Marienburg is designed as an iron bridge.[107] The piers of both bridges are designed as masonry constructions standing on concrete foundations. The lowest parts of them, from the foundations to the bridge decks, are designed as ashlar-faced walls filled with brick masonry; whereas the upper parts, over the bridge decks, are of full brick masonry and form two towers, "according to the style of the aristocratic houses of the Prussian province."[108]

Fat lime taken at the port of Danzig and trass imported from Rhineland could be used to produce hydraulic mortar, but Lentze prefers to search for local raw materials suitable for producing hydraulic cement.[109] Knowledge of English building culture probably plays a crucial role in this choice, and we cannot rule out the idea that Lentze is also inspired by Charles William Pasley's experiments about artificial cement. Lentze finds deposits of calcareous and siliceous marls about three miles away from the building site in Dirschau. He performs some experiments with burning different mixtures of both kinds of marl, until he develops a kind of cement that, when mixed with sand, hardens underwater and proves to have satisfactory compression strength.[110] Lentze moreover verifies that the kind of mortar he intends to use does not set too rapidly, so that it can be produced in appropriate plants and transported to the building sites of the different piers. Construction of the plants to produce cement, mortar and concrete is underway in 1846, shortly after the excavation work begins.

Due to financial difficulties, however, work stops in October 1847, just as Lentze goes on a second research journey to England. Once back in Prussia, he modifies the plans of both bridges, which are by then both designed as lattice truss bridges: the one in Dirschau stretching over six spans, and the one in Marienburg over two spans.[111] Nothing changes, however, for the already planned production of artificial cement and concrete, which remains as it has been conceived in 1845.

Work resumes in April 1850. The cement and concrete plants are rapidly underway. The one in Dirschau starts producing in 1850.[112] It is situated between the right bank of the Vistula and the area where the railway station is going to be built. The plant comprises a shed where calcareous and siliceous marls are mixed together (*Steinwäsche*), three further sheds where calcareous-siliceous mixtures dry, four continuous fire kilns and a steam-powered mill to grind cement and produce mortar and concrete. Calcareous and siliceous

[105] See SAS, P 1.590, n. fol.; De Villiers du Terrage, *Description du Canal de Saint-Denis*...cit., pp. 38–41. It is intriguing to observe that according to Rondelet the *Machine française de M. Saint-Léger* derives from the *Trybomilus*, i.e. a grind mill illustrated in the treatise *Theatrum Machinarum* by Georg Andreas Böckler. See, in this regard, Jean-Baptiste Rondelet, *Traité théorique et pratique de l'art de bâtir*, vol. I, Paris, Firmin Didot Frères, 1834, p. 349; Georg Andreas Böckler, *Theatrum machinarum novum*, Nürnberg, Christoff Gerhard, 1661, pl. 52.

[106] Danzig is today Gdańsk.

[107] The number of spans for the bridge in Dirschau is drastically reduced in comparison with the estimation by the Committee, in order to let ice blocks better flow during spring; see Mehrtens, *Zur Baugeschichte der alten Eisenbahnbrücken*... cit., col. 101.

[108] See Anon., *Die Eisenbahn zwischen Berlin und Königsberg*, in, "Illustrierte Zeitung," vol. VI, 1846, p. 218.

[109] See Johann Carl Wilhelm Lentze, *Beschreibung einiger Hülfseinrichtungen von dem Bau der Weichsel- und Nogat-Brücken. II. Die Cementhütte und die Mörtelmühle für den Bau der Weichselbrücke*, in, "Zeitschrift für Bauwesen," XII, 1861, (coll. 375–380), coll. 375–377.

[110] Lentze's cement contains 64% calcium carbonate, 30% silica and 4% clay in addition to small quantities of iron oxide. The mortar is made of one part cement and one and 1/7 part sand. Strength tests consist of regularly charging samples of mortar on stone over the course of a year; this way, Lentze ascertains that the mortar can withstand 1800 Pfund

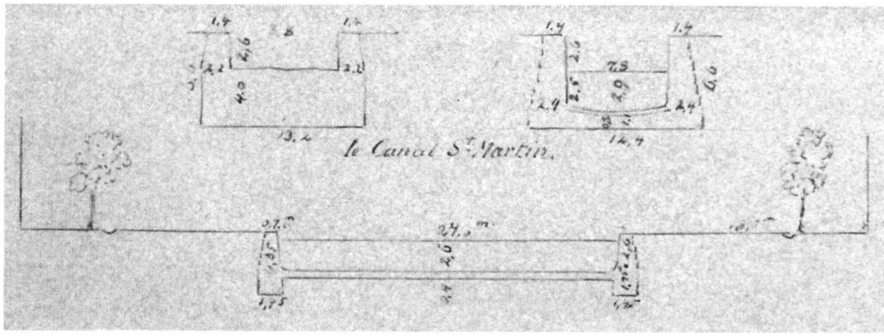

103. J. C. W. Lentze, sketch of the *Canal de St. Martin*, Paris (SAS, P.1.590).

next pages

104-106. J. C. W. Lentze, Bridge over the River Vistula, Dirschau, 1846–1855.

104. Perspective view (J. C. W. Lentze, *Die im Bau begriffenen Brücken über die Weichsel bei Dirschau und über die Nogat bei Marienburg*, in, "Zeitschrift für Bauwesen," V, 1855).

105. Elevation and plan (J. C. W. Lentze, *Die im Bau begriffenen Brücken über die Weichsel bei Dirschau und über die Nogat bei Marienburg*, in, "Zeitschrift für Bauwesen," V, 1855).

106. Plan of the site (J. C. W. Lentze, *Die im Bau begriffenen Brücken über die Weichsel bei Dirschau und über die Nogat bei Marienburg*, in, "Zeitschrift für Bauwesen," V, 1855).

107-109. J. C. W. Lentze, Bridge over the River Nogat, Marienburg, 1846–1856.

107. Perspective view, (J. C. W. Lentze, *Die im Bau begriffenen Brücken über die Weichsel bei Dirschau und über die Nogat bei Marienburg*, in, "Zeitschrift für Bauwesen," V, 1855).

108. Elevation with sections of the concrete foundations (J. C. W. Lentze, *Die im Bau begriffenen Brücken über die Weichsel bei Dirschau und über die Nogat bei Marienburg*, in, "Zeitschrift für Bauwesen," V, 1855).

109. Plan of the site (J. C. W. Lentze, *Die im Bau begriffenen Brücken über die Weichsel bei Dirschau und über die Nogat bei Marienburg*, in, "Zeitschrift für Bauwesen," V, 1855).

marls are first plunged into water, transformed into mush and mixed together with a horse-powered clay-cutting tool. Mixtures coming from the clay-cutting tool are moulded as little cubic blocks with each side measuring 2.5 inches; these are first dried in the appropriate sheds and then brought to kilns to be fired. Each kiln is provided with six gates at the base – three to charge fuel and three to take out burnt cement.[113] They are set in a square; two roofs shelter their lowest parts from the rain, whereas two bridges and a lift made of ropes and pulleys placed at the top make it possible to charge calcareous-siliceous blocs. The kilns are charged every six hours and each one produces thirty barrels (120 Sheffels) of cement per day, burning a barrel (1 and 1/5 cubic foot) of coal from Newcastle.

Burnt cement blocs are checked after having cooled and only the ones that are perfectly burnt are brought to the mill, which is installed in a three-floored building. Eight grind mills, similar to the ones that Lentze observed in Liverpool, are placed in line at the highest floor, over a longitudinal camshaft that is connected to the flywheel of a steam engine. Being used both as a mill and as mortar mixers, they are first charged with burnt cement blocks and, once these are ground into powder, workers add sand and water to produce mortar. This is then transported by means of wheelbarrows to a trap door placed in the floor and thrown into wagons on rails to be transported to the building site. To produce concrete, mortar is thrown together with some gravel through a second trap door into a sloping and steam-powered concrete drum on the second floor of the building, where the two different materials are mixed together and fall into the wagons on the ground floor.

The cement plant in Marienburg starts producing in 1851.[114] It is similar to the one in Dirschau but smaller, being equipped with a clay-cutting tool, a shed to dry marls mixtures, one kiln, six grind mills and a concrete drum. No wagons on rails are available in the Marienburg plant; mortar and concrete are transported in wheelbarrows or even in containers on shoulders, which

(about 900 Kg) of pressure. See Lentze, *Beschreibung einiger Hülfseinrichtungen*...cit., col. 380.

[111] See Albert Julius Licht, *Der Portland-Cement und der hydraulischer Kalk der Cement-Fabrik zu Dirschau*, Danzig, Schroth, 1867, p. 1; Königlich Preuzischen Ministerium der Oeffentlichen Arbeiten, ed., *Archiv für Eisenbahnwesen*, Berlin, Julius Springer, 1911, p. 911.

[112] Königlich Preuzischen Ministerium der Oeffentlichen Arbeiten, ed., *Archiv für Eisenbahnwesen*...cit., p. 105; Anon., *Auszug aus der Mittheilung des Herrn Ministers für Handel etc. an die Kamern über den Stand des Baues der Ostbahn, der Westphälischen und Saarbrücker Eisenbahn in November des Jahres 1851*, in, "Zeitschrift für Bauwesen," II, 1852, p. 126. About the development of the works, see also Anon., *Deutschland. Berlin*, in, "Frankfurter Volkszeitung," December 29, 1852, n° 311, n. p.

[113] Continuous kilns are called Rumford kilns, from the name of the inventor Sir Benjamin Thompson Count of Rumford (1753–1814), an American-born British physicist who graduates in Munich and spends the most important part of his career in Germany. The early Rumford kilns are used in Rüdersdorf. Lentze himself defines the kiln he has built in Dirschau and Marienburg as the Rumford kiln; see Lentze, *Beschreibung einiger Hülfseinrichtungen*...cit., col. 378.

[114] Anon., *Auszug aus der Mittheilung des Herrn Ministers für Handel etc. an die Kamern über den Stand des Baues der Ostbahn, der Westphälischen und Saarbrücker Eisenbahn in November des Jahres 1851*...cit., p. 124.

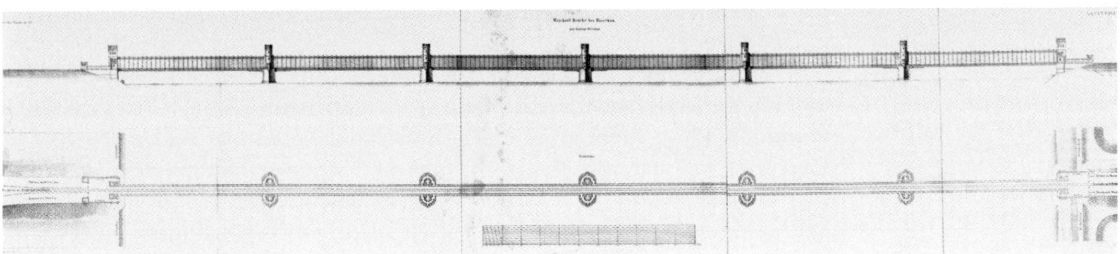

Danziger
Werder
Bahnhof
Stadt
Dirschau
Kirchhof
Weichsel
Ponton Brücke

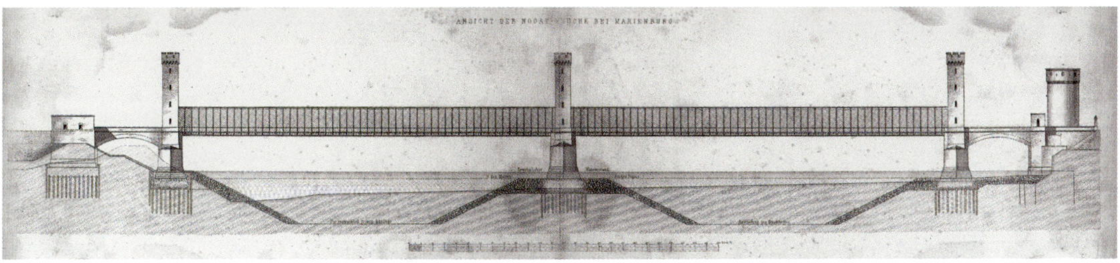

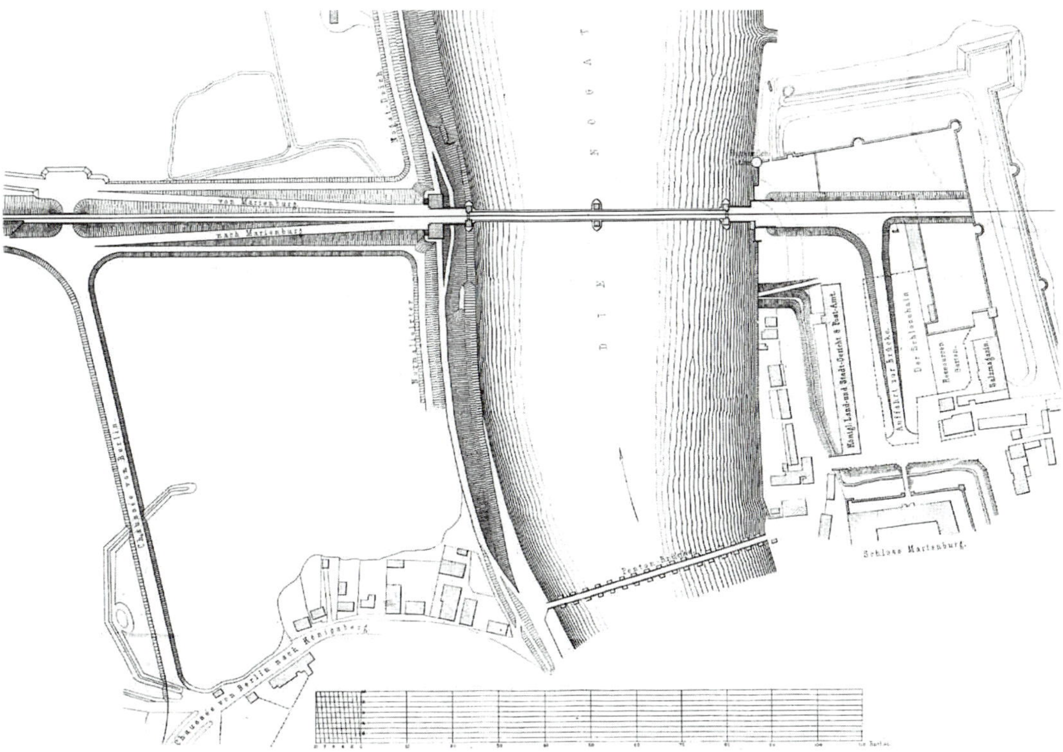
Chaussee von Berlin nach Königsberg
Schloss Marienburg.

Situations Plan der Mörtel Mühle
für den Weichsel Brücken Bau bei Dirschau
(A.)
Werkplatz für den
Mörtel Mühle
Weichsel Brücken Bau
Nach Dirschau gehö-
rige Pottlitz Gärten
Der Stadt
Dirschau gehö-
riges Weideland
Die Weichsel
7847
Berliner Hauptarchiv
Berlin-Dahlem
Allgem. Ks. 8447

110-114. J. C. W. Lentze, Bridge over the River Vistula, Dirschau, 1846–1855.

110. Plan of the building site on the link riverside (GStA PK, XI. HA AKS Allgemeine Kartensammlung G Nr. 51833).

111-112. Views of the mortar mill and of the cement plant (SAS, P.1.590).

113. The cement kilns, sections and plan (J. C. W. Lentze, *Beschreibung einiger Hülfseinrichtungen von dem Bau der Weichsel- und Nogat- Brücken. II. Die Cementhütte und die Mörtelmühle für den Bau der Weichselbrücke*, in, "Zeitschrift für Bauwesen," XII, 1861).

114. The mortar mill, section, plan and details of a single mill (J. C. W. Lentze, *Beschreibung einiger Hülfseinrichtungen von dem Bau der Weichsel- und Nogat- Brücken. II. Die Cementhütte und die Mörtelmühle für den Bau der Weichselbrücke*, in, "Zeitschrift für Bauwesen," XII, 1861).

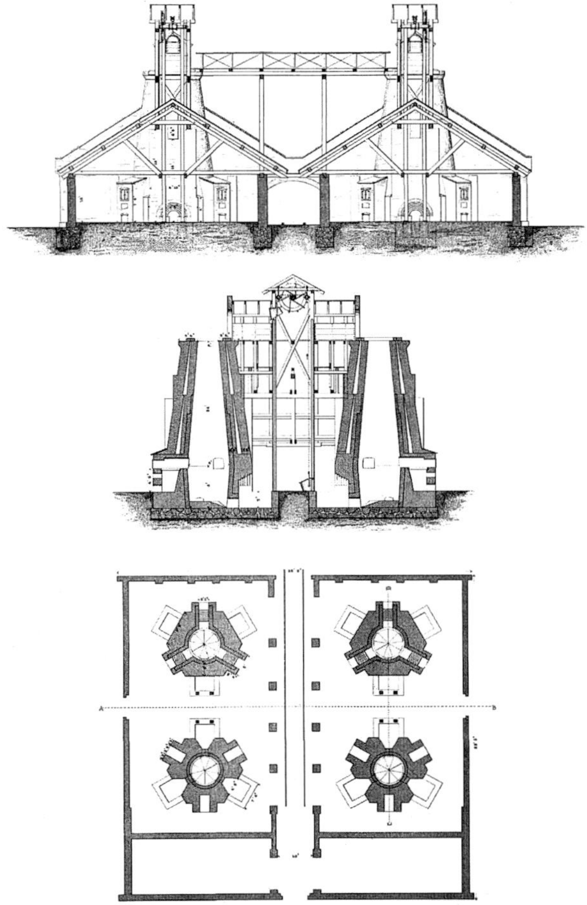

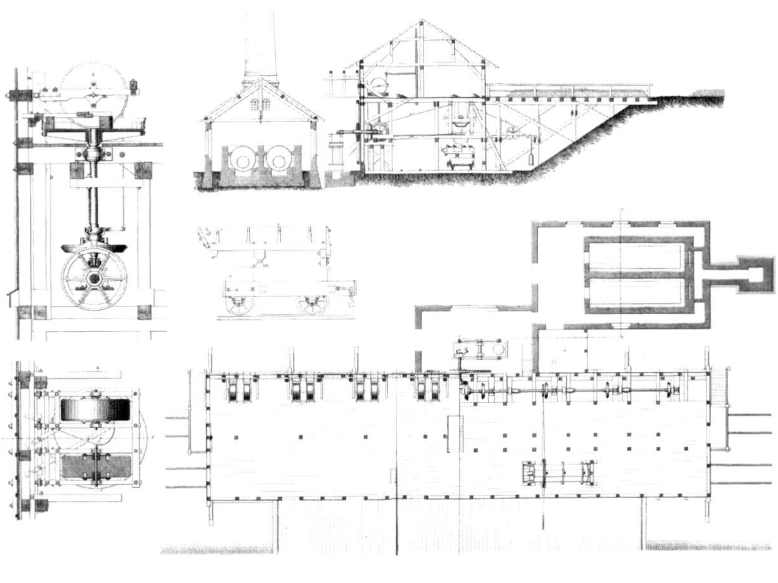

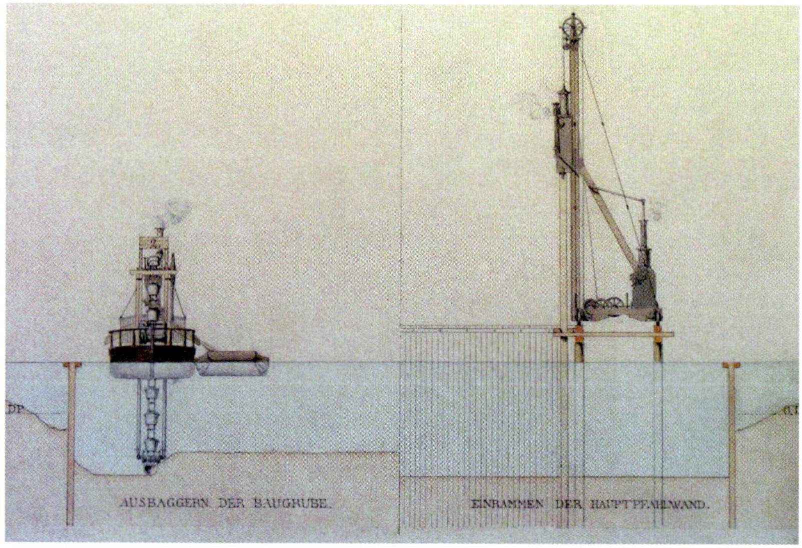

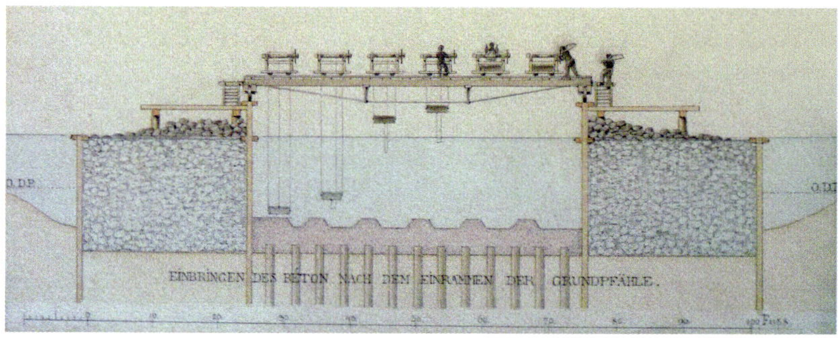

115-118. J. C. W. Lentze, Bridge over the River Vistula in Dirschau, 1846–1855. Construction of the concrete foundation for a pier (SAS, P.1.590).

are called *Vögel*. Lentze later reports that these containers weigh about 110 Pfund (55 kg) once filled, and that workers therefore would place a sort of straw pillow on their shoulders when carrying them.

Construction of the bridge in Dirschau begins in 1850, starting from the foundation of the left abutment, the closest to the cement plant.[115] The foundations of all piers are achieved in 1852, the masonry up to the level of the deck in 1853 and the towers in 1855. Construction of the bridge in Marienburg starts in 1851 from the right abutment. The foundations are completed in 1854, the lower masonry in 1855 and the towers in 1856. All piers are built the same way. An area larger than the foundation site is surrounded with sheet piles; the soil is dug up to a depth of about 10 feet under the natural riverbed and a smaller enclosure of sheet piles is rammed. Inside this second enclosure, timber piles are rammed at a depth of about 20–22 feet and their heads are cut one foot above ground. Concrete, made of one part mortar - a measure of cement and one and 1/7 of sand - and two parts of gravel, is poured over the timber piles, forming a 10 feet high base. Similar to other cases, 3 feet high concrete walls are built all around the upper surface

[115] See Johann Carl Wilhelm Lentze, *Die im Bau begriffenen Brücken über die Weichsel bei Dirschau und über die Nogat bei Marienburg*, in, "Zeitschrift für Bauwesen," V, 1855, coll. 445–458.

of the concrete bases and form a closed and watertight space where masons build the lowest masonry layers of the piers.

Concrete foundations for the abutments of the bridge in Dirschau are built directly on sand, without any timber pilings. They support structures made of about 30 feet high tapering brick pillars linked by reversed arches at the bottom and by ordinary arches at the top, which, in turn, support the slip road between the bridge deck and the land road. A roughly 3 feet high bed of lean concrete, made of one measure of mortar and six of sand, is poured over these arches to build the base for the pavement. The same kind of lean concrete is also used to partially fill the spaces between the brick pillars. As it turns out, the abutments of the bridge in Marienburg are built in a similar way to all other piers.

Besides their importance as being two pioneering examples of imposing iron constructions in Germany, the bridges in Dirschau and Marienburg are also recognized as two cases of a successful use of hydraulic mortar and concrete produced with local raw materials. In this regard, it is worth reading the words by the Prussian minister of industry, trades and public works, August von der Heydt in 1857: "The discovery and the exploitation of good materials to produce mortar and cement has already proved in France how profitable the exploration of the soil is; the same has been also proved by the brilliant success of the search for appropriate building materials that

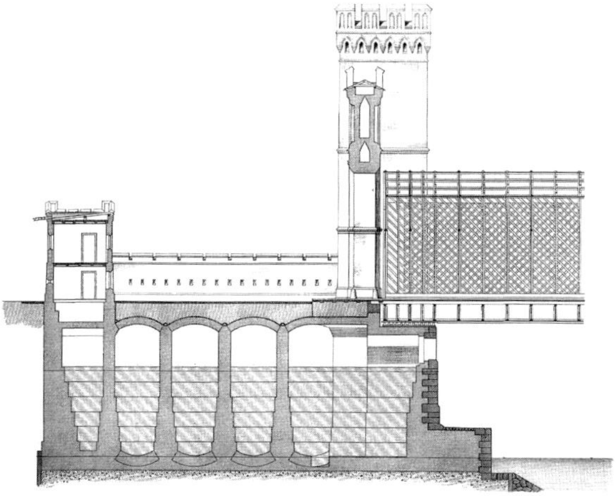

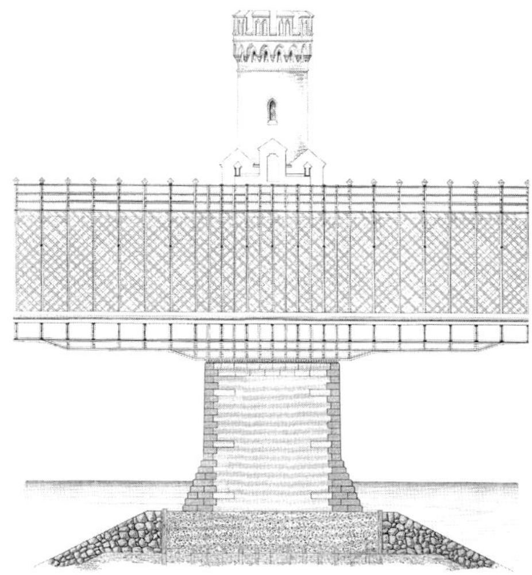

has been carried out for the construction of the bridge over the Vistula near Dirschau."[116] The renown of the cement plants in Dirschau und Marienburg is probably also due to the fact that they sell cement for other public and private works. Moreover, after the completion of the bridges, the plant in Dirschau continues to produce cement, which becomes known as the "Dirschauer Cement."[117]

Further studies about materials used to produce hydraulic mortar and concrete

The increasing need for effective and affordable hydraulic mortar and concrete requires a full accounting of all the materials that are appropriate for producing hydraulic binders. Thus, the second half of the 1840s hosts new chemical investigations. Some concern the so-called Santorin earth, a kind of volcanic sand from the Greek island of Santorini, which shows properties similar to those of pozzolana and trass. Santorin earth becomes known in Germany through accounts given by Bavarian engineers that work for the court of Otto Friedrich Ludwig von Wittelbach (1815–1867), crowned King of Greece in 1832. The engineer Wilhelm von Weiler uses Santorin earth-based concrete in the early 1840s to build, on the island of Syros, the foundations of the quays for the new port, the vaults of a military hospital and some flat terraces.[118] The

[116] "Wie lohnend die Erforschung des Bodenreichstum werden kann, haben die seit längere Zeit in Frankreich der Auffindung und Nutzbarmachung guter Materialien zur Mörtel- und Cementbereitung zugewendeten Bestrebungen dargethan; nicht minder die glänzenden Erfolge, welche die dem Bau der Weichselbrücke bei Dirschau vorangegangen Untersuchungen und die darauf gegründeten Fabricationen geeigneter Baumaterialien geliefert haben" (August von der Heydt, *Circular-Verfügung vom 19. April 1858, die Aufstellung von Baumaterialien-Verzeichnissen in den einzelnen Baukreisen betreffend*, in, "Zeitschrift für Bauwesen," VIII, 1858, (coll. 369–371), col. 369.

[117] See Lentze, *Beschreibung einiger Hülfseinrichtungen*...cit., col. 377; Licht, *Der Portland-Cement und*...cit.

[118] Von Weiler develops a plan for the town of Ermoúpoli on the island of Syros, since 1835. For more on this, see Anna Frangoudaki, Caglar Keyder, *Ways to Modernity in Greece and Turkey: Encounters with Europe, 1850–1950*, London, New York, Tauris, 2007, pp. 240–241.

[119] Karl von Körber, *Ueber den Gebrauch der Santorinerde*, in, "Allgemeine Bauzeitung," VI, 1841, pp. 266–268; see also Johann Christian Wedeke, *Handbuch der Landbaukunst und der Landwirthschaftlicher Gewerbe für Baumeister, Landwirthe und Cameralisten*, vol. I, Leipzig, Romberg, 1851, coll. 331–335.

[120] See Christian Ludwig Friedrich von Förster, *Schlussbemerkung zu diesem Jahrgange der Allgemeine Bauzeitung*, in, "Allgemeine Bauzeitung," XI, 1846, (pp. 397–398), p. 398.

[121] See Franz Stauffert, *Der jetzige Zustande der Baukunst in Griechenland*, in, "Beilage zur Allgemeinen Bauzeitung. Ephemeriden," 1844, n° 1, pp. 1–8, n° 2, pp. 17–25, n° 3, pp. 33–38, n° 4, pp. 41–58,

119-120. J. C. W. Lentze, Bridge over the River Vistula, Dirschau, 1846–1855.

119. Longitudinal section of an abutment (J. C. W. Lentze, *Die im Bau begriffenen Brücken über die Weichsel bei Dirschau und über die Nogat bei Marienburg*, in, "Zeitschrift für Bauwesen," V, 1855).

120. Cross section of a pier (J. C. W. Lentze, *Die im Bau begriffenen Brücken über die Weichsel bei Dirschau und über die Nogat bei Marienburg*, in, "Zeitschrift für Bauwesen," V, 1855).

Austrian engineer Karl von Körber reports about Von Weiler's concrete in an article published in the"Allgemeine Bauzeitung" in 1841.[119] Von Körber also reports about successful experiments with the use of Santorin earth that were carried out in the harbour of Trieste, giving rise to further and wider uses of this material along the northeast Italian coast.

Following the Athens insurgency of 1843 against king Otto, several German scholars and technicians are obligated to leave Greece. Franz Stauffert, who has worked as state architect in Athens since 1835, is among them. He probably settles in Vienna after leaving Greece, where he becomes an editor of the "Allgemeine Bauzeitung."[120] Stauffert is principally known for a long report about the Greek art of building that is published in 1844; however, he should be also remembered as the author of a report focusing on the building materials used in Greece, including the Santorin earth, which is published in 1846 in the "Journal für die Baukunst."[121] "The volcanic ash from Santorin, called ασπρόχωμα," Stauffert writes, "has such a binding force that even big buildings on the island are made of mortar prepared with this ash mixed with different sized stones – tiny, hand- and egg-sized ones."[122] In the same year, Franz Ludwig von Welden (1782–1853), a Lieutenant of the Austrian army, reports in the "Kunst- und Gewerbe-Blatt des polytechnischen Vereins für das Königreich Bayern" about the quay foundations made of alternating layers of pebbles and Santorin earth-based mortar poured into sunken caissons that he observes in Trieste.[123] Responding to such interest, the chemist Franz Carl Leonhard Elsner (1802–1874) promotes the analysis of some Santorin earth samples in the laboratory of the *Gewerbe-Institut* in Berlin.[124] He reports that the master builder Salzenberg brought samples to Berlin and wrote an article about Santorin earth. It is very probable that Elsner means the Berlin architect and Schinkel scholar Wilhelm Salzenberg (1803–1887), who travels to Constantinople in 1847 on behalf of Friedrich Wilhelm IV and makes some stops in northern Italy along the way back to Germany, evidenced by the report in the "Notiz-blatt des Architekten-Vereins zu Berlin," in 1847, about the Italian stucco he observed in Parma.[125] In the same issue, a report about Santorin earth immediately follows the one about the stucco, and one cannot rule out the idea that this is precisely the report by Salzenberg that Elsner mentions, although regrettably no author was given for it.[126] The report describes and compares the chemical contents of Santorin earth, pozzolana and trass, and furthermore states that Santorin earth "has been recently and successfully used to produce concrete for hydraulic works in Syros, Trieste and Venice, where people highly recommend it."[127] Again in 1847, a certain P. Schulze visits the building site of new quays in the harbour of Trieste,

n° 5, pp. 69–95; Id., *Über die Baumaterialien des alten und des neuen Griechenlands, und über die geognostischen Verhältnisse dieses Landes*, in, "Journal für die Baukunst," vol. XXIII, 1846, (pp. 101–130, 268–319), pp. 278–279.

[122] "Die vulcanische Asche von *Santorin*, ασπρόχωμα genannt, hat eine so bindende Kraft, dass auf der Insel grosse Gebäude aufgeführt werden, die nur aus ganz kleinen, hand- und eigrossen Steinen, in den mit der Asche angemachten Mörtel gelegt, bestehen" (Stauffert, *Über die Baumaterialien*...cit., p. 279).

[123] Franz Ludwig von Welden, *Ueber die Eigenschaften des Santorin-Mörtels und seiner Anwendung bei Bauten in und außer Wasser*, in, "Kunst- und Gewerbeblatt des polytechnischen Vereins für das Königreich Bayern," XXXII, 1846, coll. 683–693.

[124] See Franz Carl Leohard Elsner, *Ueber die Zusammensetzung des Santorins und dessen technische Verwendung*, in, "Verhandlungen des Vereins zur Beförderung des Gewerbfleißes in Preußen," XXVIII, 1849, pp. 86–87; and in, "Polytechnisches Journal," vol. CXIII, 1849, pp. 157–158.

[125] See Ottobrina Voccoli, *La rinascita dell'arte musiva in epoca moderna in Europa. La tradizione del mosaico in Italia, in Spagna e in Inghilterra*, doctoral dissertation, Universitat de Barcelona, 2009, pp. 306–307; Wilhelm Salzenberg, *Stucco a Lucido*, in, "Notiz-Blatt des Architekten-Vereins zu Berlin," I, 1847, pp. 15–16.

[126] [Wilhelm Salzenberg], *Santorin-Erde*, in, "Notiz-Blatt des Architekten-Vereins zu Berlin," I, 1847, p. 16.

[127] "Diese Erde ist nun bei den Meeresbauten zu Syra, Venedig und Triest mit sehr günstigem Erfolge zur Betonbereitung angewendet worden und wird von dort aus lebhaft empfohlen" (Anon. *Santorin-Erde*...cit.).

where he examines the same kind of concrete foundations that Von Welden had already observed and writes about them in a report published by the "Notiz-blatt des Architekten-Vereins zu Berlin" in 1848.[128] Between November 1848 and Mai 1849, the Prussian *Kriegs- und Handels-Ministerium* promotes several tests to compare the hardening of concrete produced with trass and with Santorin earth.[129] The tests take place on the building site of the Berlin cathedral. A certain quantity of Santorin earth is transported to Berlin by land, and the *Ingenieur-Major* Block provides the necessary information about how to use it, according to the published reports and his personal experience acquired during multiple journeys to Venice, Trieste and Greece. Once a supervising technical committee including the engineer Lentze has been established, the tests start. Three wooden 2.5 square feet wide boxes are assembled: the first is filled with alternating layers of trass-based mortar and pebbles, the second with layers of pure Santorin earth-based mortar and pebbles, and the third with pebbles and mortar made of Santorin earth and sand. The hardness is tested after 7, 11 and 42 days, thereby acknowledging that the quality of Santorin earth does not equal that of trass. Moreover, it is also established that concrete made of alternate layers of mortar and pebbles is not as suitable as mixtures of mortar and stones. In the context of the interest in Santorin earth, it still receives attention in a report dating back to 1851, written by Xaver Johannes Landerer, a chemist at the Bavarian court who relates having successfully filled dental cavities with mixtures of lime, Santorin earth and water.[130] This peculiar use of Santorin earth-based mortar seems a late legacy of the closeness between the domain of building and the one of medicine that characterizes the eighteenth-century studies on lime, before medicine and chemistry develop as two separate disciplines.[131]

Approximately in the same years as the studies about Santorin earth, a scientific interest in iron slags arises, probably urged by the increasing availability of such materials as a consequence of the developments in iron industry. Von Bauernfeind analyses iron slags in the context of the aforementioned trials ordered by the Nurnberg Railway-Committee, and, a few years later, Elsner promotes some further chemical investigations in the laboratory of the *Gewerbe-Institut* in Berlin. "It is well known," he asserts in 1847, "that iron slags milled into fine powder and added to lime produce a good kind of hydraulic mortar. Since the origin of this property must be due only to their chemical composition, I found important, from a technical-chemical point of view, […], to undertake some analyses."[132]

Interest in Santorin earth and iron slags does not, in any case, produce the same important consequences as the emergence of Portland cement from England. In 1847, the Bavarian chemist Max Joseph von Pettenkofer (1818–1901) conducts investigations into the chemical features of Portland cement and compares them to those of Bavarian hydraulic lime. The architect Leo von Klenze

[128] See P. Schulze, *Anwendung von Santorin-Erde und Puzzolane*, in, "Notiz-Blatt des Architekten-Vereins zu Berlin," vol. II, 1848, pp. 25–26.

[129] See Anon., *Ueber die Anwendung von Santorin-Erde in Preussen gemachten Versuche*, in, "Zeitschrift für die Bauwesen," I, 1851, coll. 293–302, 347–356.

[130] See Xaver Johannes Landerer, *Ueber die Santorinerde und deren Gebrauch als Zahnkitt*, in, "Polytechnisches Journal," vol. CXXII, 1851, p. 468.

[131] The reference is principally to the crucial studies by Joseph Black (1728–1799). See Joseph Black, *Experiments upon Magnesia alba, Quicklime and some other Alcaline Substances*, in, *Essays and observations, physical and literary*, vol. II, Edinburgh, 1756, (pp. 157–225), pp. 184–225.

[132] "Es ist bekannt, daß Hohofenschlaken [sic !], im fein gepülverten Zustande, als Kieselcemente dem gebrannten Kalk hinzugesetzt, einen sehr guten hydraulischen Mörtel liefern. Da diese Eigenschaft der Hohofenschlacken einzig und allein nur in deren bestimmter chemischen Zusammensetzung ihren Grund haben kann, so schien es mir in technisch-chemischer Beziehung von Wichtigkeit, […], einer chemischen Analyse zu unterwerfen" (Franz Carl Leohard Elsner, *Ueber Hohofenschlacken [sic!] als hydraulische Cämente*, in, "Verhandlungen des Vereins zur Beförderung des Gewerbfleißes in Preußen," XXVI, 1847, (pp. 89–90), p. 89; in, "Polytechnisches Journal," vol. CVI, 1847, pp. 312–322, and

(1784–1864) asks him to perform this research, after having verified that Portland cement is more effective than Bavarian hydraulic lime in producing mortar and plaster. He even provides Von Pettenkofer with samples of Portland cement and Bavarian hydraulic lime taken from different producers.[133] Von Pettenkofer's scholar Anton Hopfgartner undertakes early analyses, while his mentor carries them through to their conclusion. Both then share the results of their work in 1849.[134] Von Pettenkofer realizes that the higher quality of English Portland cement is essentially due to the use of more appropriate raw materials and to a stronger firing. He verifies that English cement contains some alkali that he considers to be substances facilitating the melting of clay, a process through which silica are freed from silicates (as previously argued by Fuchs), while clay covers lime particles and prevents slaking reaction. Von Pettenkofer reports having received from Fuchs a sample of marl coming from Sheppey Island. He verifies that this kind of stone produces cement that hardens very fast underwater if burnt at a moderate fire, whereas it produces cement that hardens slowly but strongly, if burnt with a strong fire. He furthermore confirms at the microscope that, whereas Bavarian hydraulic lime particles are spherical, the ones of Portland cement are scale-shaped and therefore facilitate cohesion. The example that Von Pettenkofer gives to explain such a phenomenon pertains solely to the domain of building: "To compare the two kinds of masses, one should imagine a first wall made of ashlars, and a second one made of layer of boulders; in the first case surfaces come in touch, whereas in the second case just points touch each other, leaving much more in-between empty spaces than in the first case."[135]

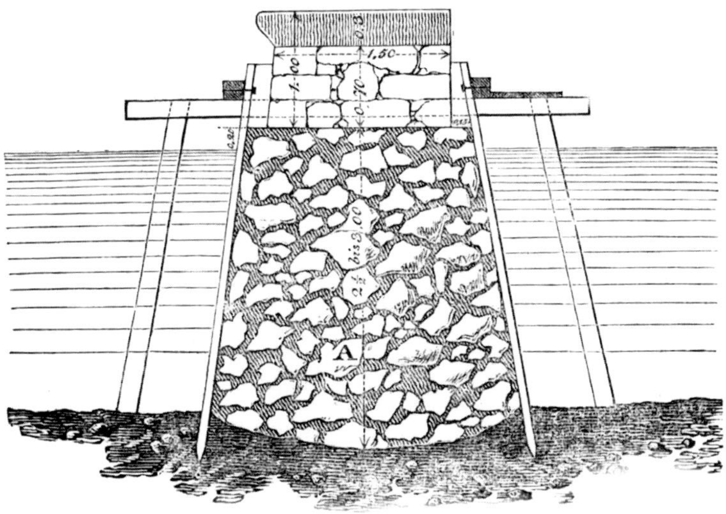

121. K. von Körber, drawing of concrete foundation of a seawall in Syros, Greece. Cross section (K. von Körber, *Ueber den Gebrauch der Santorinerde*, in, "Allgemeine Bauzeitung," VI, 1841).

in, "Kunst- und Gewerbe-Blatt des polytechnischen Vereins für das Königreich Bayern," XXXIII, 1847, coll. 686–688).

[133] See Carl von Voit, *Max von Pettenkofer zum Gedächtniss*, München, Verlag der k. b. Akademie in Commission des G. Franz'schen Verlags, 1902, pp. 26–27; Otto Neustätter, *Max Pettenkofer*, Wien, Julius Springer, 1925, p. 23; Harold Breyer, *Max von Pettenkofer Arzt im Vorfeld der Krankheit*, Leipzig, Hirzel, Teubner, 1980, pp. 44–45; Karl Wieninger, *Max von Pettenkofer. Das Leben eines Wohltäters*, München, Hugendubel, 1987, pp. 117–121.

[134] See Max Joseph von Pettenkofer, *Bemerkungen zu Hopfgartner's Analyse eines englischen und eines deutschen hydraulischen Kalkes*, in, "Polytechnisches Journal," vol. CXIII, 1849, pp. 357–371; see also Anton Hopfgartner, *Analyse zweier hydraulischen Kalkes, eines englischen (Portland-Cement) und eines deutschen hydraulischen Kalkes*, in, "Polytechnisches Journal," vol. CXIII, 1849, pp. 351–357.

[135] "Denkt man sich beide in auseinanderliegenden Massen, so läßt sich das eine einer Mauer von Quadern, das andere einer Schichte von Rollsteinen vergleichen; beim ersteren berühren sich Flächen, beim letzteren Punkte, daher so viele Zwischenräume mehr, als beim ersten" (Von Pettenkofer, *Bemerkungen zu Hopfgartner's*...cit., p. 368).

Oder-Brücke.

Gerüst zum Senken des Drehpfeilers.

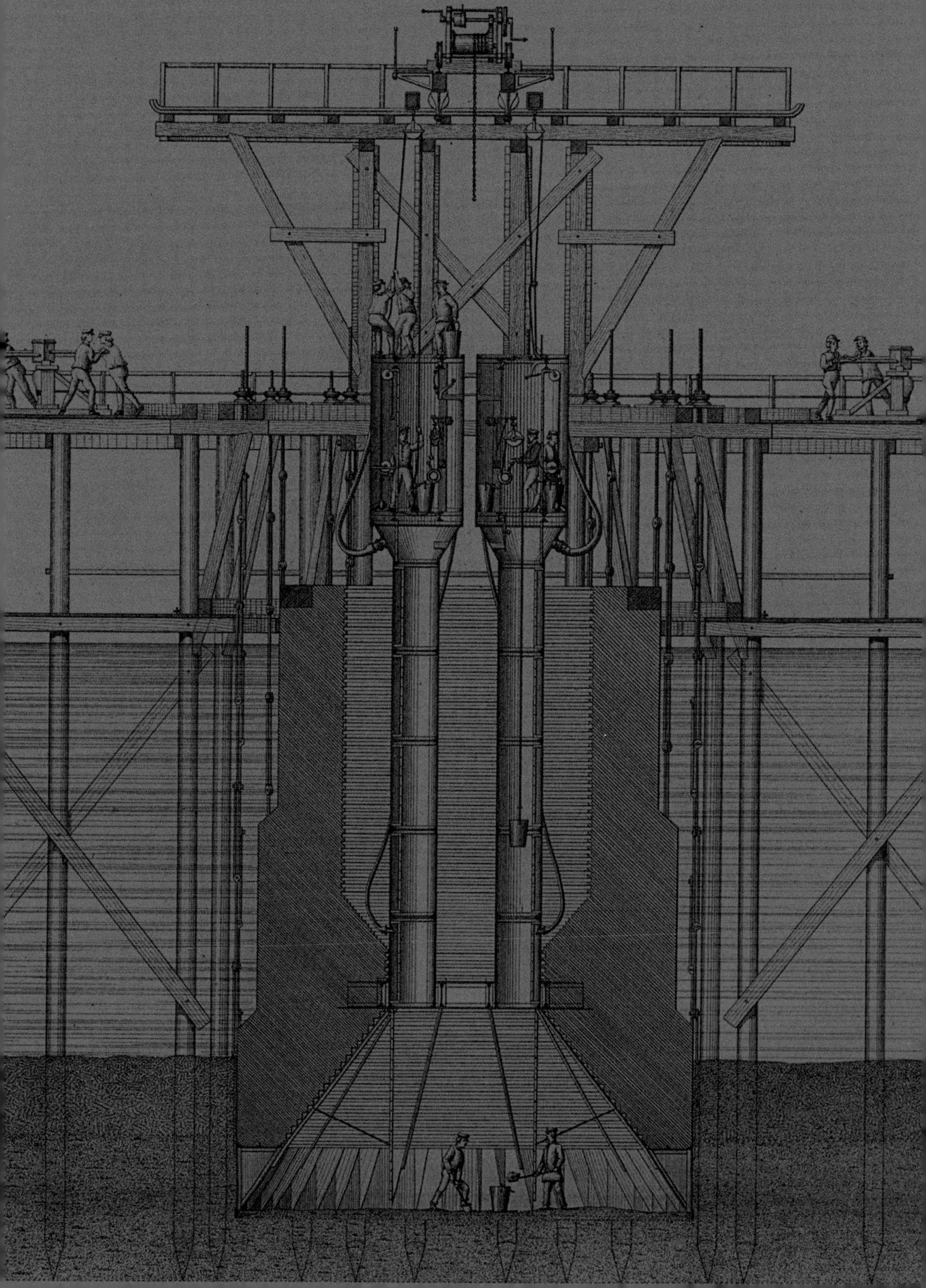

chapter five
portland cement and concrete construction, from foundations to roofs

Portland cement and its prevalent use in binding brickwork

At the turn of the 1850s, Portland cement is gaining attention. Its use gradually increases and Prussia plays a leading role in this. Between 1849 and 1850, the Berlin master builder [Wilhelm Andreas] Becker experiments with this binder during the construction of a manufacturing plant powered by a water mill in Spandau; he notably applies a kind of pure Portland cement-based mortar to fill the empty spaces that mistakenly arise when sheet piles are rammed into the soil.[1] This is only a minor example, especially when compared to the foundations made of Portland cement-based concrete that are realized in 1851 in Berlin to build the bridge over the *Luisenstädtischer Kanal*, close to the *Wassertorplatz*, along the so-called *Verbindungsbahn*.[2] This railway is built between December 1850 and October 1851, following the plans drawn by the aforementioned engineer Neuhaus, and the use of Portland cement-based concrete can probably be attributed to Neuhaus' previous experience with concrete building in northern Prussia, for the construction of the Eastern Railway branch between Berlin and Stettin.[3] A further pioneering use of Portland cement in concrete foundations is in Königsberg, where a retaining wall along the River Pregel, close to the end station of the Eastern Railway, realized be-

[1] See [Wilhelm Andreas] Becker, *Erfahrungen über den Portland-Cement*...cit., p. 1; Id., *Praktische Anleitung zur Anwendung der Cemente zu baulichen, gewerblichen, landwirthschaftlichen und Kunst-Gegenständen*, Berlin, Nicolaische Verlagsbuchhandlung, 1860, pp. 25–26; the second edition of this book is published in 1869. About the hypothesis of Becker's first name, see *Verzeichniss derjenigen Baumeister des Preussischen Staats, welche nicht im Staatsdienste als Baubeamte angestellt sind*, in, "Beilage zur Zeitschrift für Bauwesen," III, 1853, n. p.

[2] See [Wilhelm Andreas] Becker, *Der feuerfeste Treppenbau von natürlichen und künstlichen Steinen. Nach den neuesten Erfindungen und Ausführungen mit besonderer Berücksichtigung der Constructionen zum praktischen Gebrauche bearbeitet*, Berlin, Ernst & Korn, 1857, p. 56, n. 3. The foundations are 4 ½ feet high and concrete is composed of one measure of Portland cement, two measures of sand and two measures of equal amounts of limestone and brick rubble. It is mixed in wooden drums and poured by means of buckets into pits surrounded by sheet piles; it hardens in eight days.

[3] See Königlich Preussischen Ministers der Öffentlichen Arbeiten, ed., *Berlin und seine Eisenbahnen 1846–1896*, Berlin, Julius Springer, 1896, vol. I, pp. 236–257; Anon., *Friedrich Neuhaus - Ein vergessener Eisenbahnpionier*... cit., p. 110.

122. W. A. Becker, timber sheet piles sealed with cement-based mortar (W. A. Becker, *Praktische Anleitung zur Anwendung der Cemente zu baulichen, gewerblichen, landwirthschaftlichen und Kunst-Gegenständen*, Berlin, Nicolaische Verlagsbuchhandlung, 1860).

tween 1851 and 1853, is built on a foundation made of timber piles and Portland cement-based concrete, poured among the pile heads.[4]

Apart from these examples of concrete foundations, in the very early 1850s, the use of Portland cement mainly concerns the production of mortar to bind and plaster brickwork. In 1851, the master mason [Carl Ludwig] Schüttler builds a balcony on the main façade of his own house in the Berlin neighbourhood of Moabit, using bricks bound with Portland cement-based mortar.[5] Eight pillars and six beams, two longitudinal and four transversal, support the balcony. Each beam consists of a shallow-arched row of bricks arranged on edge and three further rows of flat bricks on each side of the arched one. The floor lying over the beams consists of two layers of bricks bound with Portland cement-based mortar; the ones of the lowest layer are arranged diagonally, whereas those that form the upper layer are arranged perpendicularly to the façade. The whole is finally plastered with Portland cement-based mortar. This kind of construction can be seen as the paroxysm of the belief in the feasibility of monolithic building elements made of bricks and cement-based mortar, a sentiment echoed in the words of the master builder Paul Emanuel Spieker (1826–1896) a few year later, as he describes the compression strength test that he carries out on a pillar-specimen made of bricks and Portland cement-based mortar, in the context of the construction of the Evangelical Church of Wiesbaden (1852–1862). "One can be sure of building an artificial monolith," Spieker asserts, "if Portland cement, which rapidly binds and gets the hardness of a stone, is used instead of ordinary mortar."[6]

A number of publications about English cement in the early 1850s indicate the increasing general interest in Portland cement. In 1851, the savant Karl Emil Franz von Schafhäutl (1803–1890) publishes a survey about the history of Roman and Portland cements in the "Polytechnisches Journal."[7] He traces a path from Smeaton's studies about hydraulic lime to Von Pettenkofer's analyses of Portland cement, mentioning, inter alias, Pasley's trials for the production of artificial cement and the compressive strength tests on specimens of cement-based concrete that the English cement company White & Sons carries out in 1847.[8] The master builders Johann Christian Wedeke and Johann Andreas Romberg (1806–1868) give a comparative account of Roman and Portland cement in the first volume of the *Handbuch der Landbaukunst*, which they write together and publish in 1851, asserting that the use of Roman cement is decreasing as a consequence of the spread of Portland cement.[9] They also mention the strength tests that another important English cement company, the Robins, Aspdin & Company, carries out in 1848 on beams made of bricks bound with cement-based mortar (for flexural strength tests) and artificial stones made of cement-based mortar (for compressive

[4] See Becker, *Der feuerfeste Treppenbau*...cit., p. 56, n. 3.
[5] See Becker, *Praktische Anleitung*...cit., p. 33.
[6] "Verwendet man namentlich, statt gewöhnlichen Mörtel, einen raschen bindenden, etwa den in Kurzem zu Steinhärte gedeihenden Portland-Cement, so kann man gewiss sein, einen künstlichen Monolith hergestellt zu haben" (Paul Emanuel Spieker, *Bericht über eine Probe auf Tragvermögen von Backsteinen aus geschlemmter Erde*, in, "Zeitschrift für Bauwesen," V, 1855, (coll. 474–79), col. 478).
[7] Karl Emil Franz Schafhäutl, *Das Portland- und Roman-Cement. Ein Beitrag zur Geschichte der Cemente oder hydraulischen Mörtel in England, nebst einem Anhange über die Theorie der Erstarrung der Mörtel und über den glänzenden Stucco der Alten*, in, "Polytechnisches Journal," vol. CXXII, 1851, pp. 186–208, 267–293.
[8] *Ibid.*, pp. 202–208, 267–279.
[9] See Johann Christian Wedeke, Johann Andreas Romberg, *Handbuch der Landbaukunst und der landwirthschaftlichen Gewerbe*...cit., vol. I, coll. 328–329. The dates of birth and death of Wedeke remain unknown, whereas the ascertained period of activity is to be placed between 1835 and 1858; see Katalog der Deutschen Nationalbibliothek.
[10] *Ibid.*, coll. 329–330.
[11] Johann Caspar Garthe, *Die neuesten englischen Versuch über Vergleichung des Portland- und Roman-Cements*, in, "Polytechnisches Journal," vol. CXXIV, 1852, pp. 25–48.
[12] See Karl Wilhelm Drewitz, *Casernement für das Königliche*

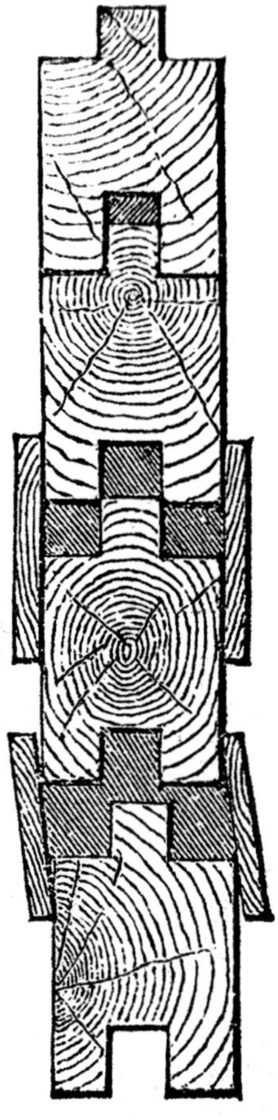

strength tests). It is then probably in thinking of Roman cement alone that Wedeke and Romberg define modern cements as unsuitable for producing concrete. In fact, Roman cement is thought to set too rapidly, which prevents workers from appropriately levelling and tamping the mass.[10] As a matter of fact, the property of setting rapidly, which allowed the building of extremely demanding masonry works like the vaults of the Thames Tunnel in London, is considered disadvantageous as the use of concrete spreads. Concrete for important hydraulic works needs to be produced in huge amounts, in appropriate plants, to then be transported to the places where it will be poured and tamped. This working sequence requires time and urges the development of artificial cements with a slower setting time. Lentze already realized this when he developed the cement for the construction of the bridges in Dirschau and Marienburg. A detailed account of the 1847–1848 tests by White & Sons and Robins, Aspdin & Company is then given in a report by the naturalist Johann Caspar Garthe (1796–1876), which is published in 1852.[11]

Some simple rudimentary experiments to test the quality of Portland cement also take place in Berlin between 1851 and 1852, on the initiative of the master builder August Ferdinand von Fleischinger (1804–1885), who works as building councillor for the Ministry of War and has planned the new barracks near *Hallesches Tor* in Berlin together with Karl Wilhelm Drewitz (1806–1888).[12] At the end of the summer of 1851, Von Fleischinger orders Becker, who has been appointed clerk of the barracks construction, to carry out experiments with Portland cement. At that time, two lime kilns produce Portland cement in Germany, according to Becker; they are the aforementioned kiln of Brunkhorst and Westphalen in Buxtehude and the kiln of Adolph Jencquel in Moorrege near Uetersen.[13] They are both situated at the mouth of the River Elbe, where they benefit from English trade, and, as a matter of fact, Jencquel imports chalk and coal from England, whereas clay is widely available in Uetersen. A chalk mill is moreover available along the River Pinnau, a right tributary of the Elbe.[14] Portland cement from England is obtainable in Berlin from four dealers, among which there is the previously-mentioned hydraulic lime maker Hasler from Moabit.[15] Becker chooses to test Portland cement from the kiln of Brunkhorst and Westphalen, which he considers to be the best produced in Germany, and Portland cement from the English factories Francis & White and Robins, Aspdin & Company. Since he believes that enough tests have been carried out that focus on the behaviour of cement-based mortar placed underwater and exposed to the air, Becker decides to investigate how Portland cement-based mortar withstands fire and frost.[16] This era of industrial and urban development requires available mortar to build heat-resistant masonry for chimneys and structures supporting

Garde-Dragoner-Regiment auf dem sogenannten Upstall vor dem Halleschen Tor bei Berlin, in, "Zeitschrift für Bauwesen," V, 1855, coll. 521–542.

[13] See Becker, *Erfahrungen über den Portland-Cement*...cit., pp. 14–15. Becker also mentions a third kiln, the Zurhelle & Elster in Hamburg, which begins to produce Portland cement in the late 1840s but stops shortly after. See, in this regard, *ibid.*, p. 15; Friedrich Quietmeyer, *Zur Geschichte der Erfindung des Portlandzementes*, doctoral dissertation, Königliche Technische Hochschule zu Hannover, 1911, p. 148. Becker could probably be Quitmeyer's source.

[14] See Michael Plata, *Die Zementfabrik in Moorrege*, in, "Demokratische Geschichte," vol. XXII, 2011, (pp. 125–142), pp. 125–131.

[15] The other dealer are Schüttler & Comp, which Becker introduces as the exclusive reseller of Portland cement from the factory of Robins, Aspdin & Comp.; Woderb and Goslich, who moreover mill trass and produce hydraulic-lime and Roman cement; and Albert Reinicke, who is the reseller of Portland cement from the factory of White and Sons. See, in this respect, Becker, *Erfahrungen über den Portland-Cement*...cit., pp. 51–52).

[16] *Ibid.*, p. 20, n. 1.

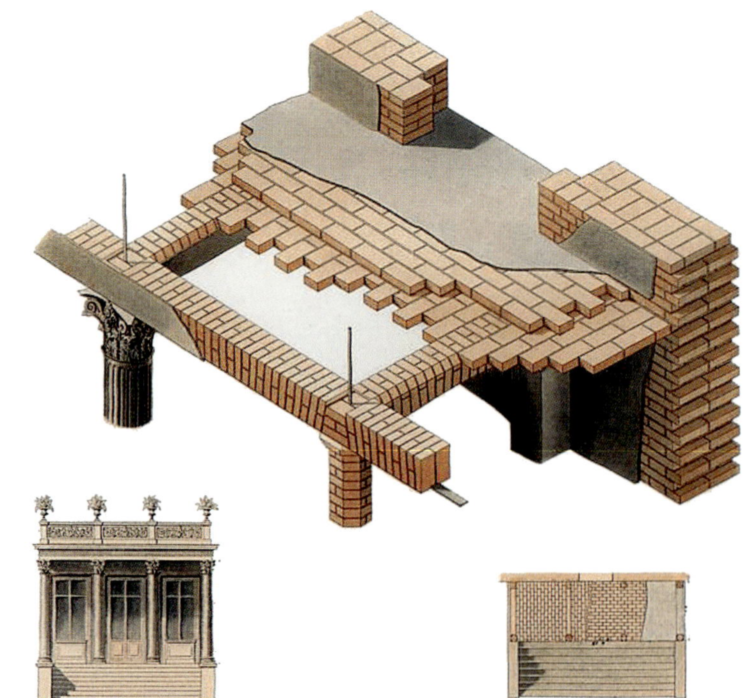

123-125. C. L. Schüttler, balcony made of bricks bound with Portland cement-based mortar, Berlin, 1851; Hahnemann, balcony made of bricks, iron bars and Portland cement-based mortar, Berlin, 1854; Riehmer, oriel windows made of bricks, iron bands and Portland cement-based mortar, Berlin, 1859 (W. A. Becker, *Praktische Anleitung zur Anwendung der Cemente zu baulichen, gewerblichen, landwirthschaftlichen und Kunst-Gegenständen*, Berlin, Nicolaische Verlagsbuchhandlung, 1860).

126. W. A. Becker, drawings of prefabricated steps made of bricks and cement-based mortar (W. A. Becker, *Der feuerfeste Treppenbau von natürlichen und künstlichen Steinen. Nach den neuesten Erfindungen und Ausführungen mit besonderer Berücksichtigung der Constructionen zum praktischen Gebrauche bearbeitet*, Berlin, Ernst & Korn, 1857).

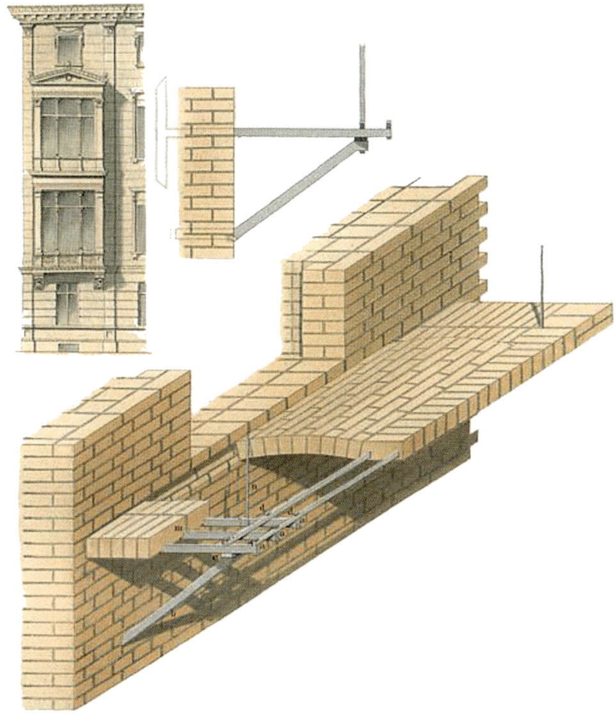

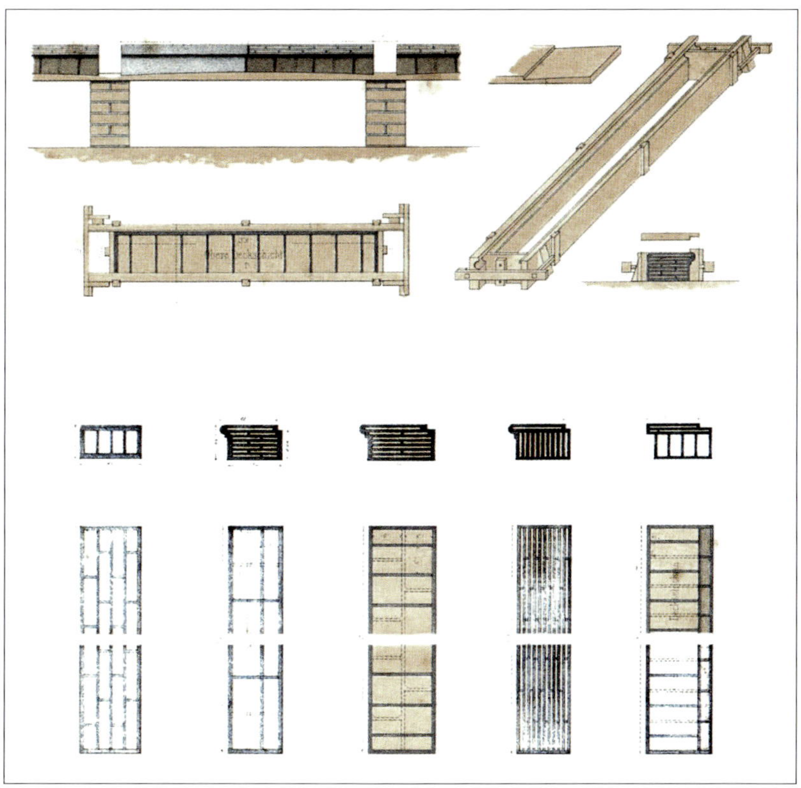

boilers for steam engines, as well as frost-resistant masonry to build water supplies, sewage networks and even public urinals, in order to improve the sanitary conditions of towns. At first, Becker burns several samples of cement-based mortar mixtures in a fireplace of the new barracks, as it is still under construction. Then, he verifies the effects of frost on some mortar specimens by soaking them in a kind of salt solution whose particles gain volume by crystallising, thereby reproducing the effect of water expansion upon freezing.[17] Becker learns this technique from the French naturalist Cyprien Prosper Brard (1786–1838), who had developed it to test the behaviour of natural stones in case of frost.[18]

While Becker's experiments continue, Portland cement is used to plaster the façades of the barracks, which look like they were made of cut stones, and to build the four staircases of the same building.[19] Three of these are entirely built on-site in brickwork bound with Portland cement-based mortar, whereas the fourth is assembled using prefabricated steps. The master mason Fehse produces these steps, arranging flat bricks and layers of Portland cement-based mortar in provisional wooden formworks, which are then removed once the mortar has set. Steps are furthermore finely plastered with the same kind of cement-based mortar that has been used to bind the bricks, and they are finally put into place as if they were made of cut stones. Fehse, Czarnikow and Schüttler are the main producers of prefabricated steps in Berlin; they actually begin to experiment with this kind of manufacturing as early as the second half of the 1840s, but it is only due to the introduction of Portland cement that they come to produce reliable steps.[20] The barracks are completed in 1854 with annexed stalls, the brick masonry vaults of which are bound with Portland cement-based mortar, after having verified that the hydraulic lime taken from the factory of Haslinger in Moabit does not have enough binding force for such a purpose.[21] Besides these examples, Portland cement is used in 1852 to plaster the interior walls of the bathrooms in the *Sommer-Lazareth* of the Charité Hospital in Berlin, which are exposed to warm and damp air, and to bind the bricks of a masonry structure supporting a boiler in the basement of the same hospital.[22] A singular experiment takes place in the same period at the Berlin *Garnison-Lazareth*, where a container made of bricks bound and plastered with Portland cement-based mortar is filled with urine and excrement and kept outside over a winter, in order to verify how Portland cement resists acids and frost.[23]

Outside Berlin, Portland cement mortar is used in a number of Prussian villages and towns for various purposes. In Königsberg, it is used with lime

[17] *Ibid.*, pp. 21–43.

[18] See Anon., *Ueber das von Brard in Vorschlag gebrachte Verfahren, um zu ermitteln, ob Steine den Wirkungen des Frostes widerstehen können, oder nicht*, in, "Verhandlungen des Vereins zur Beförderung des Gewerbfleißes in Preußen," VIII, 1829, pp. 178–180. This article is an extract from Anon., *Sur le Procédé proposé par M. Brard pour reconnaitre, immédiatement, les pierres qui ne peuvent pas résister à la gelée et que l'on désigne ordinairement par les noms de pierres gelives ou pierres gelisses*, in, "Annales de chimie et de physique," vol. XXXVIII, 1828, pp. 160–170. Further sources mentioned by Becker are Jean-Baptiste Rondelet, *Theoretisch-praktische Anleitung zur Kunst zu bauen*, ed. Carl Heinrich Distelbarth, vol. I, Leipzig und Darmstadt, Karl Wilhelm Leske, 1833, pp. 475–478; Friedrich Christian Accum, *Physische und chemische Beschaffenheit der Baumaterialien deren Wahl, Verhalten und zweckmässige Anwendung*, vol. I, Berlin, Reimer, pp. 255–260. It is furthermore worth mentioning that the master builder Johann Christian Wedeke described Brard's technique in 1842, although Becker does not mention him; see Johann Christian Wedeke, *Handbuch der bürgerlichen Baukunst: Allgemein fasslich für Maurer- und Zimmermeister und die es werden wollen*, Quedlinburg, Leipzig, Gottfried Basse, vol. I, 1842, pp. 145–147).

[19] See Becker, *Erfahrungen über den Portland-Cement*...cit., p. 20; Drewitz, *Casernement für das Königliche Garde-Dragoner-Regiment*...cit., coll. 538–541.

[20] See Becker, *Der feuerfeste Treppenbau*...cit., pp. 59–60; Id., *Praktische Anleitung*...cit., p. 34.

[21] See Drewitz, *Casernement für das Königliche Garde-Dragoner-Regiment*...cit., coll. 533–534.

127. P. J. Lenné, artificial grotto in the garden of Villa Borsig in Moabit, Berlin, approx. 1849 (W. A. Becker, *Praktische Anleitung zur Anwendung der Cemente zu baulichen, gewerblichen, landwirthschaftlichen und Kunst-Gegenständen*, Berlin, Nicolaische Verlagsbuchhandlung, 1860).

and sand to produce the mortar that binds the foundation masonry of the local railway station.[24] The cornices on the façade of the secondary school building of Anclam in Mecklenburg, which is built between 1850 and 1852, are plastered with Portland cement-based mortar, as well as the basin of the fountain that is built to decorate the space housing the industrial exhibition of Breslau in 1852. Between 1852 and 1853, for the construction of the church of the village of Brodwin, northeast of Berlin, Portland cement-based mortar is even used as a sort of glue to fix the finials (*Kreuzblumen*) on the roof.[25]

A considerable boost to the use of cement-based mortar comes indirectly from an 1853 Berlin building regulation which imposes limitations on the use of wood in order to reduce the danger of fires.[26] It establishes that jutting out balconies must be built in masonry or metal, and, following this prescription, some balconies and loggias are realized on the model of Schüttler's balcony – however, using iron bars as reinforcement. The constructions completed by the master builder Hahnemann in 1854, by the master mason Lüdke in 1858 and by the master mason Riehmer in 1858 serve as pertinent examples.[27] The same regulation from 1853 also orders that "the staircases of each building must be fireproof," consequently promoting the construction of staircases made of brick masonry bound with Portland cement-based

[22] See Becker, *Erfahrungen über den Portland-Cement...* cit., pp. 19–20; Hesse, *Das neue erbaute Sommer-Lazareth bei der Charité-Kranken-Heil-Anstalt in Berlin*, in, "Zeitschrift für Bauwesen," III, 1853, (coll. 343–350), col. 346.

[23] See Becker, *Erfahrungen über den Portland-Cement...* cit., p. 44.

[24] See Becker, *Der feuerfeste Treppenbau...*cit., p. 56, n. 2, 3; the exact mixture is composed of three measures of sand, three quarters of lime, one quarter of Portland cement. The frameworks of doors and windows of the stations along the Ostbahn are plastered with Krienberg cement-based mortar, whereas the rest of the buildings consist of exposed brick masonry; see Winterstein, *Bericht über die Reise des Architekten-Vereins und des Vereins für Eisenbahnkunde nach Königsberg*, in, "Zeitschrift für Bauwesen," IV, 1854, (coll. 543–571), col. 552.

[25] See M. Gottgetreu, *Das Gymnasium zu Anclam*, in, "Zeitschrift für Bauwesen," II, 1852, coll. 455–458; P. R. Brecht, *Kirche zu Brodwin*, in, "Zeitschrift für Bauwesen," IV, 1854, coll. 209–212; C. Studt, *Ueber die auf die Exercier-Platz zu Breslau erbaute Halle für die Schlesische Industrie-Ausstellung im Jahre 1852*, in, "Zeitschrift für Bauwesen," II, 1852, (coll. 533–538), col. 536.

[26] See Königliches Polizei Präsidium, Karl Ludwig Friedrich von Hinckeldey, *Neue Bau-Ordnung für Berlin*, in, "Zeitschrift für praktische Baukunst," XIII, 1853, coll. 133–152.

[27] See Königliches Polizei Präsidium - Karl Ludwig Friedrich von Hinckeldey, *Neue Bau-ordnung für Berlin*...cit., col. 143; Becker, *Praktische Anleitung*...cit., pp. 32–34.

mortar, as well as the manufacturing of prefabricated steps.[28] Drewitz builds a fireproof staircase made of bricks and Portland cement mortar in the guard building that he constructs in 1853 in Berlin's west end.[29]

The technique used to produce prefabricated steps is also applied to manufacturing a wide range of objects and decorations, such as containers, vases and even artificial rocks for decorating gardens, in the wake of the attempt at Glienicke in the early 1840s. Around 1849, Fehse produces such artificial rocks for the garden that Peter Joseph Lenné (1789–1866) designs for the villa of Johann Friedrich August Borsig (1804–1854), which is built in the neighbourhood of Moabit according to the plans by the architect Johann Heinrich Strack (1805–1880).[30] The technique of producing ready-made decorations and objects made of bricks and cement-based mortar spreads so widely in Berlin that it becomes known in the early 20th-century historiography as the "Berlin System."[31]

The manufacturing of Portland cement by Bleibtreu

Although, as previously noted, some lime kilns at the mouth of the Elbe are already producing Portland cement starting the late 1840s, the actual German manufacturing of Portland cement is thought to date back to the mid-1850s, when the entrepreneur and savant Hermann Bleibtreu (1821–1881) undertakes the production of cement from inland raw materials in a plant that he founds in Stettin, after having carried out studies and trials based on his knowledge and skills in chemistry. Using the expression "authentic Portland cement from inland raw materials" in 1860, Becker soon recognizes Bleibtreu as the initiator of the German manufacturing of Portland cement, in accordance with the general belief in the importance of scientific knowledge in developing manufacturing practices and the broad effort to exploit inland raw materials.[32]

Bleibtreu is born into an entrepreneurial family based near Bonn. At a young age, he begins to study natural sciences and geology, but he is forced to stop due to the death of his father in 1839, when he needs to take over the ownership of the family mine of alum, together with his brother.[33] As of 1842, he can once again pursue his passion for chemistry and attends the lectures that Von Liebig and Hermann Franz Moritz Kopp (1817–1892) give at the University of Gießen. In 1845 he becomes Liebig's assistant and moves to the London Royal College, where von Liebig's previous assistant August Wilhelm von Hoffmann (1818–1892) has been charged with the task of setting up a chemical laboratory. According to Karl Goslich, who has been the technical director of the Stettin cement plant for several years, Bleibtreu tries to get information about Portland cement during his stay in London, at the suggestion of Ernst Friedrich Zwirner (1802–1861), the master builder of the

[28] See Königliches Polizei Präsidium - Karl Ludwig Friedrich von Hinckeldey, *Neue Bau-ordnung für Berlin*...cit., col. 140; Becker, *Der feuerfeste Treppenbau*...cit., p. 2.

[29] See Karl Wilhelm Drewitz, *Das neue Wachtgebäude am Unterbaum zu Berlin*, in, "Zeitschrift für Bauwesen," V, 1855, (coll. 467–472), col. 470.

[30] See Becker, *Praktische Anleitung*...cit., pp. 27–29.

[31] See Knut Stegmann, *Das Bauunternehmen Dyckerhoff Et Widmann. Zu den Anfängen des Betonbaus in Deutschland 1865–1918*, Tübingen, Berlin, Ernst Wasmuth, 2014, p. 39. See also Wilhelm Petry, *Betonwerkststein*, München, Deutscher Beton-Verein, 1913; Adolph Leonhardt, *Von der Cementware zum Konstruktiven Stahlbetonfertigteil*, Wiesbaden, Berlin, Bauverlag, 1964, pp. 1–10.

[32] See Becker, *Praktische Anleitung*...cit., p. 66.

[33] See Wilhelm Petry, *Zum 50 jährigen Bestehen des Vereins Deutscher Portlandzement-Fabrikanten. 75 Jahre deutscher Portlandzement. Ein Rückblick auf die Anfänge der deutschen Portlandzementindustrie und auf das Schaffen von Hermann Bleibtreu*, in, "Der Bauingenieur," VIII, 1927, (pp. 643–645), p. 643. The mine was in Hardt near Pützchen, which is today a neighbourhood of Bonn.

[34] See Karl Goslich, *Stettiner Portland–Cement–Fabrick 1855–1905*, Stettin, H. Susenbeth, 1905, esp. p. 1,

Cologne Cathedral. He is then thought to come back to Bonn in 1852, taking the text of Aspdin's patent from England, and a little information about the raw materials necessary to produce Portland cement, whereas the manufacturing process of it is kept secret by the British cement makers.[34] Based on a piece of informal family writing that Bleibtreu's son Leopold made available, the engineer Wilhelm Petry (1883–1936) maintains, instead, that Bleibtreu comes back to Bonn in 1846, without having studied cement during his stay in London, and that around 1852, a certain architect named Becker pushes him to study Portland cement. The architect that Petry mentions is probably Becker from Berlin, considering moreover that Bleibtreu later sends a sample of the very early Portland cement he burns specifically to him.[35]

Apart from such biographical speculations, it is a matter of fact that Bleibtreu studies cement in 1852 and, making reference to Von Pettenkofer's explanations, he experiments with different mixtures of lime, clay and alkaline substances. He develops a procedure to burn a kind of artificial cement and applies for a patent in October 1852.[36] During the winter between 1852 and 1853, while the release of the patent is pending, Bleibtreu considers the use of chalk from Wollin and from Rügen, in Pomerania, to produce cement.[37] He therefore moves to Züllchow near Stettin, where the relatives of his wife Dorothea Rosalie Sadée (1823–1899) live, and undertakes a series of trials burning mixtures of chalk from Rügen, trachyte (which contains silica) and phonolite (which contains alkaline feldspar).[38] Petry assumes that Bleibtreu moves to Pomerania because he fears the competition of trass in Bonn, once again confirming how deep-rooted the use of this material is in Rhineland.[39] In Züllchow, Bleibtreu meets the entrepreneur Paul Gutike, who agrees to invest money to install a cement plant on the site of an old brickyard that he expressly buys in December 1852. The results of early trials, however, are not encouraging and, on February 1853, Bleibtreu goes back to Bonn, where, by reading a few books, he learns about deposits of septarian clayish stones in Pomerania, which he supposes to be similar to the ones that English cement makers use. He returns to Züllchow and, after having successfully verified what he has learned from the books in Bonn, he goes on a journey to England with the aim of learning how to build an appropriate cement kiln. In September 1853, he moves to Züllchow with his family and begins to build the first nucleus of the future cement plant. The first burning test in a proper kiln is successfully carried out in December 1853.[40]

The raw materials are chalk from Wollin and septarian clay from Züllchow; they are soaked in a container full of water, mixed together according to a ratio of 4.5:1, and shaped in moulds as little bricks called *Koksen*, which are first dried and then fired. Well-burnt *Koksen* are ground into high-quality cement powder, a barrel of which is sent as a sample to the Stettin artificial-

See also Id., *Einhundert Jahre Portlandzement*, in, "Zeitschrift für angewandte Chemie," XXXVII, 1924, (pp. 265–267, 297–298, 504–505), p. 265.

[35] See Petry, *Zum 50 jährigen Bestehen*...cit., p. 643; Hermann Bleibtreu, *Der erste Zementbrand der Stettiner Portland-Cementfabrik im Dezember 1853*, in, "Zement," XIV, 1925, n° 10, (pp. 212–213), p. 213.

[36] See Hermann Bleibtreu, "Beschreibung eines neuen Verfahrens zur Darstellung hydraulischer Cement," 1852 (GStA, I HA Rep. 120 TD.D n. 206, foll. 1–4 r/v).

[37] The patent is released on March 1853.

[38] See Goslich, *Stettiner Portland–Cement–Fabrick*... cit., pp. 1–2. See also Historische Kommission bei der Bayerischen Akademie der Wissenschaften, ed., Otto Stolberg-Wernigerode, *Neue Deutsche Biographie*, vol. II, Berlin, Dunker & Humblot, 1955, entry "Bleibtreu, Hermann," pp. 297–298.

[39] English Portland cement produced by the factory Robins Aspdin & Goodwin is moreover available in Köln, since 1848, from the dealer J. Simonis; see Haegermann, *Vom cœmentum*...cit., p. 49, image caption 111.

[40] See Petry, *Zum 50 jährigen Bestehen*...cit., p. 644; Bleibtreu, *Der erste Zementbrand*...cit.

stone maker Philipp Loewer, who is based in the village of Grabow. The remaining part is mixed with a quarter of powder obtained by grinding the partially burnt *Koksen*; samples of this lower quality cement are sent to Becker in Berlin, as previously mentioned, and to Loewer as well. The remaining cement-powder obtained from the partially burnt *Koksen* is finally packaged in nine casks. We know from Bleibtreu that two are sent to the nearby village of Anklam, two others to Berlin, and one is kept on the site of the plant to be then used, in February 1854, to plaster some provisional buildings and to test the construction of beams made of bricks and cement-based mortar.[41] In 1854, Bleibtreu's cement plant is transformed into a public limited company named *Stettiner Portland Cement Fabrik*. The British prohibition to export Portland cement and Blue Lias limestones, from December 1854 until the end of the Crimean War in 1856, probably supports the early development of this pioneering German cement company.[42] In 1855, the *Stettiner Portland Cement Fabrik* is awarded a gold medal at the Paris industrial exhibition, while the plant in Züllchow is enlarged.[43] Bleibtreu heads the company until April 1856, when he goes back to Bonn and founds a second cement factory, the *Bonner Portland Cement Fabrik*, which starts producing in 1858.

In the wake of Bleibtreu's success, in 1855, the Stettin entrepreneur Johannes Heinrich Quistorp (1822–1899) buys a lime kiln in the village of Lebbin, on the Wollin Island, and transforms it into a Portland cement factory, for which he first exploits local raw materials and later stones taken from Rügen.[44] Existing documents provide evidence that, between 1856 and 1858, Quistorp sends samples of the cement he produce to several master builders based in the towns of Swinemünde, Naugard, Stettin, Magdeburg and Halle, and that, as of 1858, he has a dealer in Berlin.[45]

The acknowledgment of the monolithic behaviour of concrete and the further spread of concrete foundations

The acknowledgement of the critical role that concrete plays as an appropriate material for building foundations, both underwater and in wet soil, further develops as of the beginning of the 1850s as did, simultaneously, the awareness that the construction of monolithic building elements is possible by using concrete, rather than ordinary masonry bound with cement-based mortar. William Löbe (1805–1891), a writer specialized in agricultural topics, expresses such a conviction in 1850. "Concrete forms dense uniform masses," he writes, "which, in a short time, become as solid and resisting as medium-hard rocks, so that a layer of concrete can be considered as a single piece of

[41] Bleibtreu, *Der erste Zementbrand*...cit., p. 213.
[42] See Becker, *Praktische Anleitung*...cit., p. 3.
[43] See Julius Manger, *Der Stettiner Portland-Cement in Versuche und Erfahrungen dargestellt und beleuchtet*, Stettin, F. Heffenland, 1860, p. 3.
[44] It is the village Lubin Insel Wollin of today. About Quistorp, see Rudolf Vierhaus, ed., *Deutsche Biographische Enzyklopädie*, vol. VIII, München, K. G. Sauer, 2007, entry "Quistorp, Johannes (Heinrich)," p. 130; Anon., "Beiträge zur Genealogie und Geschichte der Familie Quistorp. Der Stettiner Zweig," 2008, in, www.quistorp.de, pp. 111–112.
[45] See Carl Görcke, model of commercial letter with attachments, April 1, 1858 (BLHA, A Rep. 010–01–01, n° 321, foll. 1,2 r/v).
[46] "Die vorzüglichste Eigenschaft des Gußmauerwerks besteht darin, dichte und gleichförmige Massen zu bilden, welche in kurzer Zeit die Festigkeit und die Widerstandsfähigkeit von Steinen mittler Härte annehmen, so daß eine Schicht Gußmauerwerk wie ein Stein aus einem einzigen Stück angesehen werden kann. Dieser Eigenschaft wegen eignet sich das Gußmauerwerk vorzugsweise zu Grundmauern, da kein anderes Mauerwerk größere Sicherheit gegen die ungleiche Zusammenpressung des Bodens darbietet; die Ursache davon ist, daß das

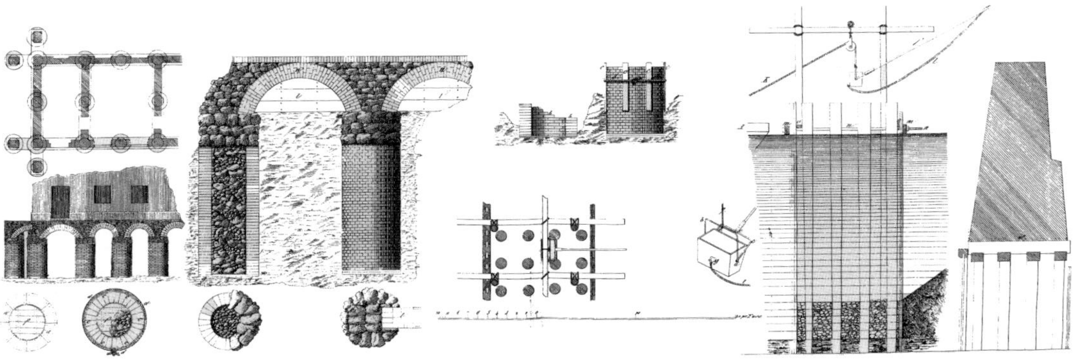

128. J. Ch. Wedeke, reproduction of Gilly's drawings of well foundations – on the left – and Winckelmann's drawings of the concrete foundation for the dock along the River Oder in Stettin – on the right (J. Ch. Wedeke, *Handbuch der bürgerliche Baukunst: Allgemein fasslich für Maurer- und Zimmermeister und die es werden wollen*, Quedlinburg, Leipzig, Gottfried Basse, vol. II, 1849).

stone. This is the excellent property of concrete that makes it appropriate to build foundations, since no other kind of masonry is more reliable against uneven pressure on the soil. This happens exactly because concrete lies homogeneously on large soil surfaces, as a uniform and stiff mass, and cannot settle partially, as it happens with ordinary stone masonry."[46] The master builders Wedeke and Romberg express the same conviction in 1851, in the first volume of the *Handbuch der Landbaukunst*, using almost the same words that Löbe used one year before.[47]

How to prepare hydraulic mortar and concrete, and how to build concrete foundations are moreover described in a number of construction manuals published in the early 1850s. Wedeke describes how to build foundations made of sunken wells filled with concrete and foundations made of concrete poured underwater in enclosure of sheet piles, straight on the soil or on timber piles, in the second volume of the *Handbuch der bürgerlichen Baukunst*, mentioning two German hydraulic works from the 1840s as pertinent and exemplary references: the lock at the Bughof in Bamberg and the dock along the River Oder in Stettin.[48] How to build foundations for bridge piers by pouring concrete in enclosures of sheet piles is also explained in the huge manual about the art of building bridges that the Bremen building superintendent Heinrich Müller (1819–1890) publishes in four volumes between 1850 and 1854.[49] Max Becker (1817–1884), who is the responsible master builder for hydraulic engineering in the Grand Duchy of Baden and professor in the School of Engineering in Karlsruhe, deals with the way of building concrete foundations for bridge piers and for locks in the *Baukunde des Ingenieurs*,

Gußmauerwerk, welches eine gleichförmige starre Masse bildet, und mit einer breiten Fläche auf dem Boden aufliegt, sich nicht theilweise setzen kann, wie dies bei den Steinen des gewöhnlichen Mauerwerks möglich ist" (Willliam Löbe, *Encyclopädie der gesammten Landwirthschaft, der Staats-, Haus- und Forthwirthschaft und der in die Landwirthschaft einschlagenden technischen Gewerbe und Hülfswissenschaften*, vol. I, Leipzig, Otto Wigand, 1850, pp. 204–205).

[47] See Wedeke, Romberg, *Handbuch der Landbaukunst und der landwirthschaftlichen Gewerbe*...cit., vol. I, col. 330.

[48] See Johann Christian Wedeke, *Handbuch der bürgerlichen Baukunst*...cit., vol. II, 1849, pp. 60–62.

[49] See Heinrich Müller, *Die Brückenbaukunde in ihrem ganzen Umfange. Ein Handbuch für Ingenieure und Baumeister*, Leipzig, Romberg, vol. I–II, 1850, vol. III, 1852, vol. IV, 1853, esp. vol. I, pp. 33–43 about mortar and concrete, pp. 84–87 about the mortar tests published by the master builder Zimmermann in 1829, pp. 225–238 about concrete poured underwater in enclosure of sheet piles, and vol. III, pp. 66–84 about foundations of bridge piers.

129-130. Ch. A. E. Pötzsch, warehouse, Leipzig, 1850. Plan; cross section (Anon., *Die neuen Lagerhäuser in Leipzig*, in, "Zeitschrift für praktischen Baukunst," XIII, 1853).

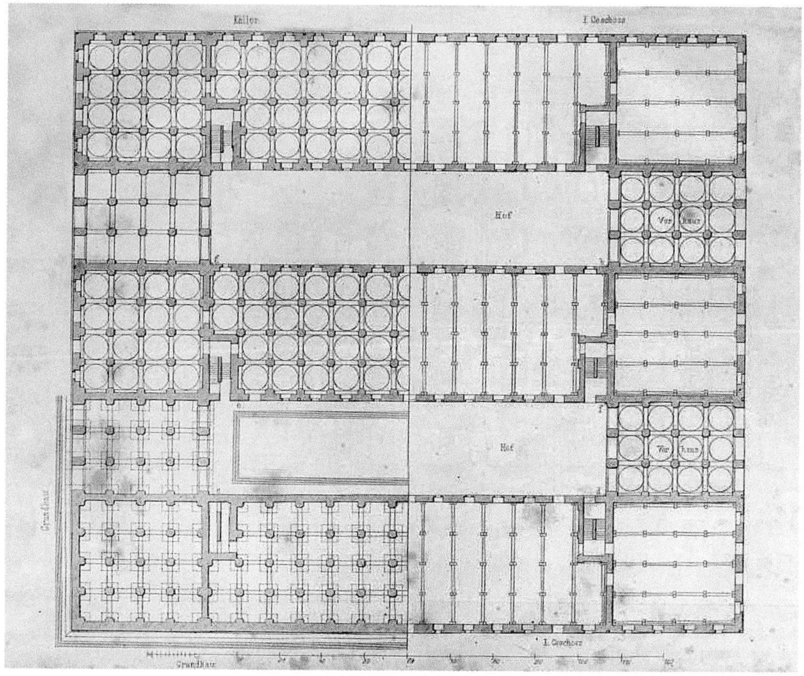

while further indications about concrete as an appropriate material to build foundations underwater, as well as in wet and unsteady soils, are also provided in the building lexicon by the master builder Leo Bergmann.[50]

Beyond such literary sources, the adoption of concrete foundations in building practice gradually increases as of the early 1850s, driven by the development of the railways, which gives rise to the need for different buildings and constructions. The previously described concrete foundations of the bridges in Dirschau and Marienburg, which are completed in the early 1850s although they are conceived in the second half of the 1840s, as well as the foundations of the railway bridge near the *Wasserthorplatz* in Berlin and of the retaining wall along the River Pregel near the Königsberg railway station, serve as examples of this increasing practice, in addition to being pioneering cases of artificial and Portland cement use.

The now well-known technique of pouring and tamping concrete in simple enclosures of sheet piles remains predominant for a little over a decade. A pertinent example is the foundation of the warehouses that the architect Christian August Eduard Pötzsch (1803–1889) conceives as an ensemble of different contiguous buildings to construct in different phases on a large site situated in Leipzig, close to the railway stations of the lines to Magdeburg and to Dresden.[51] The construction begins in 1850, and it is the presence of

[50] Max Becker, *Der Brückenbau in seinem ganzen Umfange, und mit besonderer Rücksicht auf die neusten Constructionen*, Stuttgart, Carl Mäcken, 1854, esp. p. 87 about the use of layers of hydraulic mortar and concrete as a waterproof filling over vaults, pp. 97, 100 about the concrete foundations and the use of concrete on the deck of the *Nydeckbrücke* in Bern, p. 131 about the concrete foundations of the *Pont du Carroussel* in Paris, p. 221 about the concrete foundations of the piers of the suspension bridge in Budapest; Id., *Der Wasserbau in seinem ganzen Umfange*, Stuttgart, Carl Mäcken, 1856, esp. p. 15 about sunken wells filled with concrete, p. 136 about the concrete foundation of the connection lock between the River Elz and the Elz-Dreisam Canal, p. 191 again about concrete foundations for locks, p. 267 about the waterproofing of canal bridges by means of hydraulic mortar or concrete, p. 280 about

ground water close to the surface (between 7 ½ and 11 feet underground), due the proximity of the River Parthe, that leads to the adoption of a concrete foundation. The concrete is composed of slaked lime, trass from Brohl, sand, bricks and stone rubble. It is poured and tamped in five layers all over the building site, reaching a thickness up to 6 feet, while the pit is kept dry by means of pumps. The basement, which is built straight on the concrete foundation, consists of brick masonry walls and vaults that are bound with Roman cement-based mortar, while trass-based mortar is used to build a 2 ½ feet high plinth of unrefined stones supporting the outer walls.[52]

More demanding concrete foundations are constructed so as to build, between 1854 and 1855, two towers housing mechanical elevators for freight wagons in Ruhrort and Homberg near Duisburg, on the two opposite sides of the Rhine. The elevators are intended to move the wagons from the ordinary level of the embankments, onto an appropriate steamboat that crosses the river. They are to replace a previous and inefficient system of ramps that was put into service in 1852, whereas the construction of a bridge is not taken into account for military reasons. With British models as a reference, the master builder Emil Hermann Hartwich (1802–1879) proposes the construction of the elevator towers system to the Cöln-Mindener and to the Ruhrort-Crefeld-Kreis Gladbacher railway companies.[53] The foundations must be particularly

concrete foundations of river embankments and locks at the mouths into the sea. See also Leo Bergman, *Baulexicon oder Realencyclopedie des gesammten Bauwesens*, vol. I, Leipzig, Ernst Schäfer, 1855, pp. 224–225.

[51] See Anon., *Die neuen Lagerhäuser in Leipzig*, in, "Zeitschrift für praktischen Baukunst," XIII, 1853, coll. 97–100.

[52] *Ibid.*

[53] The hydraulic mechanisms powered by a steam engine are delivered by British engineer and industrialist William George Armstrong, who is also the owner of a patent for such elevator system. About the conception and building of the elevator towers in Ruhrort and Homberg, see Theodor Weishaupt, *Die Homberg-Ruhrorter Rheintraject-Anstalt*, in, "Zeitschrift für Bauwesen," VII, 1857, coll. 347–390.

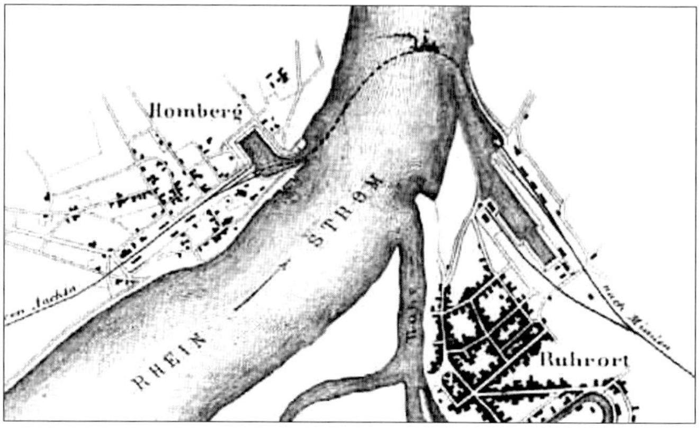

131-139. E. H. Hartwich, elevator towers for freight wagons, Duisburg, 1854–1856. Plans of the site; plan of the concrete foundation; longitudinal section of the concrete foundation; front elevation of the tower from the earth; front elevation of the tower from the river; longitudinal section of the tower; view; side elevation of the tower; cross section of the tower (Th. Weishaupt, *Die Homberg-Ruhrorter Rheintraject-Anstalt*, in, “Zeitschrift für Bauwesen,” VII, 1857).

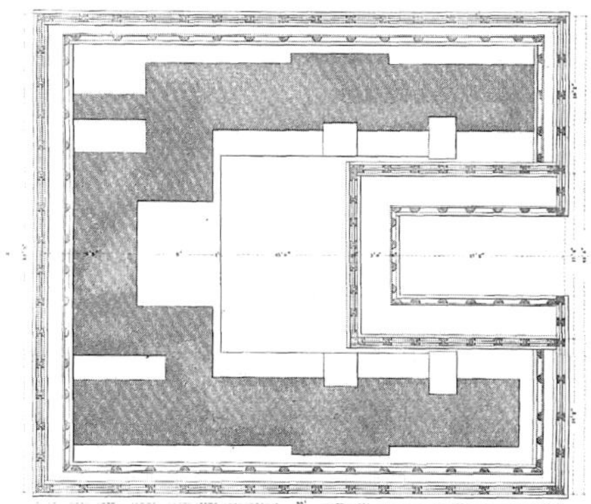

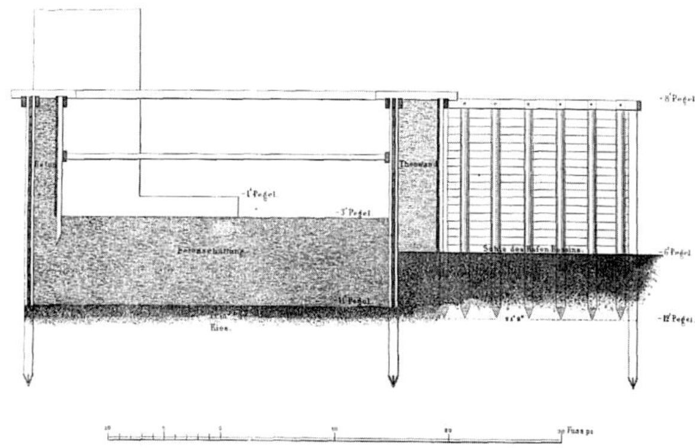

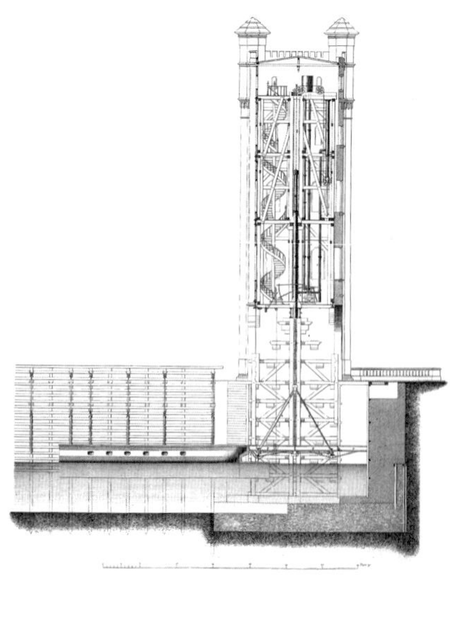

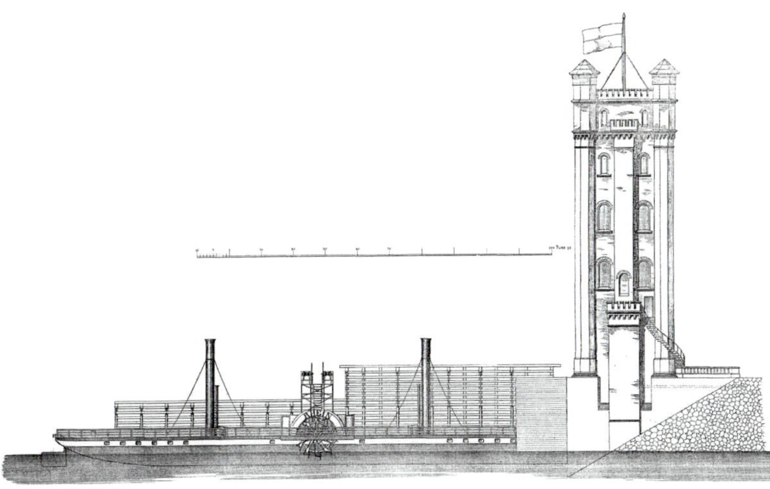

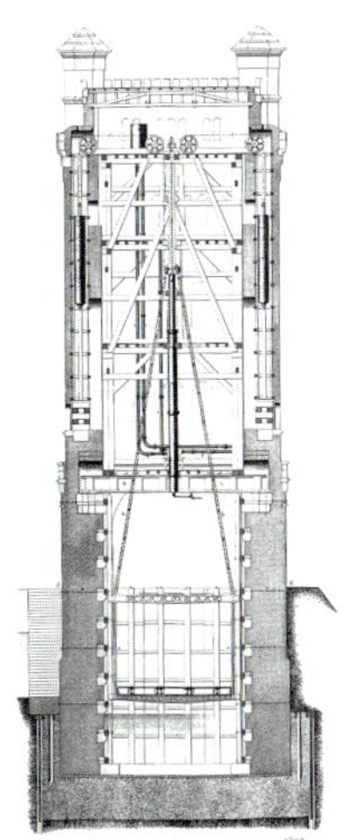

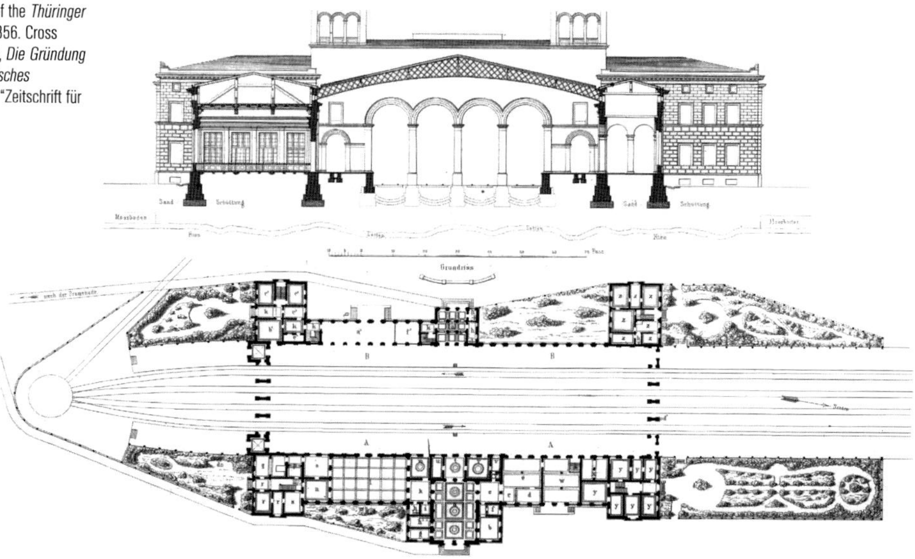

140. Witzeck, drawings of the *Thüringer Bahnhof*, Leipzig, 1855–1856. Cross section and plan (Witzeck, *Die Gründung der Gebäude des Thüringisches Bahnhofes bei Leipzig*, in, "Zeitschrift für Bauwesen," X, 1860).

solid, considering that they need to support both the weight of the towers, which are about 100 feet high, and the stress produced by the machineries. Each foundation consists of an 8 feet high concrete base surrounded by three concrete retaining walls on the landside, whereas the waterside remains open to allow the connection between the steamboat and the elevator platform, after having been delimited by means of a provisional cofferdam during the construction. Concrete is composed of hydraulic lime from the nearby town of Enningerloh, trass from Brohl and sandstone rubble. The use of trass is considered unavoidable after a preliminary hardening test of hydraulic lime-based mortar fails. The masonry of the towers partially consists of ashlars and partially of bricks; the outer joints are made of Portland cement-based mortar, whereas the inner ones consist of trass-based mortar.

The old way of pouring and tamping concrete in superimposed layers of rubble and mortar is applied in 1855, in Leipzig, to construct the foundations of the buildings that compose the *Thüringer Bahnhof*, namely a locomotive shed, a gasworks, a shed for coal, a building for workers and the passenger hall, which stand on extremely marshy soil.[54] As the construction of each building begins, the marshy soil is removed, over a depth of about 4 or 5 feet, and replaced with rammed sand, while different pumps keep the pits dry. Despite being a better kind of bearing soil, such a basement of rammed sand is not yet solid and compact enough to support the footings made of cut stones that have been planned as the foundations of the above-ground walls. Therefore, about one foot-thick bases of concrete made of layers of rubble and mortar are poured and tamped onto the sandy soil, so as to better distribute the weight of the masonry footings over the sand, and prevent stones from detaching from each other. For the central role that the rammed sand plays, this kind of foundation is called *Sandfundirung*.

[54] See Witzeck, *Die Gründung der Gebäude des Thüringisches Bahnhofes bei Leipzig*, in, "Zeitschrift für Bauwesen," X, 1860, coll. 213–222. The replacement of marshy soil with sand was tried in Berlin to build the foundation of the *Hamburger Bahnhof*; however, the attempt failed and the foundation was built with sunken wells and timber framework. See, Friedrich Hoffmann, *Der Bahnhof der Berlin-Hamburger Eisenbahn in Berlin*, in, "Zeitschrift für Bauwesen," VI, 1856, (coll. 487–496), coll. 494–496.

141-145. F. W. Wallbaum, J. H. Strack, Chatedral Bridge over the River Rhine, Cologne, 1855–1859.

141-144. Plan of the site; sections of the abutments and of a pier (Lohse, *Die Rhein-Brücke bei Cöln*, in, "Zeitschrift für Bauwesen," XIII, 1863).

145. The bridge under construction, photograph, 1857 approx. (personal collection).

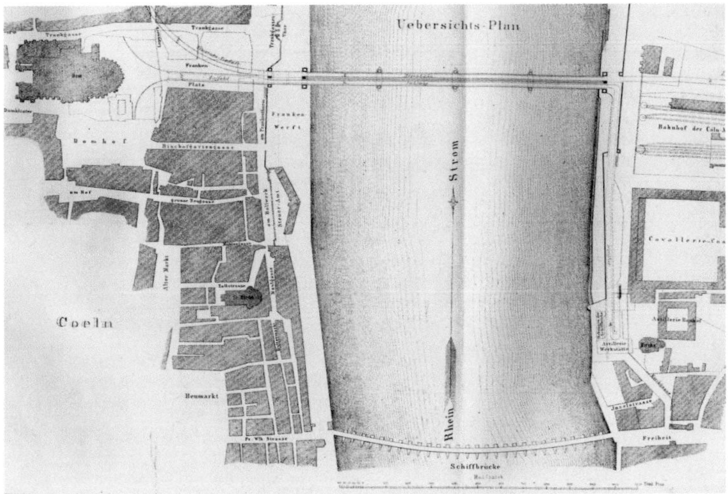

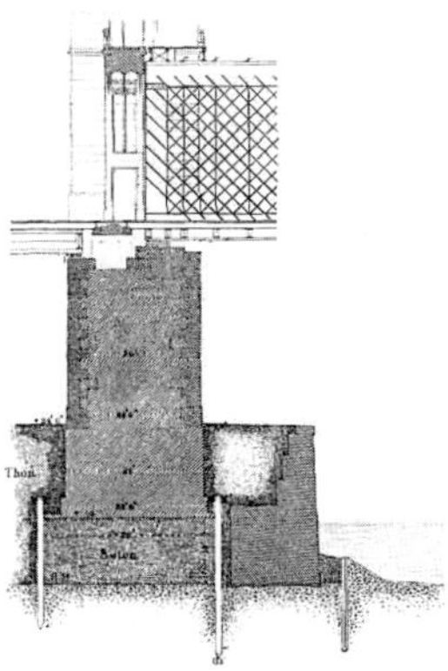

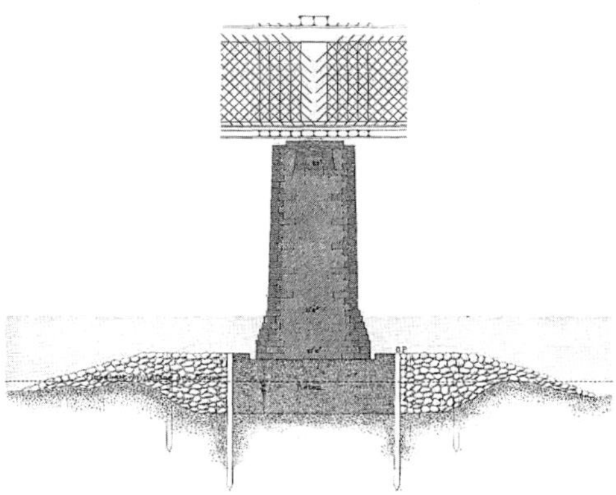

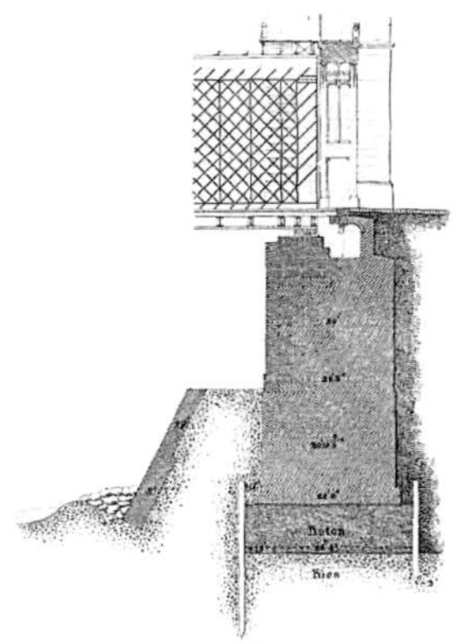

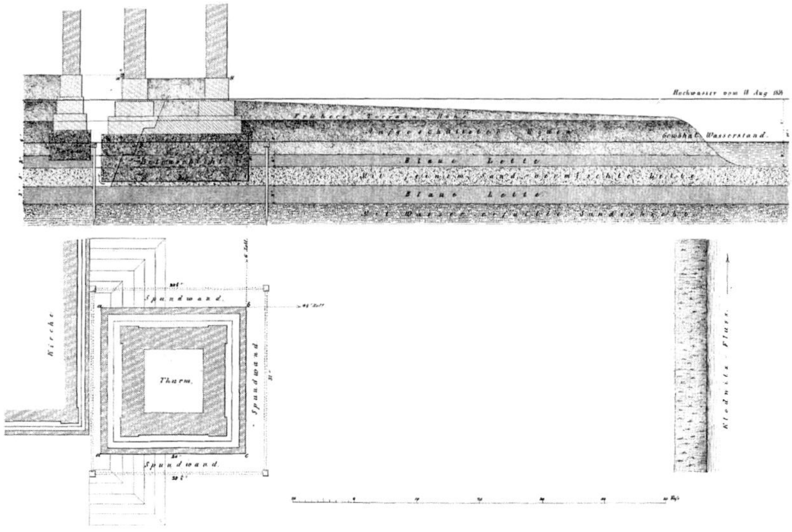

146. F. A. Stüler, Church of Saint Barbara, Gleiwitz, Silesia, 1855–1859. Cross section and plan of the concrete foundation of the tower (Assmann, *Verfahren bei Geraderichtung des Thurmes an der evangelischen Kirche in Gleiwitz*, in, "Zeitschrift für Bauwesen," XIII, 1863).

Concrete, instead, plays the main role in the foundations that support the piers and the abutments of the bridge over the Rhine in Cologne, known as *Dombrücke* (Cathedral Bridge), which is built between 1855 and 1859.[55] The foundations of the piers are built directly on a kind of gravely soil that proves to be solid enough to render timber piles unnecessary. The concrete is made of trass-based mortar mixed with pieces of basalt-lava stones and the melted remainders from brick production.[56] It is mixed in a concrete-mill powered by a steam engine and is then transported on the river by means of an appropriately-sized steamboat, to finally be poured underwater, inside enclosures of sheet piles, by means of boxes.[57] The masonry of the piers consists of basalt-lava stone faced walls with brick masonry filling. Pure trass-based mortar is used to bind the masonry underwater, whereas for the rest of it, some sand is added to the mortar.[58] The foundation of the left abutment partially consists of a 10 feet high concrete base standing on gravel, and partially of a 14 feet high masonry footing made of bricks bound in pure trass-based mortar. A similar kind of construction is adopted for the foundation of the right abutment, the concrete base being 7 ½ feet high, and the masonry footing 26 ¾ feet high.[59]

[55] About the conception and the building of the *Dombrücke* in Cologne, see Hermann Lohse, *Die Rheinbrücke bei Cöln*, in, "Zeitschrift für Bauwesen," VII, 1857, coll. 307–314; Id., *Die Rhein-Brücke bei Cöln*, in, "Zeitschrift für Bauwesen," XIII, 1863, (coll. 175–196, 335–370), coll. 178–192, 189–196. The hydraulic master builder Friedrich Wilhelm Wallbaum drew the initial plans, which are approved in 1854; work on the foundations of the piers start in the spring of 1855 under the supervision of the master builder Hermann Wilhelm Lohse (1815–1893), who simultaneously modifies the plans, according to new requests from the railway company, until 1856; the architect Johann Heinrich Strack (1805–1880) develops the architectural plans, and the bridge is completed in 1859.

[56] *Ibid.*, col. 190.

[57] Lohse, *Die Rheinbrücke bei Cöln*, cit., 1857, col. 312.

[58] Lohse, *Die Rheinbrücke bei Cöln*, cit., 1863, coll. 189–191.

[59] *Ibid.*, coll. 191–194.

[60] The bridges near Küstrin are the *Warthe-Brücke* (bridge over the River Warthe), the *Fluth-Brücke der Warthe* (bridge over the tidal land next to the River Warthe), the *Festungsgraben-Bridge* (the bridge over the moat of the Küstrin fortress), the *Oder-Brücke* (the bridge over the River Oder), and the *Vorfluth-Brücke der Oder* (bridge over the tidal land next to the River Oder). About these bridges, see Stein, *Die Bauanlagen der Kreuz-Cüstrin-Frankfurter Eisenbahn*, in, "Zeitschrift für Bauwesen," VIII, 1858, coll. 459–486. About the bridge in Alstaden, see *Mittheilungen aus Vereinen. Architekten-Verein zu Berlin. Versammlung am 31. Mai 1862*, in, "Zeitschrift für Bauwesen," XII, 1862, col. 563. For exhaustive information about the construction of this bridge, taking into account the bridge deck, see also Baeusch, *Der eiserne Ueberbau der Eisenbahnbrücke über die Ruhr bei Alstaden auf der Strecke Mülheim - Duisburg*, in, "Zeitschrift für Bauwesen," XXIII, 1873, coll. 341–344. About the bridges along the railway from Wittenberg to Halle, see *Protokoll des Vereins für Eisenbahnkunde zu Berlin*, October 9, 1860, in, "Zeitschrift für Bauwesen," XI, 1861, coll. 334–348.

147-148. G. Th. Schirrmacher, H. v. der Hude, *Kunsthalle*, Hamburg, 1864–1869. Cross section; perspective view (H. von der Hude, *Die Kunsthalle in Hamburg*, in, "Zeitschrift für Bauwesen," XVIII, 1868).

In addition to the previously described major cases, there are other noteworthy examples of railway bridges built on concrete foundations in the 1850s and 1860s, like the five ones near Küstrin, along the railway line connecting Küstrin to Frankfurt (Oder), whose construction dates back to 1856–1857; the bridge over the Ruhr in Alstaden, which is built in 1862 along the railway line connecting Mühlheim and Duisburg; and the bridge over the River Elbe and the adjacent tidal land, which is built between 1857 and 1859 near the town of Wittenberg, south-east of Berlin, along the railway line connecting this same town to Halle.[60] The concrete used in Wittenberg is produced with sand, stone rubble and Portland cement. It is mixed in drums powered by a steam engine and is transported first in wagons on rails and then on barges, close to the foundation pits, where it is poured down by means of boxes. For the construction of the bridge in Wittenberg, concrete is also used to build 3 feet thick cofferdams enclosing the foundation pits of the piers standing in the riverbed. Similar 5 feet thick concrete cofferdams are also adopted to build, between 1856 and 1857, the piers of the bridge over the Mosel near Konz, along the Saarbrücken-Trier-Luxemburger railway, whereas the founda-

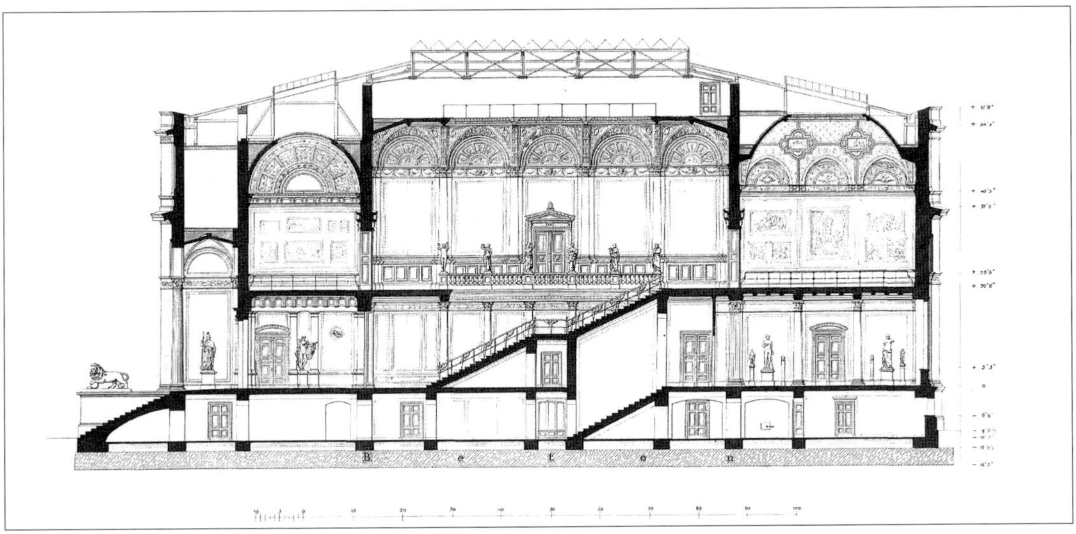

tions themselves are made, in this case, of unrefined sandstones bound with pure trass-based mortar.[61] Ordinary cofferdams filled with rammed earth are instead built to surround the pits for the concrete foundations of the piers supporting the railway bridge over the River Neckar near Jartfeld, which is built between 1867 and 1869, using Portland cement-based concrete.[62]

It is probably in light of such examples that some other kinds of buildings are built on concrete foundations. This is the case of the Church of Saint Barbara in Gleiwitz in Silesia, which is built between 1855 and 1859 according to the plans conceived by Stüler.[63] Again because of the presence of groundwater close to the surface, due to the proximity of the River Klodnitz, the walls of the church are built on about 4 ½ feet high concrete footings, and the tower on a roughly 7 feet high concrete base.[64] An unexpected and irregular subsidence of the soil produces a worrisome incline of the tower when it is still under construction and has reached the height of 60 feet. Despite the accident, the concrete foundation does not suffer any damage, the tower is charged with a counterweight of 5000 *Centner*, and it is then finally straightened up between 1860 and 1861, by digging some earth and stabilizing the soil with sheet piles.[65]

The Hamburg *Kunsthalle*, which is built between 1864 and 1869 according to the plans by the architects Georg Theodor Schirrmacher (1833–1864) and Hermann von der Hude (1830–1908), also stands on a roughly 3 ½ feet high concrete foundation, which stretches all over the building site. According to a report by Von der Hude, the foundation soil proves to be almost solid but a certain irregularity in the composition of it requires a concrete foundation.[66] The concrete is produced manually by mixing slaked lime with Portland cement, sand and crushed bricks; the building of the entire foundation takes about eight weeks.

Lean conglomerate as a German kind of concrete for above-ground constructions

Even once concrete is definitively recognized as an appropriate material for foundations, there is no agreement about the suitability of it for building above ground, despite the several examples of houses entirely constructed from poured concrete in France and in England. The architect Gustav Adolph Breymann (1807–1859), a professor of architecture at the *Polytechnische Schule* in Stuttgart, distrusts the solidity of *Gußmauern* and *Gußgewölbe*, that is, poured walls, which he depicts as walls that need to be moulded rather than built, and poured vaults.[67] He asserts the better suitability of traditional masonry and advises against the construction of walls made of cut stone-faces and concrete filling, since he believes that "the solidity of a wall chiefly

[61] Anon., *Die Bauanlagen der Saarbrücken-Trier-Luxemburger Eisenbahn*, in, "Zeitschrift für Bauwesen," XIII, 1863, (coll. 43–58), coll. 49–50.

[62] See Max Becker, *Handbuch der Ingenieur-Wissenschaft*, vol. V, *Ausgeführte Constructionen des Ingenieurs*, n° 6, *Die Kabelschifffahrt und die Eisenbahnbrücke über den Neckar bei Jartfeld*, Stuttgart, Carl Mäcken, 1870, pp. 36–38.

[63] Gleiwitz is now the Polish town of Gliwice. About the Church of Saint Barbara, see Anon., *Eine frühere Gleiwitzer St. Barbara. Nach 400 Jahren durch den Neubau der evangelischen Kirche ersetzt*, in, *Oberschlesien in Bilder*, 1934, n. p.; Jolanta Rusinowska-Trojca, *Städtebau und Wohnarchitektur des 19. Jahrhunderts in Gleiwitz (Gliwice)*, doctoral dissertation, Rheinische Friedrich-Wilhelms-Universität, Bonn, 2005, p. 55.

[64] About the concrete foundation of the tower, see Assmann, *Verfahren bei Geraderichtung des Thurmes an der evangelischen Kirche in Gleiwitz*, in, "Zeitschrift für Bauwesen," XIII, 1863, coll. 67–70.

[65] 250 tonnes (1*Centner* = 100 *Pfund* = 50 kg).

[66] See Hermann von der Hude, *Die Kunsthalle in Hamburg*, in, "Zeitschrift für Bauwesen," XVIII, 1868, coll. 3–8.

[67] See Gustav Adolph Breymann, *Allgemeine Bau-Constructions-Lehre, mit besonderer Beziehung auf Hochbauwesen*, Stuttgart, Hoffmann, 1849, pp. 25–26, 82-83.

[68] "Die Stärke einer Mauer nicht nur von ihrer Dicke, sondern besonderrs auch von einem guten Verbande abhängig ist" (*ibid*. p. 26).

[69] See Löbe, *Encyclopädie der gesammten Landwirthschaft...* cit., p. 205.

depends on the quality of the masonry bound, not only on the width of the wall itself."[68] In accordance with this belief, he also suggests limiting the construction of freestanding concrete walls to fences, which do not need to bear any load other than their own, and warns about a supposedly lesser reliability of concrete vaults in comparison to masonry ones, although he seems to be aware of the solidity of the ancient vaults that Romans used to build with pozzolana-based concrete.

Unlike Breymann, Löbe makes a brief mention of the suitability of concrete for building above ground, using Lebrun as a main reference.[69] However, he gives more emphasis to the use of lean conglomerate rather than to concrete, and stresses that the two materials can be applied using the same technique, namely by pouring and tamping the conglomerate masses into formworks. In this way, Löbe hints at a certain closeness between lean conglomerate and concrete.[70] Such closeness becomes acknowledged equivalence in Wedeke's description of basements made of hydraulic lean conglomerate, which he calls "rammed mortar". "It is recommended," he writes in 1850, "not to build the walls of basements in rammed mortar, because it does not harden in places where the air is wet, nor does it become solid enough to support the weight of the standing walls above. If someone wants to build basements in rammed mortar, he has to use fast hardening hydraulic lime mixed with much stone rubble."[71] He then explains that such a mortar is called concrete, and invites readers to refer to the *Bürglerliche Baukunst*, which he has written together with Romberg, to find out how to produce and apply concrete. In a similar way to Wedeke, [Friedrich Conrad Theodor] Krause lists concrete among the materials that can be used to build the foundations of walls in lean conglomerate.[72]

The recognition of the closeness between concrete and lean conglomerate develops further during the first half of the 1850s, and it is even a widespread belief that lean conglomerate has prevented the spread of concrete use for above-ground construction. In this regard, Wedeke asserts that "rumours coming from France seven or eight years ago, about the replacement of bricks and stones with concrete to build walls, seem to have passed unheeded," whereas "the attempts at building freestanding walls, and even entire buildings made of rammed mortar, are giving satisfactory results."[73] In 1851, Engel agrees with Wedeke, asserting: "The technique of building with rammed lime and sand [...] has almost suppressed the use of concrete."[74]

In the wake of such opinions, lean conglomerate comes to be considered a sort of more affordable German version of concrete for building above ground. Leuchs gives evidence of this approach, reacting to early news about

[70] *Ibid.*, pp. 202–204.

[71] "Kellerwände aber mit diesem Mörtel aufzustampfen, kann nicht angerathen werden, weil der Mörtel in der feuchten Luft nie erhärten, also auch niemals die genügende Festigkeit erlangen würde, die Wände der Etagen mit vollkommener Sicherheit zu tragen. Will man aber auch das Souterrain in dieser Art aufführen, so muß hierzu schnellbindender hydraulischer Kalk verwendet werden" (Wedeke, *Der Bau mit gestampftem Mörtel*...cit., p. 11). It is, however, almost contradictory on the part of Wedeke to advice against the use of formworks considered a waste of wood; see Johann Christian Wedeke, Johann Andreas Romberg, *Handbuch der Landbaukunst und der Landwirthschaftlichen Gewebe für Baumeister, Landwirte und Cameralisten*, vol. III, Leipzig, Romberg, 1858, col. 148.

[72] Krause, *Anleitung zur Kalk-Sand-Bau-Kunst*...cit., pp. 64–65.

[73] "Es sind etwa sieben oder acht Jahre her, seit aus dem südlichen Frankreich sich eine Stimme vernehmen ließ, die das dahin übliche Material zur Aufführung von Wänden, nähmlich künstlich und natürliche Steine, durch Beton ersetzt wissen wollte; ihr Ruf scheint in Deutschland verhallt zu sein [...]. Versuche, um freistehende Mauern und ganze Gebäude von reinem Mörtel zu erbauen, die nach mehrjährigen Bemühungen her genügende Erfolge lieferten" (Wedeke, *Der Bau mit gestampftem Mörtel*...cit., p. 3).

[74] "Die Methode, Gebäude aus Kalk und Sand auszustampfen [...] hat die Verwendung des Betons fast ganz verdrängt" (Johann Daniel Friedrich Engel, *Der Kalk-Sand-Pisébau. Nachtrag*, Charlottenburg, Huber, 1856, p. 2).

149-150. J. D. F. Engel, barn with walls in lean conglomerate and interior timber framing structure, Stechow, Brandenburg, 1853–1854. Longitudinal section and plan; side elevation, cross sections and detail of the truss construction (J. D. F. Engel, *Sammlung landwirthschaftlicher und ländlicher Bauausführungen*, Berlin, Ernst & Korn, 1864).

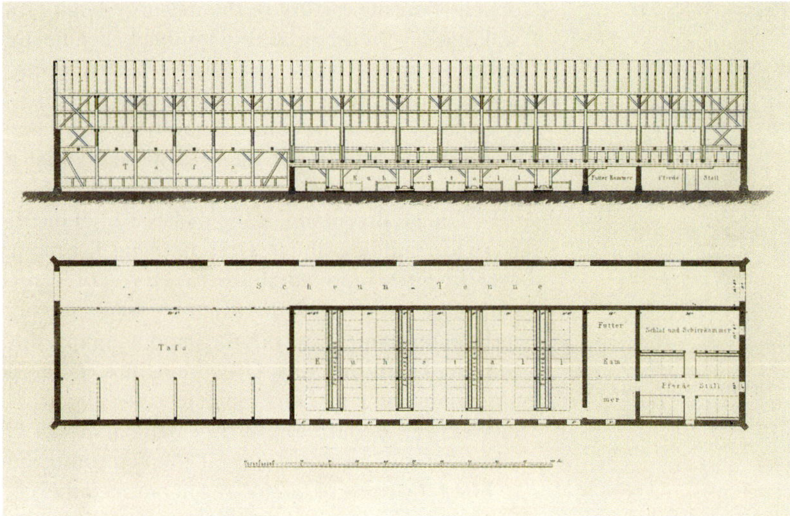

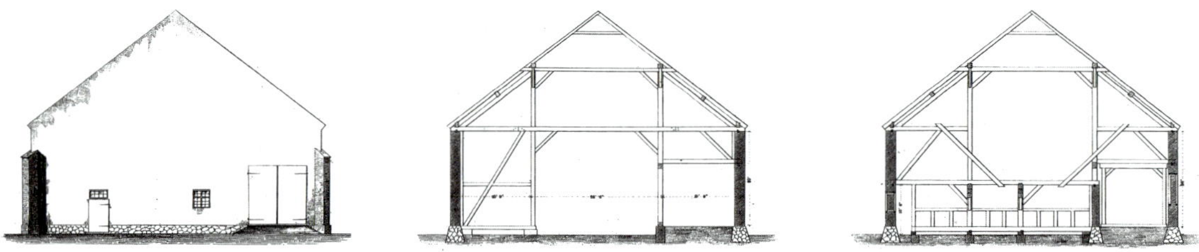

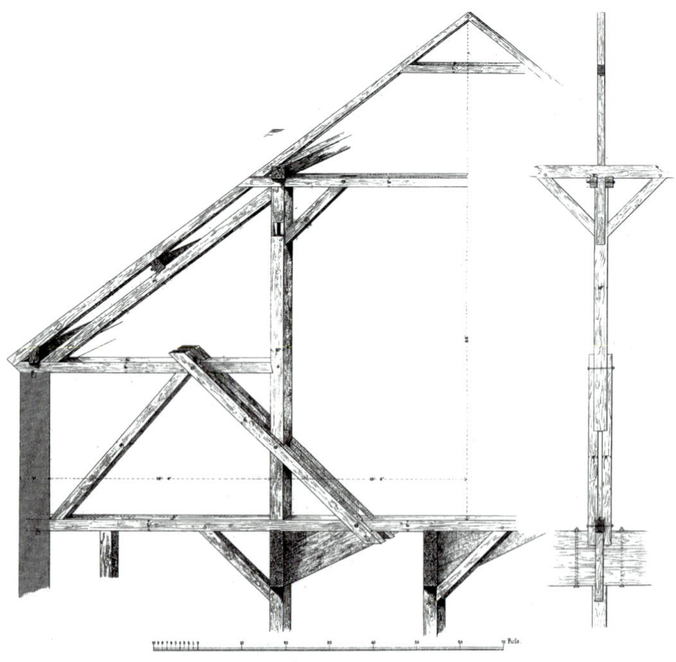

François Coignet's buildings made of "mortar pressed into moulds," which appear in German periodicals around about 1855.[75] Leuchs asserts the German primacy over the building technique that Coignet promotes. "For a long time," he writes, "we have been constructing buildings of artificial stone in Germany using a more simple mixture, which consists of twelve parts of sand and one part of lime."[76] Engel also maintains a sort of actual analogy between Coignet's concrete and the German lean conglomerate. Dealing with the possibility of adding ash to the composition of lean conglomerate, in a short report that he writes in 1856 as a supplement to the book about *Kalksand* from 1851, Engel asserts that "the company Coignet Père & Fils & Co. in Paris produces a kind of compressed lean conglomerate (compressed concrete) made of one measure of lime, one measure of coal ash and between six and eight measures of coarse sand, which provides very good results."[77] Engel even suggests adding some Portland cement into lean conglomerate mixtures – to build foundations or just to improve the quality of the conglomerate.[78]

In such a scenario, further buildings are constructed with walls in lean conglomerate. Engel considers the one built in 1853 in the Brandenburg village of Stechow to house a barn and a stall particularly worth mentioning, because of its impressive dimensions. It measures 250 feet by 60 feet, and the walls are 20 feet high. All outer ones, and even a few inside, are made of lean conglomerate. A timber framework structure, however, still plays a fundamental static role, as does the four brick masonry buttresses at the four corners of the building.[79] For an administration building of a quarry in Rüdersdorf, lean conglomerate is used to fill the fields of the exposed timber framing structure.[80] Further examples of lean conglomerate use exist near Hamburg, where, as of 1856, the engineer Scheer undertakes experiments to find an appropriate mixture for building the outer walls of the signalman's houses along the railway line between Altona and Kiel.[81] He finally develops a mixture consisting of one measure of slaked lime and eight measures of gravel containing walnut-sized stones. About ten years later, the walls of the signalman's houses along the Ostholsteinische railway are also built using a similar kind of lean conglomerate.[82]

During the second half of the 1850s, the technique of building in lean conglomerate goes through some further experiments concerning the construction of vaults. The master builder E. H. Hoffmann (1822–1896) experiments with the construction of sixteen small vaults in lean conglomerate, even

[75] See François Coignet, *Benutzung des geformten und zusammengepressten Mörtels*, in, "Polytechnisches Journal," XXXVII, CXL, pp. 101–104, and in, "Kunst- und Gewerbe-Blatt des polytechnischen Vereins für das Königreich Bayern," XLII, 1856, coll. 404–408 (Id., *Emploi des bétons moulés et comprimés*, in, "Le Génie industriel," vol. X, 1855, pp. 218–220).

[76] "In Deutschland schon längst Gebäude aus künstlichen Steinen von einer viel einfacheren Zusammensetzung, nämlich aus 12 Sand und 1 Kalk hergestellt werden" (Anon., *Ueber Häuserbau mit Mörtel*...cit.).

[77] "So fertigt gegenwärtig das Haus Coignet Père & Fils & Comp. in Paris einen Kalksandpisé (béton comprimé) aus 1 Theil Kalk, 1 Theil Steinkohlenasche und 6 bis 8 Theilen grobem Sand an, welcher sehr gute Resultate liefern soll" (Engel, *Der Kalk-Sand-Pisébau. Nachtrag*...cit., p. 6).

[78] *Ibid.*, pp. 5–6.

[79] Johann Daniel Friedrich Engel, *Sammlung landwirthschaftlicher und ländlicher Bauausführungen*, Berlin, Ernst & Korn, 1864, foll. XXII-XXIII.

[80] See Engel, *Der Kalk-Sand-Pisébau. Nachtrag*... cit., p. 15; Id., *Sammlung landwirthschaftlicher und ländlicher*...cit., foll. XXIV.

[81] H. Tellkampf, *Ueber den Bau von Wärterhäusern mit Umfassungsmauern von Kalkpisé*, in, "Zeitschrift für praktische Baukunst," XXVI, 1865, coll. 113–122.

[82] *Ibid.*

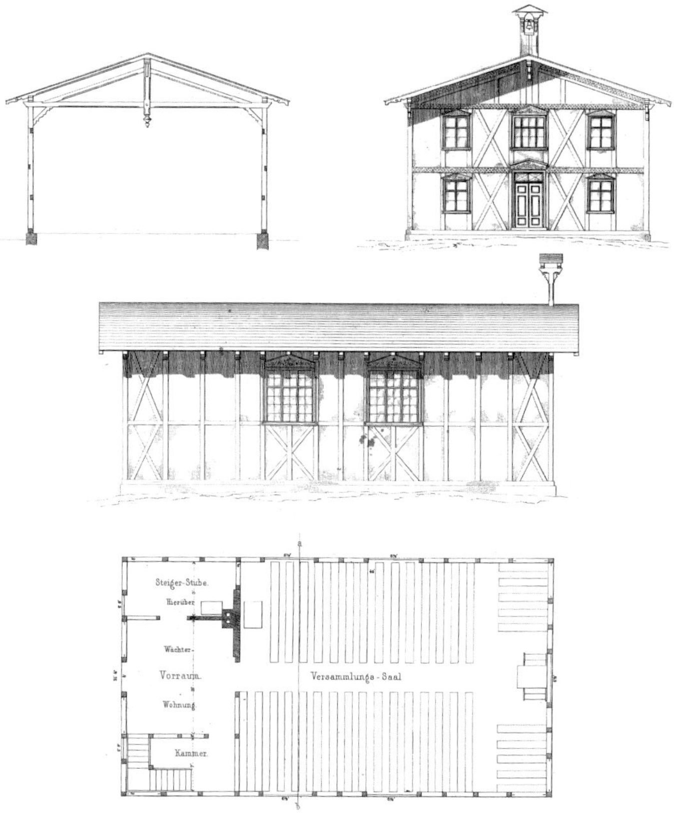

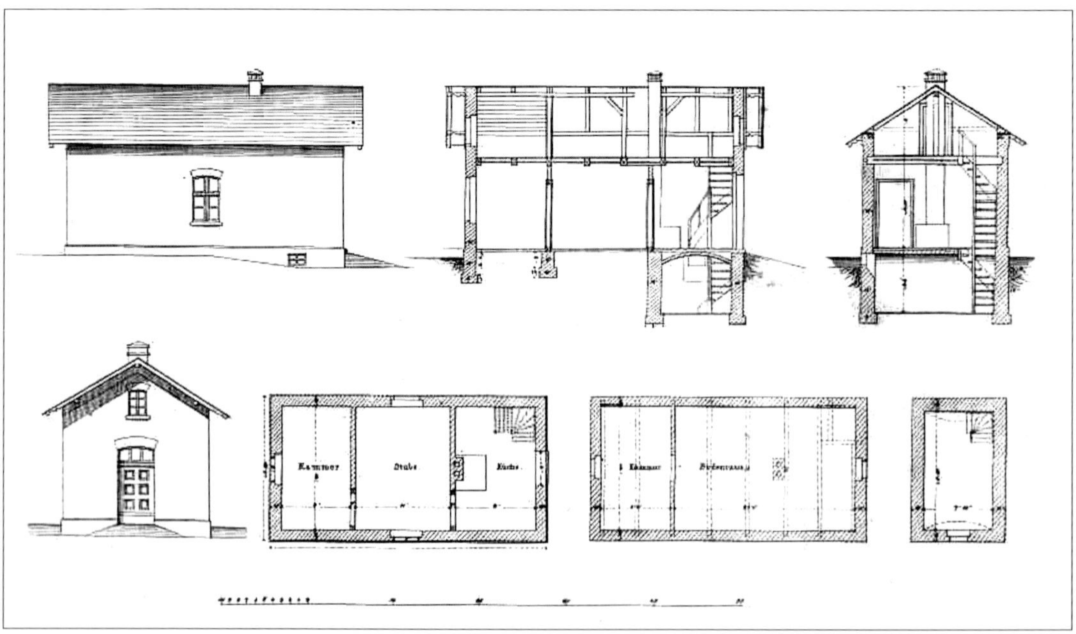

151. J. D. F. Engel, drawings of an administration building of a quarry, Rüdersdorf, n. d. Cross section, front and side elevations, plan (J. D. F. Engel, *Sammlung landwirthschaftlicher und ländlicher Bauausführungen*, Berlin, Ernst & Korn, 1864).

152. Scheer, signalman's house along the railway line Altona-Kiel, approx. 1856. Side elevation, longitudinal and cross sections, front elevation, plans (H. Tellkampf, *Ueber den Bau von Wärterhäusern mit Umfassungsmauern von Kalkpisé*, in, "Zeitschrift für praktische Baukunst," XXVI, 1865).

153. W. Salzenberg, draft of the groin vaults of the Baths of Diocletian (W. Salzenberg, *Einige Bemerkungen über die Gewölbeconstruction des antiken Rom*, in "Zeitschrift für Bauwesen," VII, 1857).

154. E. H. Hoffmann, barn with lean conglomerate walls and discharging arches, western Prussia, approx. 1855. Side and front views, longitudinal adn cross sections (E. H. Hoffmann, *Mittheilungen über Gewölbe aus Stampfmörtel*, in, "Zeitschrift für Bauwesen," VIII, 1858).

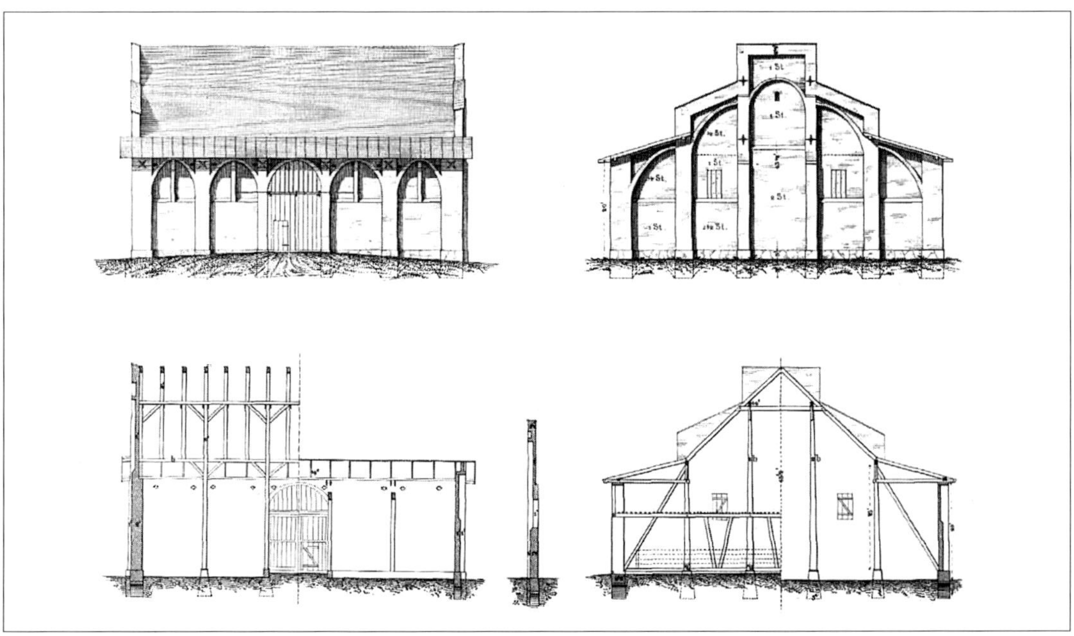

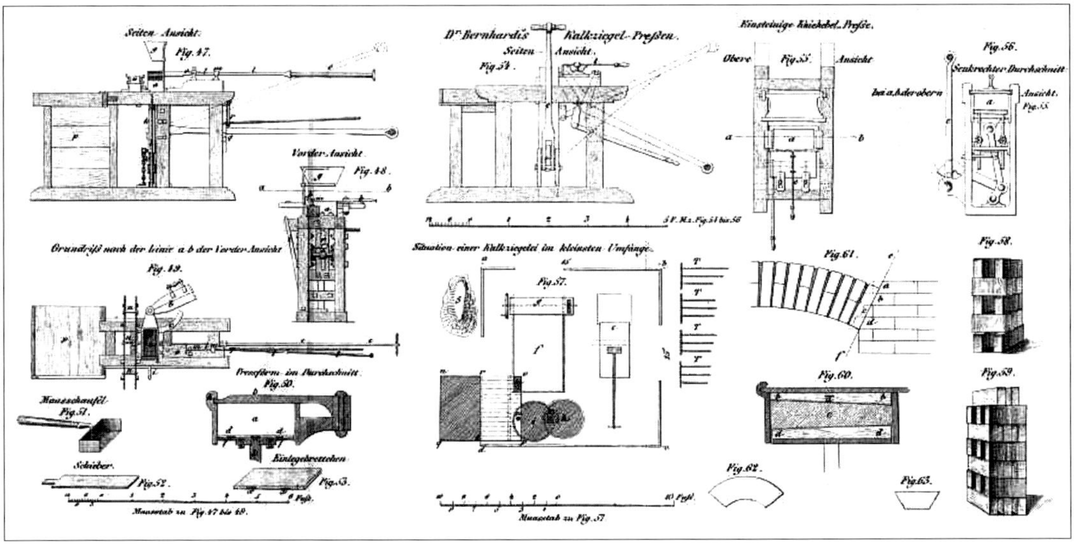

155. F. A. A Bernhardi, cast iron press for sand-lime bricks patented in 1856 (J. D. F. Engel, *Der Kalk-Sand-Pisébau und die Kalksand-Ziegelfabrikation*, Leipzig, C. H. Seemann, 1865).

using small amounts of Portland cement in four cases, as he builds his own house in 1856.[83] Most of these vaults are a failure and need to be demolished and rebuilt, but, according to Hoffmann, the reason for the failure is found in geometrical mistakes, rather than in the unsuitability of lean conglomerate. In the same year, the previously mentioned master builder Salzenberg writes a report about the ancient Roman art of building conglomerate vaults; he mentions the ones of the Baths of Diocletian and those of the Basilica of Constantine in Rome as eminent examples, and provides the description with drafts.[84] He ascribes the solidity of ancient Roman vaults to the use of pozzolana-based mortar, which he believes to harden well, rapidly and, above all, without shrinking. At the same time, he advises against the construction of new conglomerate vaults, since he believes that the kind of mortar that is usually employed at his time shrinks too much and causes the collapse of the vaults. "The ordinary mortar used in our brick masonry buildings has not these properties," Salzenberg writes referring to ancient Roman constructions, "it hardens slowly, shrinks and should therefore be used in as little amount as possible; while the use of conglomerate to build vaults should be absolutely ruled out."[85] In 1858, reporting about the lean conglomerate vaults he has built in his own house, Hoffmann replies to Salzenberg's observations about the problem of mortar shrinking and maintains that the small amount of lime that is used in lean conglomerate produces modest shrinking effects,

[83] See E. H. Hoffmann, *Mittheilungen über Gewölbe aus Stampfmörtel*, in, "Zeitschrift für Bauwesen," VIII, 1858, coll. 453–456. Hoffman published a book about vaults in 1853; see E. H. Hoffmann, *Über Form und Stärke gewölbter Bögen*, Berlin, Nauck, 1853.
[84] Wilhelm Salzenberg, *Einige Bemerkungen über die Gewölbeconstruction des antiken Rom*, in, "Zeitschrift für Bauwesen," VII, 1857, coll. 424–429.
[85] "Dem gewöhnlichen Mörtel unserer neueren Backsteinbauten mangeln diese Eigenschaften, er erhärtet langsam, schwindet beim Erhärten und darf daher nur in geringer Quantität angewendet werden; Gusswerk in Gewölben muss bei diesem Mörtel ganz ausgeschlossen bleiben" (Salzenberg, *Einige Bemerkungen über die Gewölbeconstruction*...cit., col. 425).
[86] "So ist doch zu erwähnen, dass wir ein jenem Mörtelguss ganz ähnliches Material besitzen, nämlich das Stampfmörtelwerk. Bei diesem fällt die getadelte und tadelnswerthe Eigenschaft unseres gewöhnlichen Mörtels, das Schwinden, fort, indem dieses, wie es scheint, in dem zu grossen Kalk- Gehalt des gewöhnlichen Mörtels begründet ist, welchen der Stampfmörtel nicht besitzt" (Hoffmann, *Mittheilungen über Gewölbe*...cit., col. 453).
[87] See E. H. Hoffmann, *Ueber Anwendung von Stampfmörtel bei kleinen Brückengewölben*, in, "Zeitschrift für Bauwesen," X, 1860, coll. 263–266; Engel, *Der Kalk-Sand-Pisébau und die Kalksand-*

therefore making lean conglomerate suitable to vault construction. "It is worth mentioning that we have a material similar to [Roman] poured mortar, it is the rammed lean conglomerate," Hoffmann writes, "which does not have the blameworthy property of shrinking because it has no large amounts of lime as ordinary mortar."[86] In the same report from 1858, Hoffmann also describes a barn he conceived for a private builder in western Prussia as a construction with walls in lean conglomerate, which, in the interest of savings, are composed of pillars, discharging arches and thin filling-panels. According to Hoffman's report, Stüler visits this building and offers hardly any criticism.

In 1857 Hoffmann builds a thin (7 inches thick at the key and 8 inches at the imposts), 5 feet span vault in lean conglomerate for a small road bridge, whereas, in 1858, to build three further bridges, he resorts to a kind of raw stone masonry with quite wide tamped mortar joints.[87] He first pours and tamps a roughly 3-inch thick layer of mortar on the centrings, then he places voussoirs at a distance of 2 inches from each other, and finally covers and tamps the whole construction with mortar. The mortar mixtures he adopts to build the joints of these bridges are almost the same as those commonly used for masonry, being composed of one part of cement and five parts of sand, in one case, and one measure of hydraulic lime and six measures of sand, in the two other cases.

A corollary of the evolution of lean conglomerate applications during the second half of the 1850s concerns the serial manufacturing of artificial stones called *Kalksandziegeln* (sand-lime bricks), based on the use of a cast iron press, which Friedrich August Anton Bernhardi (1813–1889), a physician based in Eilenburg in Saxony, develops from 1850 to 1854 and patents in 1856.[88] The essential concept of sand-lime bricks goes back to Prochnow and to Sachs' experiments with lean conglomerate-based artificial stones, whereas the ameliorated technique benefits from improved expertise in producing mechanical cast iron machines.[89] The pressure produced with the long levers of the press makes bricks more compact and resistant than the ones that are produced by just tamping mortar into moulds with a pestle.[90] Bernhardi himself starts manufacturing presses in Eilenburg around 1854, and, in the following years, other entrepreneurs use his idea to develop similar machines, as demonstrated by the fact that the *Ekart Machinenbau-Anstalt* in Berlin produces a kind of press for sand-lime bricks since at least the late 1850s. It is documented that an Ekart press is used on the manor of Lietzow near Nauen, north-west of Berlin, to construct, probably after the fire of 1859, a number of buildings, which are partially made of ordinary bricks, partially of poured lean conglomerate and partially of sand-lime bricks.[91]

The development of sand-lime brick manufacturing is also connected to the production of artificial concrete ashlars that is spreading in France, in the

Ziegelfabrikation...cit., pp. 85–86.

[88] Bernhardi is politically engaged and fights to improve the conditions of the working class. It is probably in the context of this engagement that he develops the idea of producing affordable quality bricks, in order to allow the building of hygienic and affordable flats for workers. It is interesting how Bernhardi studies the social disadvantages of mechanical manufacturing (he publishes in 1848 a study entitled *Ueber die sozialen Nachtheile des gewerblichen Maschinenwesens*), but, at the same time, exploits the principle of mechanical manufacturing to improve workers' living conditions. Bernhardi himself starts the production of the press he invents, in 1854. About this, see Anon., *Kalksand-Mauerstein - vom Arzt entwickelt*, in, "Die Baustelle," 2014, n° 2, pp. 26–27.

[89] See Engel, *Der Kalk-Sand-Pisébau und die Kalksand-Ziegelfabrikation*...cit,. p. 4.

[90] Bernhardi claims to have also invented a mechanical mixer to prepare the mortar necessary to produce the bricks; according to Engel, this mixer is similar to the lean conglomerate mixer already developed in Fiddikow.

[91] See Anon., *Steinpresse zur Fabrikation von Ziegel- und Pisé-steinen*, in, "Zeitschrift für Praktischen Baukunst," XXI, 1861, coll. 127–128; Anon., *Ueber Kalkpisébau*, in, "Zeitschrift für praktische Baukunst," XXVII, 1867, coll. 162–163; Engel, *Der Kalk-Sand-Pisébau und die Kalksand-Ziegelfabrikation*... cit., pp. 104–105.

156-158. Anon., houses with walls of sand-lime bricks, Eilenburg, Jeseswitz [sic.], Lübchow (J. D. F. Engel, *Der Kalk-Sand-Pisébau und die Kalksand-Ziegelfabrikation*, Leipzig, C. H. Seemann, 1865).

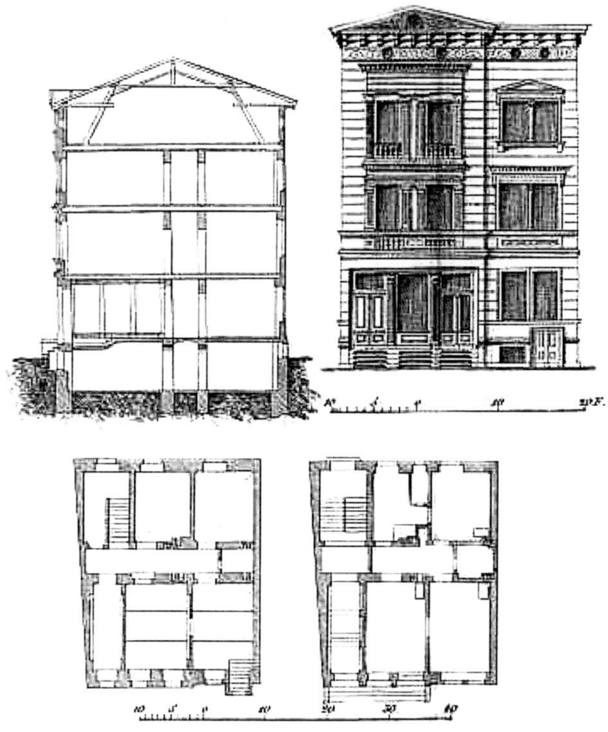

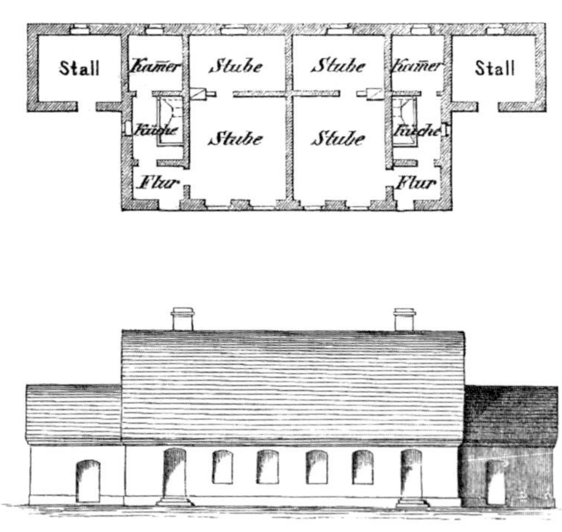

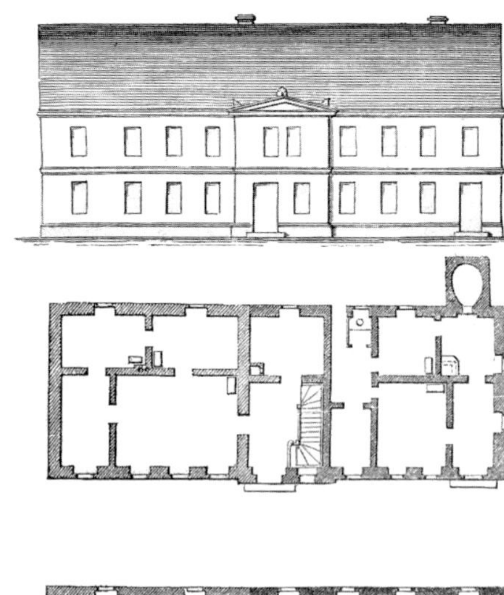

United Kingdom and in Germany as well. As Engel observes, the use of artificial concrete stones, like the ones produced by the Czarnikow factory in Berlin, develop in Germany slower than in other countries, since bricks remain more affordable. Sand-lime bricks could therefore be seen as a further adjustment of a technique that is spreading abroad, to local and limited economic conditions, as if sand-lime bricks were a sort of more affordable version of artificial concrete stones, just like poured lean conglomerate is considered a more affordable version of concrete. Documented examples of the use of sand-lime bricks in the years immediately after the invention of Bernhardi's press are evident in Eilenburg, in the nearby village of Jeseswitz [sic.], and in Lübchow near Cöslin in Pomerania. Among them, the building realized in Jeseswitz is particularly noteworthy, since it is a sort of experiment, being partially built in ordinary brick masonry, and partially with sand-lime bricks; according to Engel, this experiment proves that sand-lime bricks dry sooner than the ordinary ones.[92]

The manufacturing of *Cementwaaren*

As of roughly the turn of the 1850s, on the basis of the availability of high-quality cements and of the increased skills in producing mortar and concrete, the old ambition of producing artificial stones, architectural decorations and various other objects by moulding cement-based compounds gives rise to a veritable branch in manufacturing, and objects made of moulded cement-based compounds become known as *Cementwaaren*. The aforementioned Czarnikow factory develops a mixture composed of Portland cement, gravel and natron, by which it produces moulded decorations for important buildings constructed during the 1850s and the 1860s, in Germany and abroad, like the *Gildengebäude* of Riga, which is built between 1853 and 1860, and the *Stadttheater* of the same town, which is built between 1861 and 1863 according to the plans by the architect Ludwig Bohnstedt (1822–1885).[93] The decorations are transported from Berlin to Stettin by barges along the River Oder and, later, by ships on the Baltic sea to Riga.[94] Among them, the bas-relief of the pediment realized by the sculptor Friedrich August Wittig (1823 or 1826–1893)[95] according to drawings by Bohnstedt, which shows "Poetry and her influence on life," becomes particularly known.[96] Czarnikow also produces all the window cornices of the psychiatric hospital in Eberswalde, in northeast Berlin, which is designed by the architect Martin Carl Philipp Gropius (1824–1880) and built between 1862 and 1865.[97] Gropius praises the durability of Czarnikow moulded cornices, along with the aesthetic effect that the grey colour of cement produces in combination with the colour of bricks. "The intention to produce a slightly more agreeable effect of colours on the façades pushed to provide windows with moulded, cement-based

[92] *Ibid.*, pp. 115–118.

[93] See Anon., *Die Kunststeingießerei im Dienste der modernen Baukunst*, in, "Zeitschrift für praktische Baukunst," XXVI, 1866, coll. 23–28; Anon., *Der Werth des künstlichen Baumaterials*, in, "Zeitschrift für praktische Baukunst," XXVI, 1866, (coll. 117–122), col. 120; Leonhardt, *Von der Cementware zum Konstruktiven Stahlbetonfertigteil*...cit., p. 3. About the new Gildengebäude in Riga, see Arnold von Tideböhl, *Fürst Alexander Suworow, General-Gouverneur von Liv-, Esth- und Kurland, 1848–1861*, Riga, W. F. Häcker, 1862, pp. 45–47; about the Stadttheater, see *ibid.*, pp. 96–98.

[94] Anon., *Der Werth des künstlichen Baumaterials*...cit.

[95] See Hermann Alexander Müller, *Biographisches Künstler-Lexikon der Gegenwarth*, Leipzig, Bibliographische Institut, 1882, p. 563; Daelen, Eduard, "Wittig, August," in, *Allgemeine Deutsche Biographie*, vol. XLIII, Leipzig, Duncker & Humblot, 1898, pp. 638–639.

[96] "Die Poesie in ihrer Einwirkung auf das leben" (Anon., *Die Kunststeingießerei im Dienste der moderne Baukunst*...cit., col. 27).

[97] See *ibid.*, coll. 27–28; Martin Carl Philipp Gropius, *Die Provinzial-Irren-Anstalt zu Neustadt-Eberswalde*, in, "Zeitschrift für Bauwesen," XIX, 1869, (coll. 147–194), coll. 182–183.

159. Beyne, *Gildengebäude*, Riga, 1853–1860. Entrance tower (Anon., *Die Kunststeingießerei im Dienste der moderne Baukunst*, in, "Zeitschrift für praktische Baukunst," XXVI, 1866).

cornices. [...]. The manufacturer Czarnikow in Berlin assumed a ten-years long durability guarantee against weather effects. Until now, such cement-based moulded works, which are produced on the building site, have proven to be durable, and their colour suits that of bricks."[98] The arch-shaped lintels consist of single pieces, whereas the windowsills and the jambs each consist of two joined pieces. Some difficulties arise while finishing the joints, since the hue of poured cement mortar is slightly different from the one of the precast elements. Moreover, several sills prove not to be perfectly sealed and cause water seepage. Such complications still do not affect Gropius' confidence in the opportunity to decorate façades with cement-based moulded elements. "Such minor complications," Gropius writes in this regard, "may not detract from such experiments, since the development of imperishable artificial stones would be a great advantage for architecture. Nothing else could be more appropriate to monumental buildings. It would be possible to reproduce ancient architectural elements like plates, pillars, abacus, plinths, cornices, etc., in desired sizes and without joints, which affects the ideal forms of such elements."[99] As a matter of fact, an increasing demand for such decorative elements, especially on the part of the rising middle-class, supports the production of cement-based mouldings.[100] The apostles' figures, as well as the pinnacles and the stone fretworks of the altar that the aforementioned Loewer from Grabow constructs in 1858 for the *Stephanskirche* in Gartz, south of Stettin, are made of moulded, Portland cement-based mortar, whereas the load-bearing parts are made of bricks, bound and plastered with the same kind of Portland cement-based mortar.[101] The altar of the *Marienkirche* of Strasburg in the Uckermark is produced in a similar way by Czarnikow around about 1865.[102] Czarnikow also moulds the high reliefs and the bas-reliefs inside the pediments and over the central windows of the façade of the *Stadttheater* in Leipzig, which is built between 1864 and 1868 according to the plans by the architect Carl Ferdinand Langhans (1781–1869).[103]

Along with architectural decorations, the manufacturing of moulded cement-based pipes for water supply and drainage systems develops significantly. The brothers Born from Erfurt in Thüringen criticize the method patented by Karlinger in 1848 for producing circular pipes, which they consider too complex. Instead, they believe in the greater effectiveness of a manufacturing process similar to the one that Fleuret had developed at the turn of the century, which consists in pouring and tamping concrete between timber boards, all around a pole, the surface of which has been previously treated with some grease, so that it can be extracted once the concrete has set.[104] The brothers Born report having successfully built a 700 feet-long water pipe in 1852 for

[98] "Der Wunsch, den Façaden eine etwas freundlichere farbige Wirkung zu verleiben, führte auf die Anwendung der in Cementguss ausgeführten Fenstereinfassungen für die Aussenfronten. [...]. Der Fabrikant, Herr Czarnikow in Berlin, hat eine zehnjährige Garantie für die Wetterbeständigkeit übernommen. Es haben sich bis jetzt diese Steingussarbeiten, die auf der Baustelle selbst angefertigt werden, in Bezug auf ihre Dauer wohl bewährt. Die Farbe steht sehr gut zu der Farbe der Ziegel" (*ibid.*, col. 182).

[99] "Diese kleinen Nachtheile können indess nicht Veranlassung sein, derartige Versuche aufzugeben. Es würde ein grosser Gewinn für die Architektur sein, wenn es gelingen sollte, einen schönfarbigen und unverwüstlich dauerhaften künstlichen Stein herzustellen. Nichts würde geeigneter für monumentale Bauwerke sein. Man würde dann die durch die antike Kunst erfundenen architektonischen Elemente: Wand, Stütze, Abakus, Plinthe, Gesims etc., nach beliebigen Verhältnissen und Dimensionen herstellen können, ohne ihre ideale Form durch die

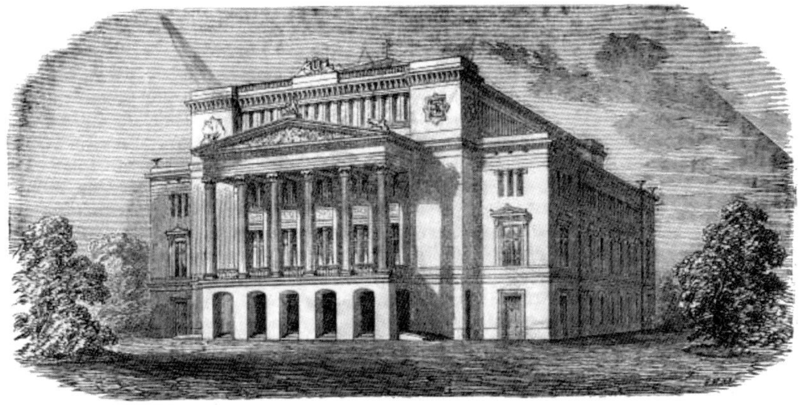

160-161. L. Bohnstedt, *Stadttheater*, Riga, 1861–1863. Perspective view; pediment (Anon., *Die Kunststeingießerei im Dienste der moderne Baukunst*, in, "Zeitschrift für praktische Baukunst," XXVI, 1866).

the town of Kölleda, northeast of Erfurt, and a 200 feet-long water pipe in 1854, in the town of Erfurt.

Some lavatories in the already mentioned *Garnison-Lazareth* in Berlin are equipped with a drainage system based on the use of pipes made of Portland cement-based mortar. Each main drain is assembled with several hollow and frustum-shaped elements, which are produced applying mortar around a wooden open mould. Some of such elements are provided with side holes to allow connections with sanitary installations at each level. They are fabricated by a master mason named Eyffert, together with the previously mentioned artificial stone maker Fehse.[105] Similar drains are also built by the master mason Lehmann in 1853 for the lavatories of the *Lazareth* in Spandau, and, in 1854, for the *Lazareth* in Prenzlau.[106] Concerning sanitary installations, Fehse moulds toilets and other sanitary basins in cement-based mortar.[107]

Small concrete open-channels are built along a number of streets in Berlin as early as 1853. They are made of concrete composed with Portland cement, sand and different kinds of stone and brick rubble. Some of them are poured on site, using a wooden profile to give the channel the desired concave shape, whereas some others consist of precast elements produced in appropri-

Roheit der Ziegelfuge zu beeinträchtigen" (*ibid.*, col. 183).
[100] See Mohamed Scharabi, *Architekturgeschichte des 19. Jahrhundert*, Tübingen, Wasmuth, 1993, p. 161.
[101] See Herrmann, *Hochaltar in der Stephanskirche zu Gartz, aus Portland-Cement erbaut*, in, "Zeitschrift für Bauwesen," X, 1860, coll. 311–316.
[102] See Anon., *Die Kunststeingießerei im Dienste der moderne Baukunst*...cit., coll. 27–28.
[103] The reliefs in the main pediment are moulded on models by the sculptor Hugo Hagen (1818–1871), and the ones in the two minor pediments are moulded on models by the sculptor August Wittig (1823–1893); see Anon., *Die Künstlerichen Verzierungen des neuen Stadttheaters zu Leipzig*, in, "Zeitschrift für praktische Baukunst," XXVIII, 1868, (coll. 309–314), col. 309; Brückwald, *Das Stadt-Theater in Leipzig von Langhans*, in, "Zeitschrift für Bauwesen," XX, 1870, coll. 291–298.
[104] See Gebrüder Born, *Ueber Verfertigung von Wasserleitung-Röhren aus Cement*, in, "Polytechnisches Journal," vol. CXXXIV, 1854, pp. 136–40. See also Fleuret, *L'art de composer*...cit., pl. 18–21.
[105] See Becker, *Praktische Anleitung*...cit., p. 42.
[106] *Ibid.*
[107] See C. Hennicke, *Ueber die Anlage von Abtritten in Wohngebäude und öffentlichen Anstalten*, in, "Zeitschrift für Bauwesen," VII, 1857, (coll. 123–144), col. 139.

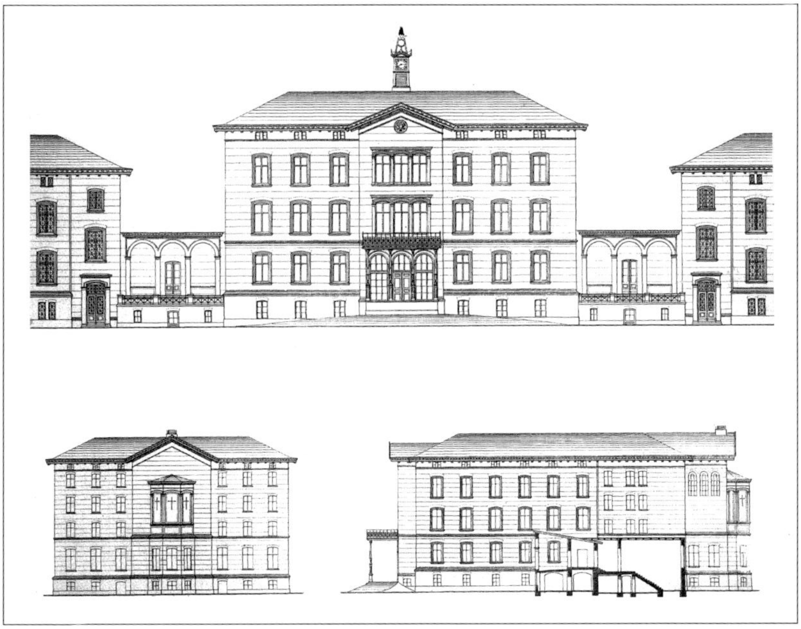

162-163. M. C. Ph. Gropius, hospital, Eberswalde, 1862–1865. Front and rear elevation, cross section through the side gallery with side elevation of the main building; elevation, plan and section of a window with cement-based moulded cornices (M. C. Ph. Gropius, *Die Provinzial-Irren-Anstalt zu Neustadt-Eberswalde*, in, "Zeitschrift für Bauwesen," XIX, 1869).

ate workshops.[108] Underground drainage pipes are built assembling elements that are similar to the ones for open-channels. They are, however, appropriately joint-shaped along the linear edges and at the curved extremities.[109] The bottom of such pipes is flat so that they may be easily placed on the ground, and the different bottom and upper elements are assembled staggered as in an ordinary masonry bond. Pipes of such a kind are used to build a number of drainage installations in Berlin around the mid-1850s, and to build road culverts in the region of Sławno in eastern Pomerania, probably using cement from the *Stettiner Portland-Cement Fabrik*.[110] In the early 1860s, they are also produced by the factory Krüger in Gdansk, in order to build the culvers along the railway connecting Bromberg and Torun, in Pomerania.[111] Moreover, the sewage system of the town of Witten along the River Ruhr is fabricated, starting in 1867, with cement-pipes that are produced using cement from the *Bonner Portland-cement Fabrik*.[112]

As of roughly 1862, the cement producers Wilhelm Gustav Dyckerhoff (1805–1894) and Carl Brentano (1833–1898), who found the company Dyckerhoff & Brentano in 1861, in Eltville am Rhein near Wiesbaden, also manufacture cement-based moulded objects, principally floor slates, which earn the company a contract for paving some stations along the Baden rail-

[108] See Becker, *Praktische Anleitung*...cit., pp. 35–37.
[109] *Ibid.*, pp. 51–53.
[110] Becker mentions a 168 feet long channel realized in 1855 by the master mason Eyfferth between the *Potsdamerstraße* to the *Leipzigerstraße* (the round shaped part of it is, however, made of masonry), a 1300 feet long channel realized in 1856 between the *Matthäusstraße* and the *Bellevuestraße*, a 200 feet long channel realized at *Kopernicher-Tor*, and a channel realized by the master mason Lehmann in 1857 at the *Humboldt-Hafen*; see Becker, *Praktische Anleitung*...cit., pp. 51–52. For the culvert in the region of Sławno, see Sanftleben, *Anfertigung und Verwendung von Portland-Cement-Röhren zu Chaussee-Durchlässen*, in, "Zeitschrift für Bauwesen," IX, 1859, coll. 417–420, and in, "Polytechnisches Journal," XL, vol. CLIV, pp. 421–423; Becker, *Praktische Anleitung*... cit., p. 51. Krüger in Gdansk also produces poured cement-mortar pipes.
[111] P. Ernst, *Ueber Cementröhren und deren Anfertigung*, in, "Zeitschrift für Praktische

ways.[113] In 1864, Brentano leaves the cement company and Dyckerhoff carries on the business with his two first sons; the company henceforth takes the name Dyckerhoff & Söhne. In 1865, Dyckerhoff, together with the merchant Heinrich Lang (1818–1887), and probably also in collaboration with the master builder Franz Erwin Serger (1817-1879), founds the company Lang & Cie in Karlsruhe, which is registered at the local Chamber of Commerce as a *Cementwaaren* company, focusing the production of "pipes, floor slabs, farming tools and architectural decorations."[114] The Dyckerhoffs devote much of their initial professional activity to the search for appropriate techniques for producing cement-based moulded objects, and they use Berlin factories like Czarnikow as models. In this way, they learn how to produce objects made of brick cores surrounded by cement-based mortar. The relation between the companies Dyckerhoff & Söhne and Lang & Cie is close, especially with the entry of Eugen Julius Richard Dyckerhoff (1844–1924), the third son of Wilhelm, into Lang & Cie in 1866. Eugen Dyckerhoff has already experimented with the manufacturing of concrete floor slabs in the family cement factory, around 1864, before leaving for a professional training in Marseille. Once in Lang & Cie, he carries out studies and experiments in the laboratory of Dyckerhoff & Söhne, focusing particularly on the composition of concrete. He

164. Ph. Loewer, altar of the *Stephankirche*, Gartz, Pomernia, n. d. Front elevation (Herrmann, *Hochaltar in der Stephanskirche zu Gartz, aus Portland-Cement erbaut*, in, "Zeitschrift für Bauwesen," X, 1860).

165. Czarnikow, altar of the *Marienkirche*, Strasbourg in the Uckermark, approx. 1865. Front elevation (Anon., *Die Künstlerichen Verzierungen des neuen Stadttheaters zu Leipzig*, in, "Zeitschrift für praktische Baukunst," XXVIII, 1868).

Baukunst," XXXI, 1871, (coll. 235–242), col. 235.

[112] See Freudenberg, *Die Canalisierung der Stadt Witten*, in, "Zeitschrift für Bauwesen," XXIII, 1873, (coll. 135–152), coll. 138, 146.

[113] In 1863, Dyckerhoff and Brentano builds a larger factory in Amöneburg am Rhein, which historically remains the main seat of the company; the headquarters of the company is still nowadays located in Amöneburg; see *ibid.*, p. 27. About the history of the Dyckerhoff & Söhne, see Curt Piorkowski, *Dyckerhoff Portland-Zementwerke A.G.: Mainz-Amöneburg*, Leipzig, Arnd, 1937, pp. 5–13.

[114] About the history of Lang & Cie, see *ibid.*, pp. 21–53, esp. p. 33 for what concerns the foundations and the registration at the Chamber of Commerce in Karlsruhe; Gert v. Klass, *Weit spannt sich der Bogen: 1865-1965: die Geschichte der Bauunternehmung Dyckerhoff & Widmann*, Wiesbaden, Bartels, 1965, pp. 7–53.

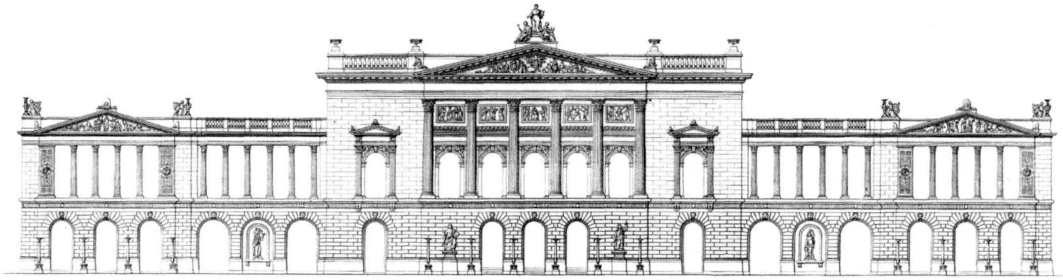

166-167. C. F. Langhans, *Stadttheater*, Leipzig, 1864–1868. Front elevation; pediment (Anon., *Die Künstlerichen Verzierungen des neuen Stadttheaters zu Leipzig*, in, "Zeitschrift für praktische Baukunst," XXVIII, 1868).

soon replaces the Berlin way of building objects containing a brick masonry core with a moulding technique consisting of pouring and tamping in moulds almost mushy cement-based concrete, containing as little water as possible. The technique is considered to be related to Coignet's *béton aggloméré*, especially on the basis of Eugen's stay in France. Nevertheless, it should not be ignored that the concept of manufacturing artificial stones and other objects by pouring and tamping conglomerate masses in formworks or moulds has already been part of German practices of building and manufacturing for at least a couple of decades. From about 1867, Lang & Cie deliver early moulded concrete pipes to the town of Karlsruhe and to a number of other towns in Baden.[115] The technique that is adopted to produce such concrete pipes is essentially similar to the ones that Karlinger had patented in 1848. Besides the search for improving concrete mixtures, Lang & Cie also engages in producing a collection of moulds for architectural decorations, working with several artists of the time, among them is the sculptor Karl Friedrich Moest (1838–1923), who later produces statues that remain representative symbols of the company, and, at the same time, evidence of talent in moulding concrete.

In 1869, following the entry of Gottlieb Widmann (1817–1894) into the business, Lang & Cie is transformed into Dyckerhoff & Widmann. The new company exhibits a range of cement-based moulded products at the *Landesgewerbeausstellung* (regional industrial exhibition) that takes place in Offenbach in 1869, taking advantage of the pavilion of Dyckerhoff & Söhne. Shortly thereafter, Dyckerhoff & Widmann entrusts Moest with the realization of a cement-based moulded statue, intended to represent the expertise that the company has achieved, at the *Allgemeine Industrieausstellung* (General Industrial Exhibition) of Kassel in 1870. The statue is the celebrated *Triumph der Galathea*, which is also known as *Galatheabrunnen*, since the figures stand in a fountain. The *Galatheabrunnen* ultimately is not exhibited in Kassel, but sold to the town of Karlsruhe, to be placed in the park called *Sallenwäldchen*. Given the presence of a wire netting inside the cement-based mortar, the statue is also related to early examples of objects made of metallic nets and cement-based mortar, which are mainly developing in France and would have important consequences in Germany around the mid-1880s.[116] A second *Galatheabrunnen* is produced to be exhibited at the International Exhibition that takes place in Vienna in 1873; whereas, for the *Kunst- und*

[115] Stegmann, *Das Bauunternehmen Dyckerhoff Et Widmann*...cit., p. 46. The celebrated expression *Stampfbeton* does not appear until the end of the 1880s; *ibid.*, p. 44, n. 218.

[116] The reference goes to the patents by the French inventor Joseph Monier (1823–1906). Between 1881 and 1886, German engineers Gustav

Gewerbeaustellung (Exhibition of Art and Industry) that takes place in Munich in 1876, Moest erects two further statues, namely two virile figures representing the construction and the functioning of railways (*Eisenbahnbau* and *Eisenbahnbetrieb*).

Improvements to scaffoldings and sheet pilings for foundations underwater, the building of the bridge over the Rhine in Koblenz

The option to pour concrete underwater in enclosures surrounded by sheet piles, instead of building cofferdams, has long been considered a major advantage of concrete foundations, since sheet pilings are more affordable and easier to build than cofferdams. Nevertheless, due to particular orographic conditions, even the construction of sheet pilings sometimes demands complex solutions needing additional attention, as is the case when the bridge over the Rhine in Koblenz is built, between 1862 and 1864, under the supervision of the previously mentioned master builder Hartwich.[117]

The bearing soil lies under deposits of gravel, the thickness of which is irregular. At the place where the left pier is to be built, the stratum of gravel is about 16–17 feet-deep; this is considered enough to build an enclosure of sheet piles fixed on scaffoldings made of poles rammed into the gravely soil, cross beams, and metal wind bracings, which are tied up at the poles by means of a metallic ring instead of screws and bolts, in order to prevent a supposed weakening of the poles. Once the scaffolding is constructed, the sheet piles are driven into the gravel, making them slip between two longitudinal beams, which are tied up at the poles of the scaffoldings by means of double ferrules. Securing the sheet piles to the poles of the scaffoldings in a very solid manner is necessary because all gravel must be removed from inside the enclosure, to reach the bearing soil, and the pressure of the exterior gravel, along with the one produced by the stream, are considered a danger for the stability of the piles.

Even more complex is the assembly of the provisional works to build the foundation of the right pier. The depth of the gravel stratum amounts to 2 or 3 feet, which is considered too shallow to drive both the poles of the scaffoldings and the sheet piles as well. Scaffoldings are therefore pre-assembled on dry land and transported on the river by means of barges and rafts, to be then sunk into the water at the desired places using stones as ballast. The longitudinal scaffoldings are built and put into place first. They are transported by

Adolf Wayss (1851–1917) and Matthias Koenen (1849–1924) move from Monier's German patent to developing an early mathematical model to calculate reinforced concrete slabs.

[117] See Emil Hermann Hartwich, *Rheinbrücke bei Coblenz*, in, "Zeitschrift für Bauwesen," XIV, 1864, (coll. 385–416, 529–579), coll. 404–409.

168-171. W. A. Becker, drawings of concrete pipes and channels. Concrete pipes poured on site; concrete sanitary drain of the *Garnison-Lazareth*, Berlin; concrete open-channel; concrete pipes for draining systems (W. A. Becker, *Praktische Anleitung zur Anwendung der Cemente zu baulichen, gewerblichen, landwirthschaftlichen und Kunst-Gegenständen*, Berlin, Nicolaische Verlagsbuchhandlung, 1860).

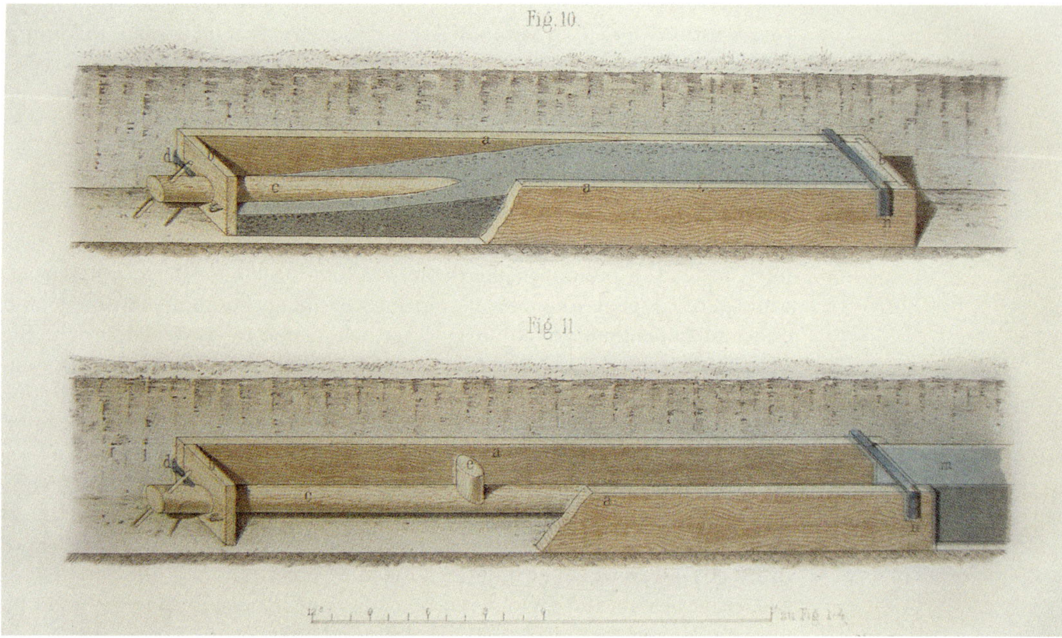

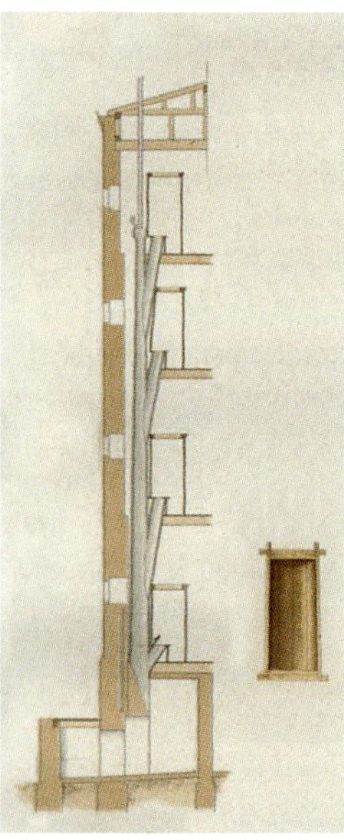

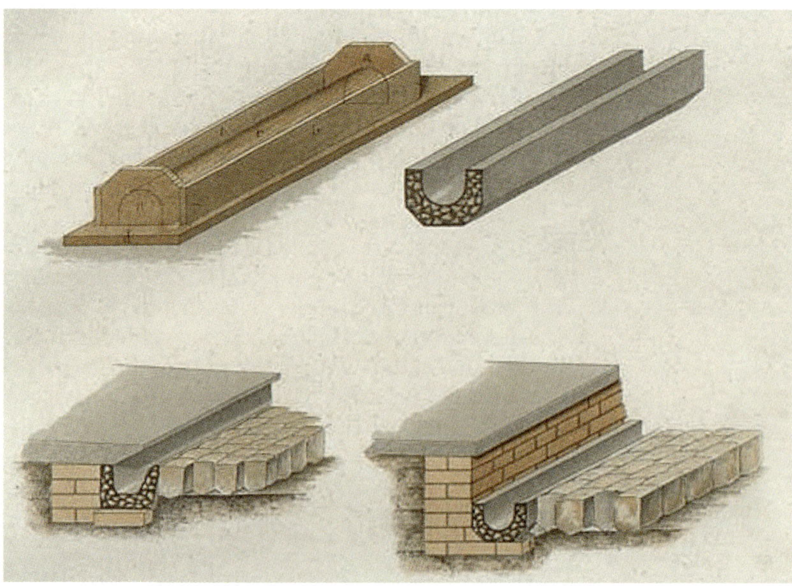

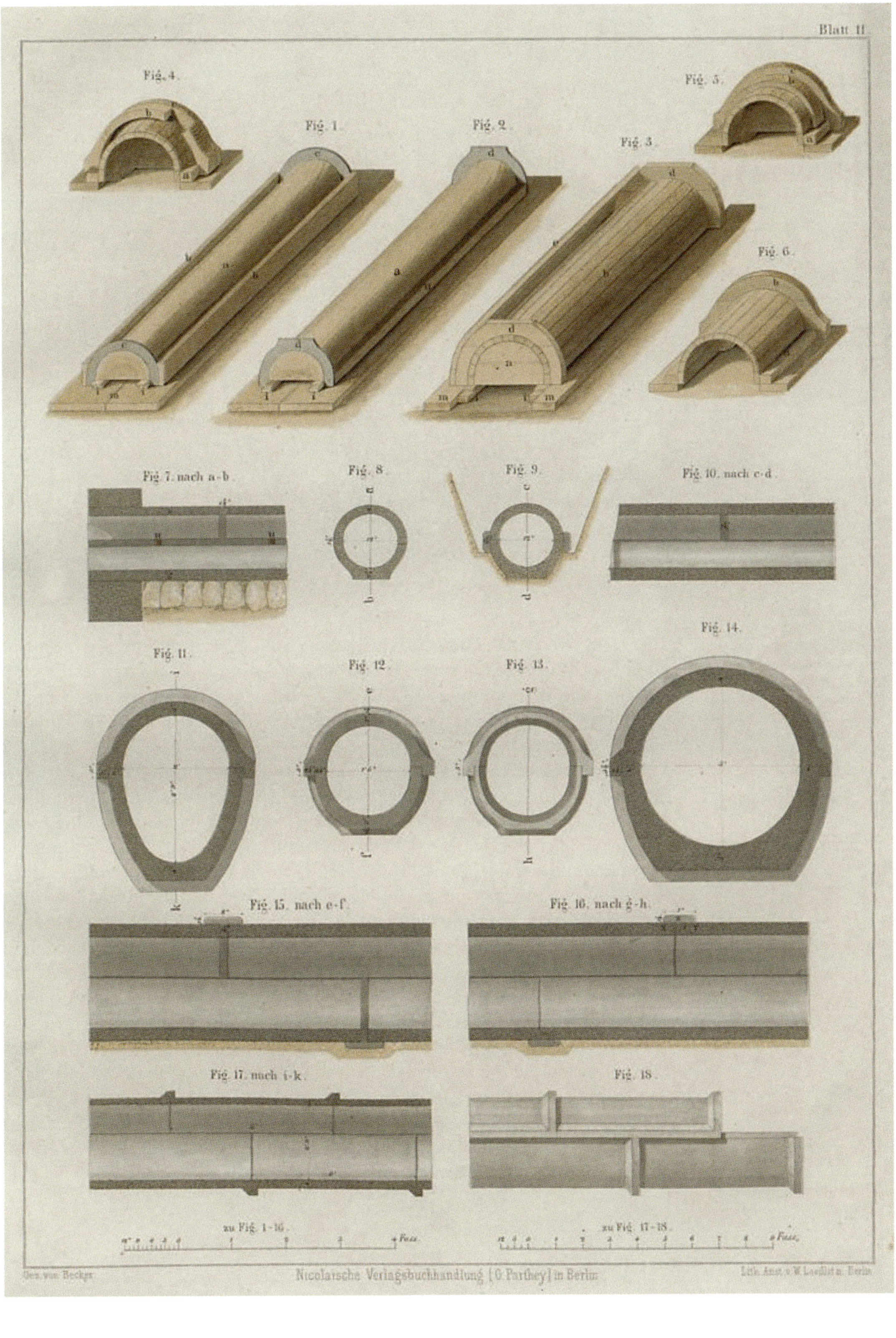
Blatt 11.
Fig. 4.
Fig. 1.
Fig. 2.
Fig. 3.
Fig. 5.
Fig. 6.
Fig. 7. nach a-b.
Fig. 8.
Fig. 9.
Fig. 10. nach c-d.
Fig. 11.
Fig. 12.
Fig. 13.
Fig. 14.
Fig. 15. nach e-f.
Fig. 16. nach g-h.
Fig. 17. nach i-k.
Fig. 18.
zu Fig. 1-16.
zu Fig. 17-18.
Fuss.
Nicolaische Verlagsbuchhandlung [G. Parthey] in Berlin

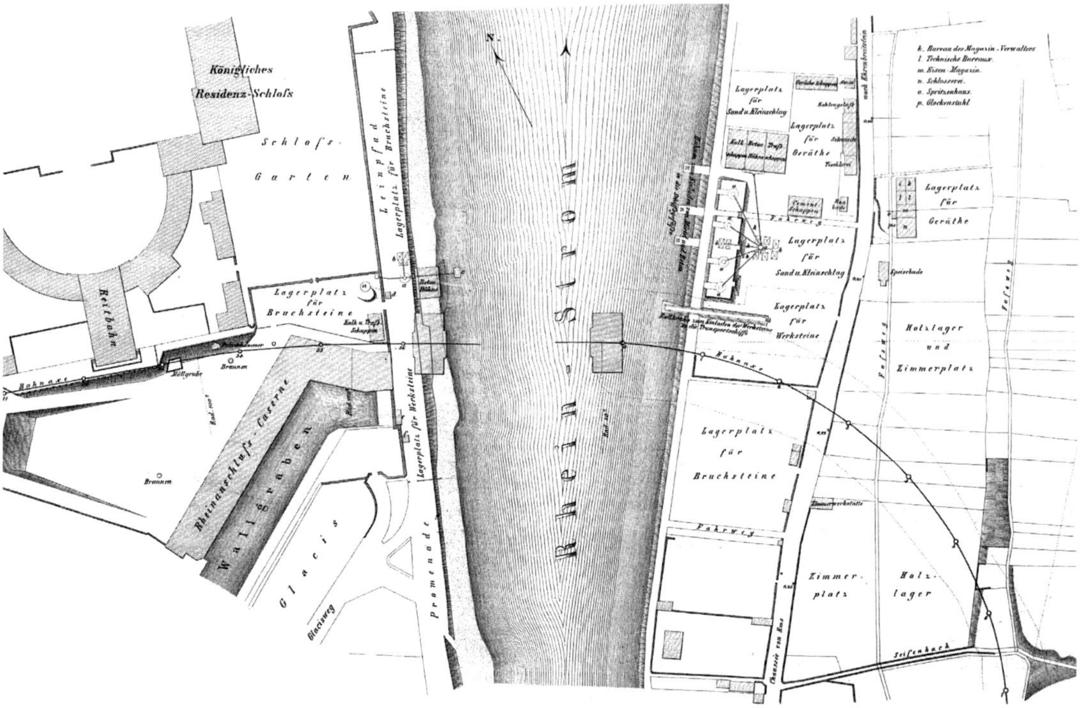

172-175. E. H. Hartwich, bridge over the River Rhine, Koblenz, 1862–1864. Plan of the site; front (south) elevation and plan; fronth (north) elevation of the western span and cross section of the western abutment; left pier under construction, cross section, plan and details of scaffoldings (E. H. Hartwich, *Rheinbrücke bei Coblenz*, in, "Zeitschrift für Bauwesen," XIV, 1864).

means of two parallel barges, which are kept at a fixed distance by means of a timber floor. This floor is provided with a central hole as large as the scaffolding, and it supports a wooden truss outfitted with pulleys and ropes to hold the scaffolding and to sink it into the water when the barges are situated over the right spot.

Due to a lack of space, each transversal scaffolding is instead put on a raft, which can partially enter into the free space between the longitudinal scaffoldings; it is fixed at these scaffoldings and kept hanging over the water by means of chains, so that the raft can be taken away, before the scaffolding is sunk into the water. Once all scaffoldings are in place, the sheet piles are installed, fixing the ones upstream with a double system of beams and ferrules, for precaution against the force of the stream.

Concrete is poured down using a pipe anchored at a bridge crane, and it forms a roughly 10 feet-high base. Following a procedure that is now common, concrete cofferdams are erected on the concrete bases, in order to form watertight caissons where workers build the lowest masonry layers of the piers, once water has been drained. Trass-based concrete is used for the foundations of the left pier, whereas, for the right one, some Portland cement is added to the concrete mixture. It is taken partially from the plant of Bleibtreu in Bonn and partially from England. Alternatively, the foundations of the two abutments of the bridge are built in enclosures made of ordinary sheet piles, using trass-based concrete.

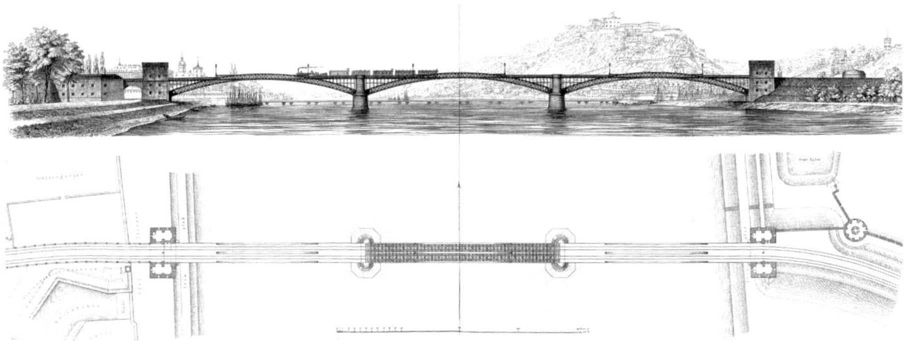

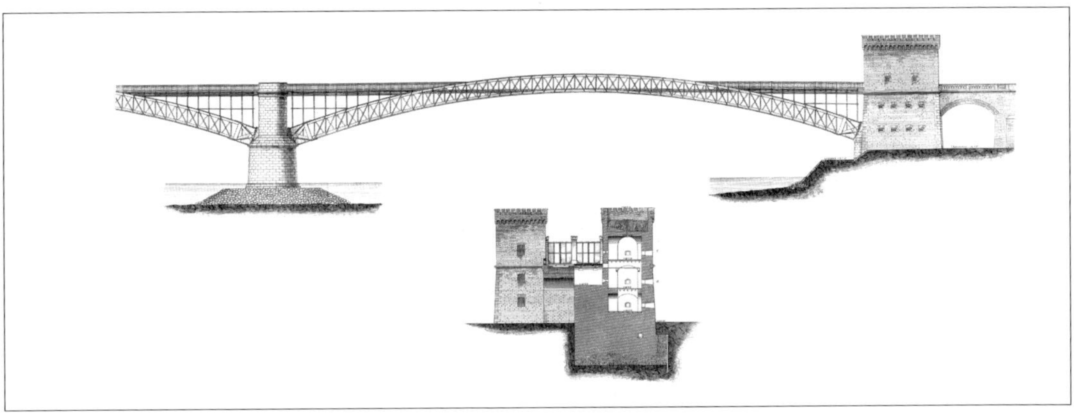

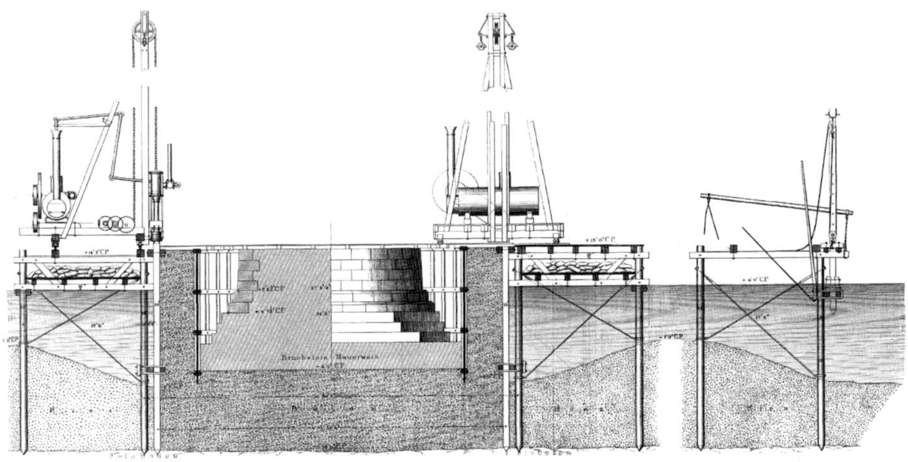

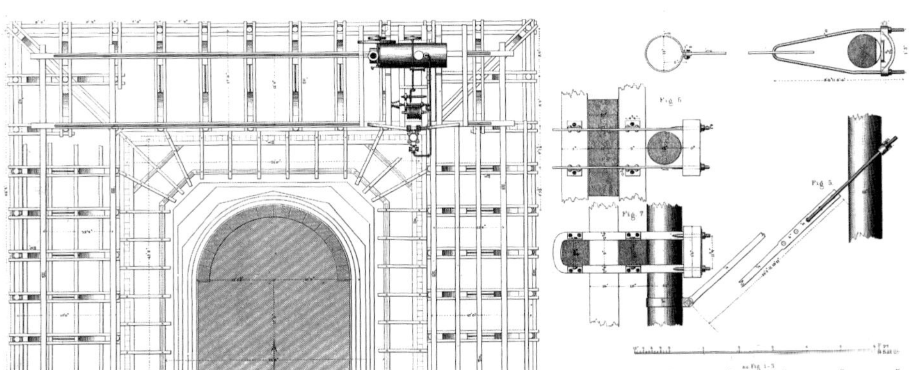

176-177. E. H. Hartwich, bridge over the River Rhine, Koblenz, 1862–1864. Scaffoldings to build the foundation of the right pier. Partial plan and side elevations, details of the system to bind the scaffoldings, the poles and the sheet piles; the transport of the scaffoldings by means of a berge and by means of a raft (E. H. Hartwich, *Rheinbrücke bei Coblenz*, in, "Zeitschrift für Bauwesen," XIV, 1864).

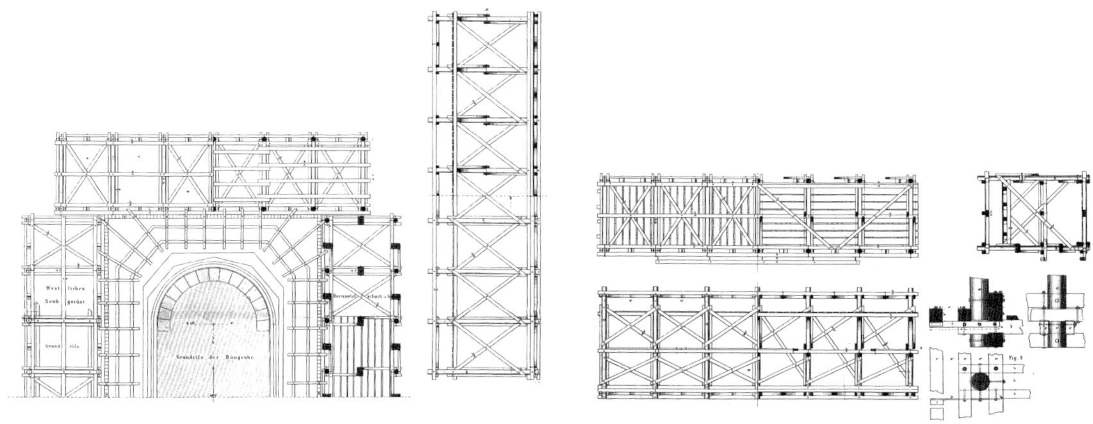

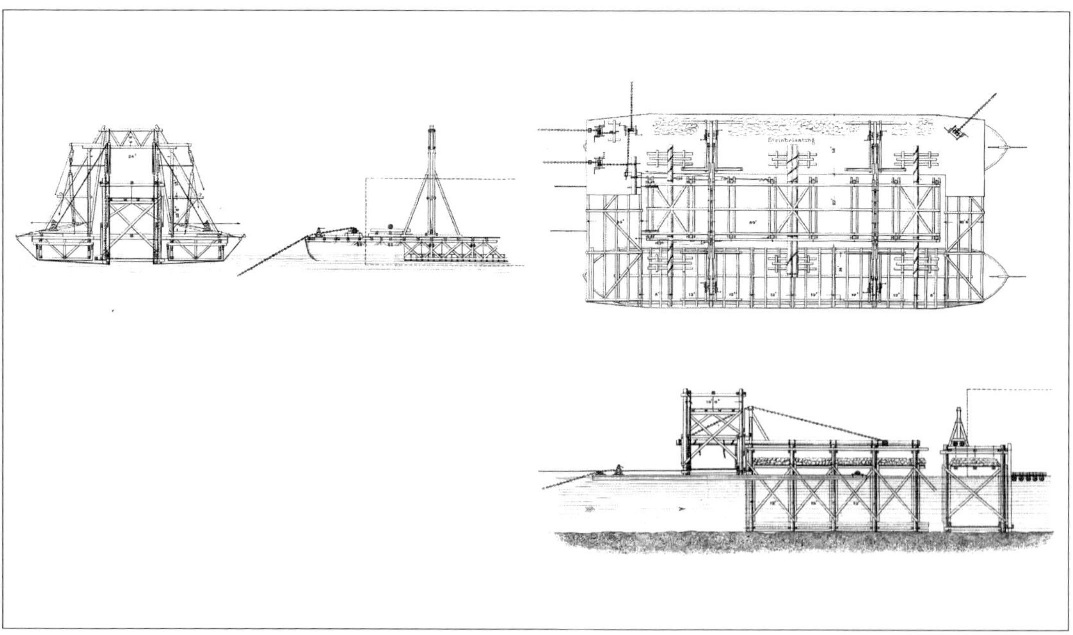

178-179. E. H. Hartwich, bridge over the River Rhine, Koblenz, 1862–1864. Construction of the left pier. The pouring process; construction of the masonry on the concrete foundation (E. H. Hartwich, *Rheinbrücke bei Coblenz*, in, "Zeitschrift für Bauwesen," XIV, 1864).

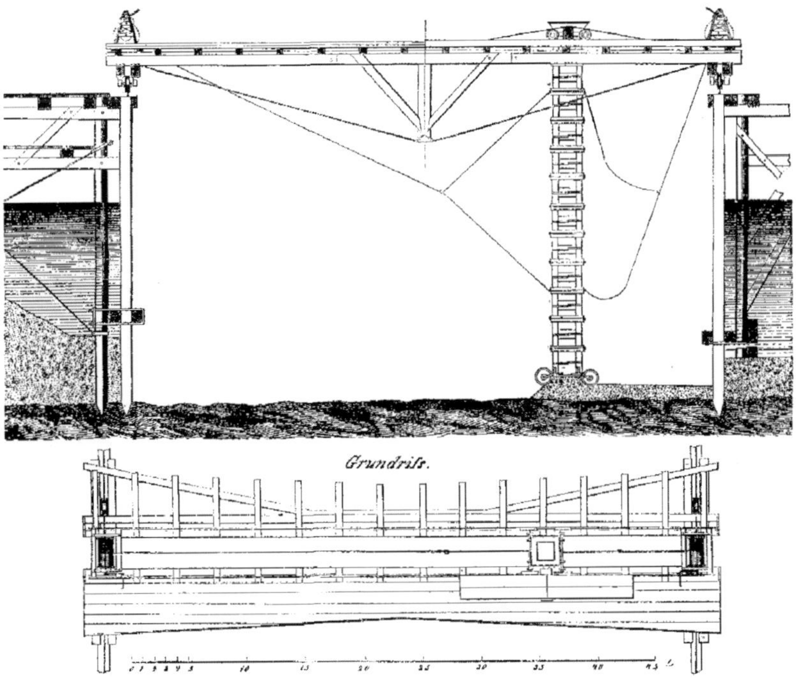

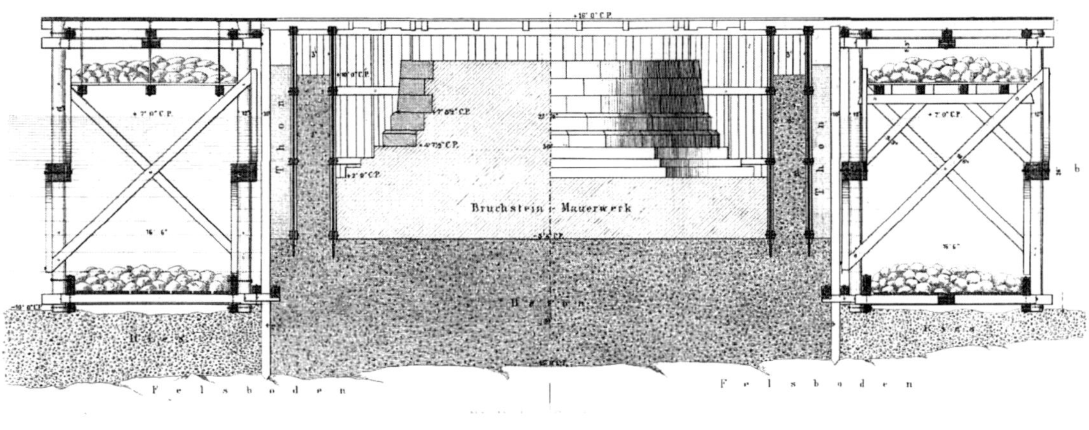

180-182. E. H. Hartwich, bridge over the River Rhine, Koblenz, 1862–1864.

180. The bridge under construction, photograph on paper (Technische Universität Berlin, Architekturmuseum, inv. n° BZ-F 01, 001).

181. The bridge under construction, view from the eastern riverside. A barge for transporting scaffoldings is visible in the foreground, photograph on paper (Technische Universität Berlin, Architekturmuseum, inv. n° BZ-F 01, 004).

182. The completed bridge, photograph on paper (Technische Universität Berlin, Architekturmuseum, inv. n° BZ-F 01,006).

Sinking wells, sinking caissons and pneumatic caissons

Starting around the mid-1860s, the ability of German master builders to build foundations made of sunken wells is confronted by further evolutions, due to the increasing need to build foundations in marshy soil and underwater – sometimes at considerable depth. This is the case in the Oder valley south of Stettin, which is marshy land, crossed by two main arms of the River Oder and by several other smaller watercourses. As of 1864, new bridges and docks need to be built there, in order to improve the previously mentioned Eastern Railway, which reaches Stettin from Berlin on the west side of the Oder and then crosses the Oder valley to extend further to Stargard.[118] The so-called *Fluthbrücke* (bridge over the tideland), which is built between 1864 and 1865, stands on thirteen piers and two abutments, the foundations of which consist of masonry sunken wells that are partially filled with concrete. Each well is made of a 1½ feet-thick circular wall standing on a timber crown. It encloses a 12 feet-wide circular space, which is gradually dug out, while water is drained by means of a pump. The tool used to dig the soil inside the well is the so-called Indian spade. It consists of a shovel hinged at a long pole and moved by means of a chain, so that workers can stand on scaffoldings placed over the well and do not need to work inside of the it while it is sinking. A 4 feet-thick bed of concrete is poured onto the bottom of each well, with the main aim of preventing rising water from the soil, whereas the rest of it is then filled with brick masonry.[119] The two abutments are each built on three sunken wells, whereas each pier stands on two wells.

The viaduct over the *Silberwiese*, which is built between 1867 and 1868 in the context of the same improvement project for the Stettin-Stargard railway line, is also initially conceived as a bridge set on sunken well foundations.

[118] See Architekten Verein zu Berlin, *Mittheilungen aus Vereinen, Versammlung am 18. November 1865*, in, "Zeitschrift für Bauwesen," XVI, 1866, col. 320; Verein für Eisenbahnkunde zu Berlin, *Protokoll vom 14. November 1865*, in, "Zeitschrift für Bauwesen," XVI, 1866, col. 330; Hagen, *Brunnen, Wasserleitung und Fundirungen*...cit., p. 365; Theodor Stein, *Erweiterungsbauten der Berlin-Stettiner Eisenbahn ausgeführt 1864–1869*, Berlin, Ernst & Korn, 1870; Wilhelm von Haselberg, *Die Brücken der Berlin-Stettiner Eisenbahn im Oderthal bei Stettin*, in, "Zeitschrift für Bauwesen," XXIV, 1879, (coll. 359–378), coll. 362–364; Teichgräber, *Die Brücken im Oderthale bei Stettin*, in, Vv. Aa., ed., *Bericht über eine Bauwissenschaftliche Studienreise nach der Pommerschen Künste*, Julius Springer, Berlin, 1887, pp. 7–13. Regarding soil composition in the Oder valley, see W. v. Haselberg, *Die Brücken der Berlin-Stettiner*... cit. col. 362.

[119] See Stein, *Erweiterungsbauten der Berlin-Stettiner Eisenbahn*...cit., pp. 2–3.

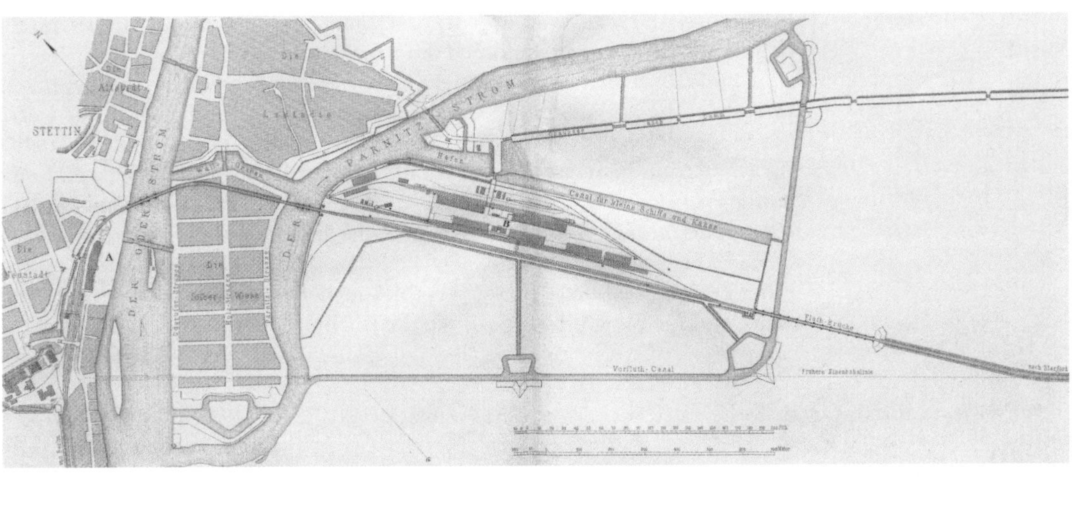

183. Th. Stein, plan to improve the Stettin-Stargard railway line, Stettin, 1864–1869 (Th. Stein, *Erweiterungsbauten der Berlin-Stettiner Eisenbahn ausgeführt 1864–1869*, Berlin, Ernst & Korn, 1870).

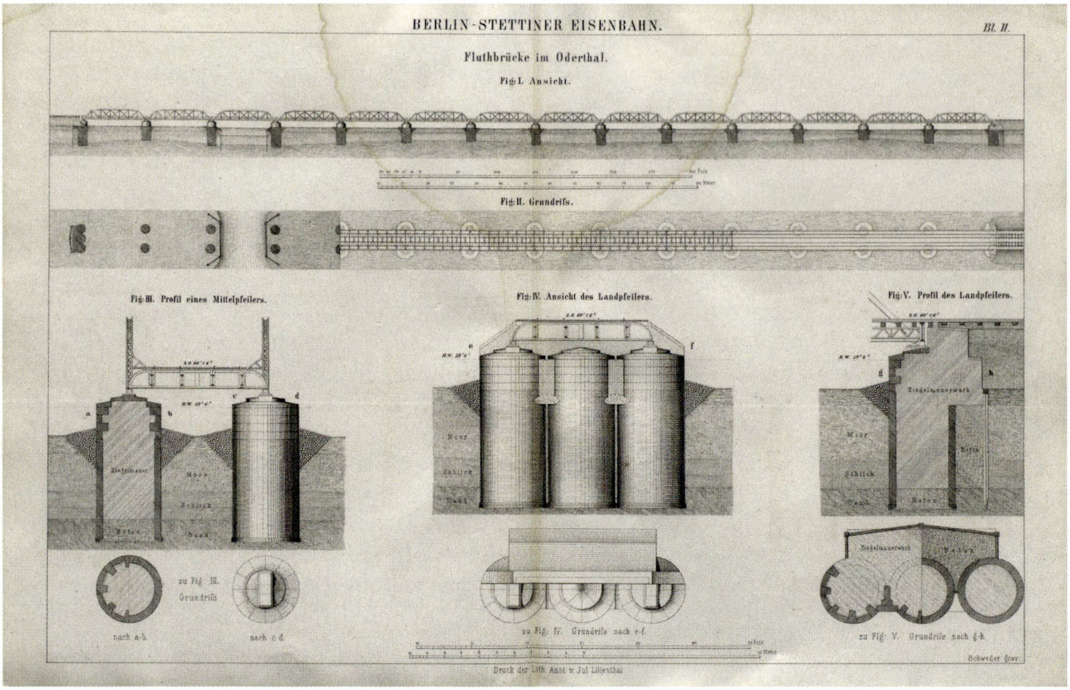

184. Th. Stein, the improvement of the Stettin-Stargard railway line, Stettin, 1864–1869. The *Flüthbrücke* (bridge over the tidal land). Front elevation and plan of the bridge, front elevations, sections and plans of a pier and of an abutment (Th. Stein, *Erweiterungsbauten der Berlin-Stettiner Eisenbahn ausgeführt 1864–1869*, Berlin, Ernst & Korn, 1870).

Nevertheless, after having successfully verified the bearing capacity of the soil by means of a test-pier standing on an ordinary masonry foundation, the plan for the construction of the sunken wells is abandoned, and the piers are actually built on ordinary masonry foundations.[120] Instead, the main pier of the swing bridge over the watercourse called *Kahnfahrt* (east of the *Fluthbrücke*) is built on a sunken well foundation. It consists of a cylindrical wall standing on a crown made of stone plates, which are kept together by means of curved iron bars. As already observed in the case of the *Fluthbrücke*, concrete is poured only at the base of the well, whereas the remaining part of the pillar is filled with brick masonry. Alternately, the right abutment of the same bridge over the *Kahnfahrt* stands, on a roughly 7 meters deep foundation made of concrete poured inside an enclosure of sheet piles, and the piers of the nearby bridge over the waterway called *Zeglin* stand on an ordinary timber piling foundation, with concrete poured among the pile heads.[121]

Building a new central freight depot is also part of the improvement project for the Berlin-Stettin-Stargard railway. Carried out between 1864 and 1869, it involves the construction of a dock along the River Parnitz, to enable the connection between water and rail transports. This dock is built on foundations consisting of square-shaped sunken wells, with each sidewall measuring 16 feet in length and 21 inches in thickness, standing 8 feet from each other.[122] They are filled with concrete and masonry, and are anchored to each other by means of rails. Two-stones-thick arches cover the space between the wells, and the remaining part of the dock is built in ordinary masonry.

[120] *Ibid.*, pp. 15–16. This change to the original project mislead Hagen, who, in 1870, asserts that the Viaduct of the *Silberwiese* is built on sunken well foundations; see Hagen, *Brunnen, Wasserleitung und Fundirungen*...cit., p. 365.

[121] Haselberg, *Die Brücken der Berlin-Stettiner Eisenbahn*... cit., coll. 362–363.

[122] See Stein,

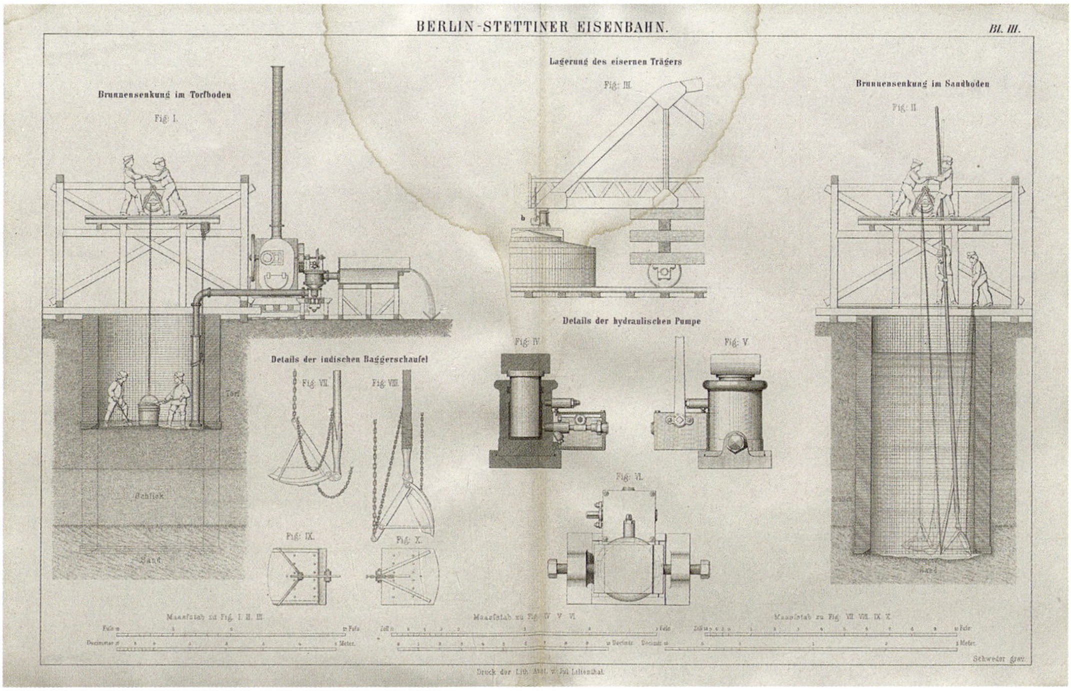

185. Th. Stein, the improvement of the Stettin-Stargard railway line, Stettin, 1864–1869. Construction drawings of a well foundation, digging tools, a hydraulic pump, beam placement (Th. Stein, *Erweiterungsbauten der Berlin-Stettiner Eisenbahn ausgeführt 1864–1869*, Berlin, Ernst & Korn, 1870).

Hagen defines the square-shaped sunken wells as *Senkkasten*, sinking caissons, which should not be confused with the caissons for underwater foundations that Bélidor describes in his *Architecture hydraulique*.[123] During the 1860s, the term *Senkkasten* is also used to describe a lighter version of the masonry square-shaped sunken wells, which actually refer to preassembled closed timber walls that are sunk into the soil as the excavation of the soil continues. They are used to build the foundations of ordinary buildings, in case of shallow bearing soil, as it happens for the locomotive shed of the Berlin-Potsdam-Magdeburg railway, which is built between 1863 and 1864 in Berlin.[124] The entire building consists of a main semi-circular shed standing along the rails and housing a turntable with thirteen places for locomotives, an octagonal tower accommodating facilities for workers, and a smaller staircase tower. The vertical bearing structures consist of masonry pillars standing on masonry footings, which, in turn, stand on concrete bases. For precautionary reasons, the foundation pits along the rails are shored up with 16 feet-high sloping sheet piles along the rails, while the other sides are provided with ordinary timber shoring works. All other foundation pits, however, are built adopting the method of the sinking caissons. At first, a kind of caisson made of poles, wind bracing and horizontal boards, is used. This proves to be weak, and so it is replaced with a stronger one, which is lower and made of 1 ½ inch-thick vertical boards held together by horizontal boards that form two bands around the entire inner surface of the caisson. The largest caisson is the octagonal-shaped one, which is assembled for the foundation of the staircase

Erweiterungsbauten der Berlin-Stettiner Eisenbahn... cit., p. 53.

[123] See Hagen, *Brunnen, Wasserleitung und Fundirungen*...cit., p. 364.

[124] See H. Weise, *Locomotiv-Haus der Berlin-Potsdam-Magdeburh Eisenbahn in Berlin*, in, "Zeitschrift für Bauwesen," XV, 1865, coll. 435–458.

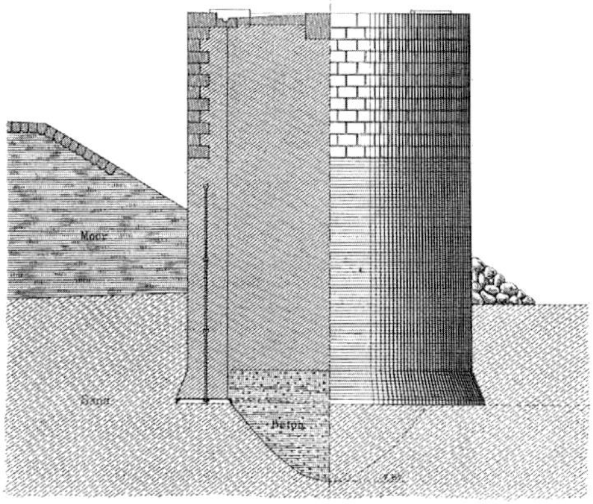
Moor
Sand
Beton

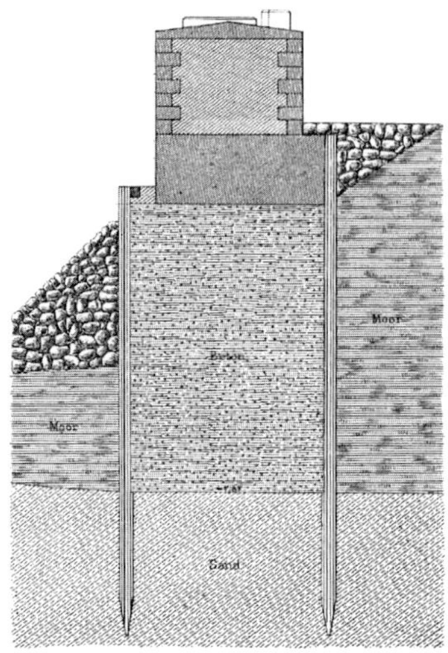
Moor
Beton
Moor
Sand

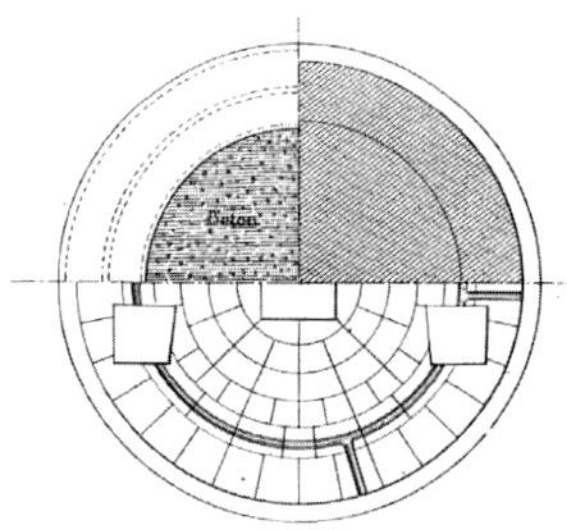
Beton

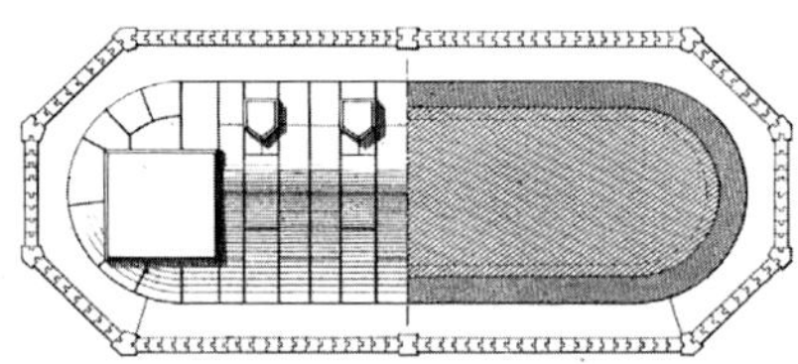

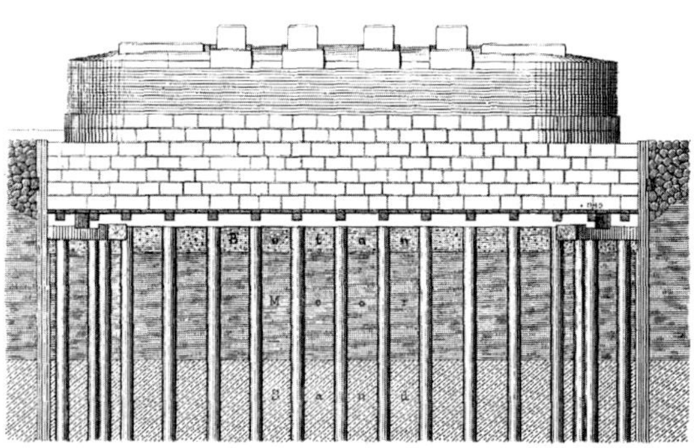
B e t o n
M o o r
S a n d

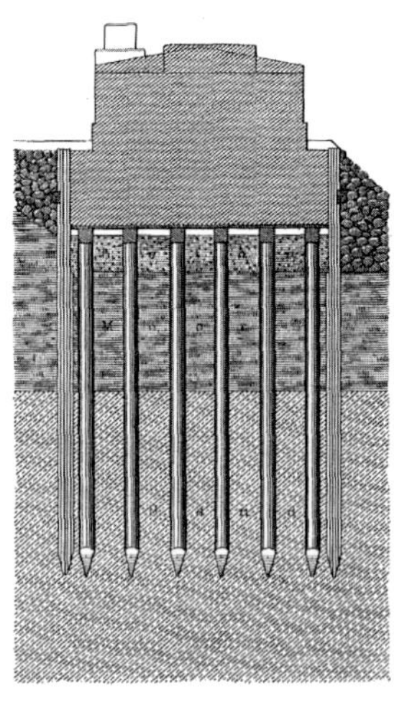

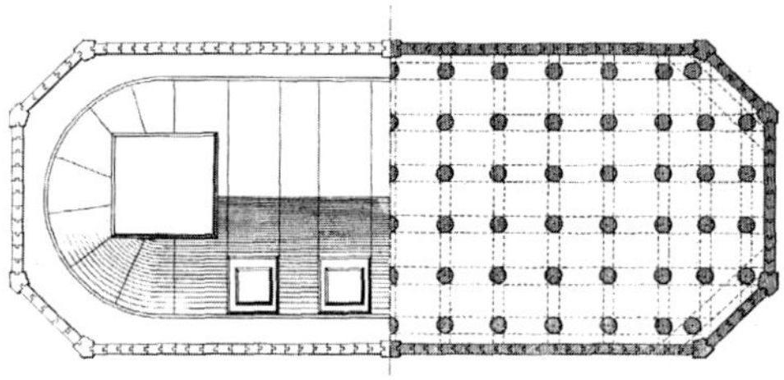

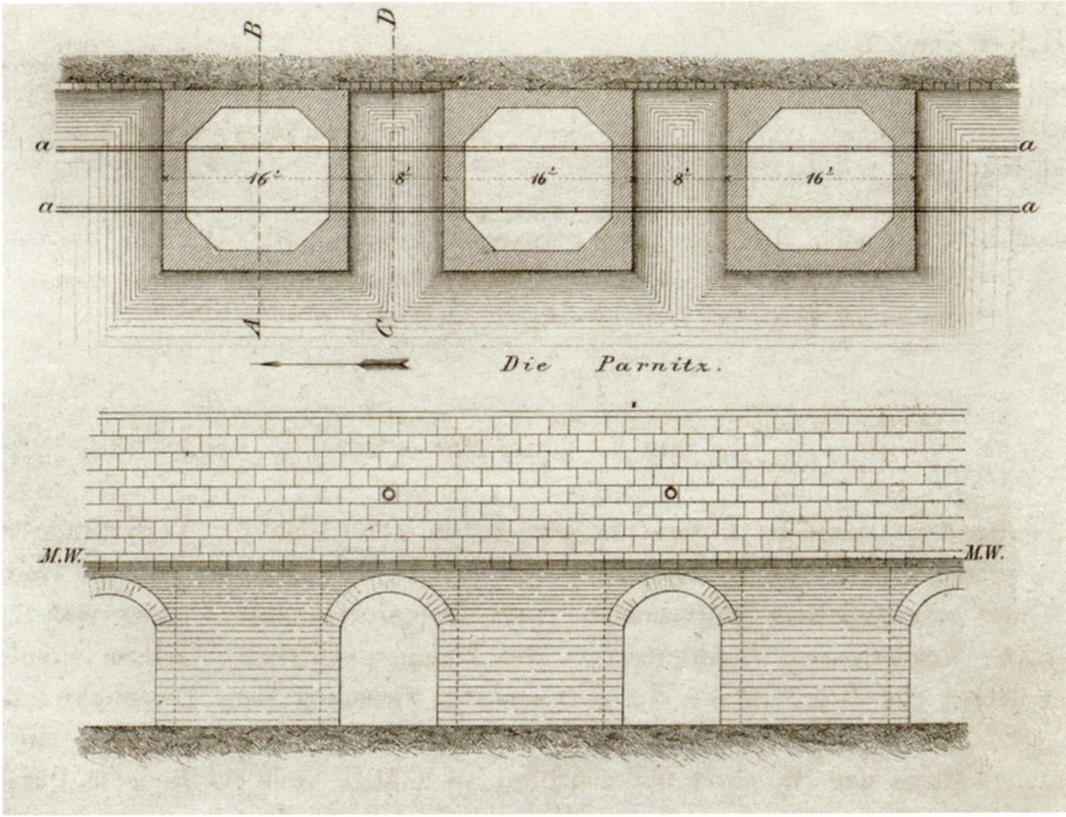

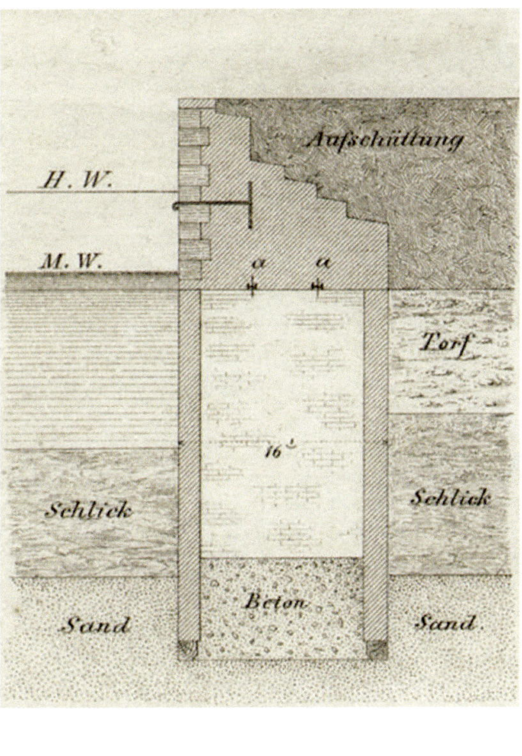

186. Th. Stein, bridge over the Watercourse *Kahfahrt*, Stettin, 1872–1878. Construction drawings of a pier, partial section and elevation, plan at different levels; construction drawings of the right abutment, section and plan at different levels (W. v. Haselberg, *Die Brücken der Berlin-Stettiner Eisenbahn im Oderthal bei Stettin*, in, "Zeitschrift für Bauwesen," XXIV, 1879).

187. Th. Stein, bridge over the Watercourse *Zeglin*, Stettin, 1872–1878. Construction drawings of a pier, longitudinal and cross section, plan at different levels (W. v. Haselberg, *Die Brücken der Berlin-Stettiner Eisenbahn im Oderthal bei Stettin*, in, "Zeitschrift für Bauwesen," XXIV, 1879).

188-189. Th. Stein, the dock along the River Parnitz, 1864–1869. A stretch of the foundation plan and elevation; cross section (Th. Stein, *Erweiterungsbauten der Berlin-Stettiner Eisenbahn ausgeführt 1864–1869*, Berlin, Ernst & Korn, 1870).

190-192. C. Hoffmann, A. Fromholz, H. Weise, the train station of the Berlin-Potsdam-Magdeburg Railway, Berlin, 1863–1864. Plan of the site; elevations, sections and plan of the locomotive shed with the tower; plan and sections of the foundations (H. Weise, *Locomotiv-Haus der Berlin-Potsdam-Magdeburh Eisenbahn in Berlin*, in, "Zeitschrift für Bauwesen," XV, 1865).

tower. Concrete is poured at the bases of the caissons; it consists of Portland cement, sand and crushed stones.

Quite large sinking caissons, entirely filled with concrete, are also used to build the foundations of the *National-Galerie* in Berlin, which is designed by Stüler from 1862 to 1865 (the year Stüler dies), and constructed between 1866 and 1876.[125] The Austro-Prussian War of 1866 stops the preliminary work for a short period, during which time some tests are done to verify the hardening of different kinds of cement underwater along with the compression strength of several specimens – some made of concrete and some others made of bricks and cement–based mortar.[126] Works resume at the end of the war and caissons are driven into the soil, which is dug out and replaced with concrete. On the basis of the outcomes of the strength tests from 1866, concrete is produced using Portland cement from the *Stettiner Portland-Cement Fabrik*, sand and stone rubble. It is mixed in hand-powered drums and is lowered down into the caissons by means of usual boxes. The entire foundation is achieved in the summer of 1867.

Huge sunk caissons made of masonry and entirely filled with concrete are conceived by the Hamburg engineer Johannes Christian Wilhelm Dalmann (1823–1875), in order to build the masonry dock on the southern side of the *Sandthor* dock, between 1866 and 1872, after having observed a similar kind of foundation while visiting the port of Saint–Nazaire in France in 1854.[127] Although they are square-shaped, Dalmann uses the expression *Senkbrunnen*, sunken wells, to indicate such constructions. The building of a masonry dock in the port of Hamburg is unusual at that time, as most constructions are made of timber. Dolphins, which consist of a number of piles driven into the riverbed and connected above the water level, are the most typical structure to provide a fixing point for ships, while barges transport goods to and from the riverbanks. Dalmann himself has built the northern quay of the *Sandthor* dock, which is

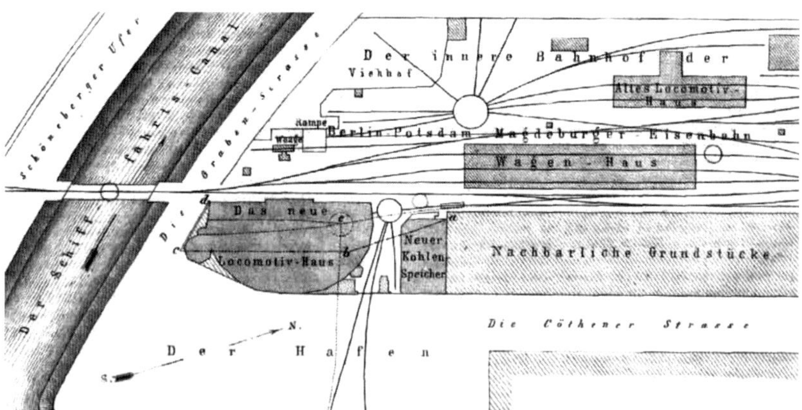

[125] See Georg Gustav Erbkam, *Die Königliche National-Galerie*, in, "Zeitschrift für Bauwesen," XIX, 1869, coll. 265–282; Hermann Zwick, ed., *Jahrbuch über die Leistungen und Fortschritte auf dem Gebiet der practischen Baugewerbe*, vol. I, Leipzig, Carl Scholtze, pp. 253–254. The gallery is built under the supervision of the master builders Carl Ferdinand Busse (1802–1868), Rienicke and Johann Heinrich Strack (1805–1880).

[126] See Erbkam, *Die Königliche National-Galerie*…cit., coll. 270–276.

[127] See Johannes Christian Wilhelm Dalmann, *Hafen-Anlagen in Frankreich und Holland*, in, "Zeitschrift für Bauwesen," (VI, 1856, coll. 361–392, VII, 1857, coll. 313–348), VII, 1857, coll. 329–339.

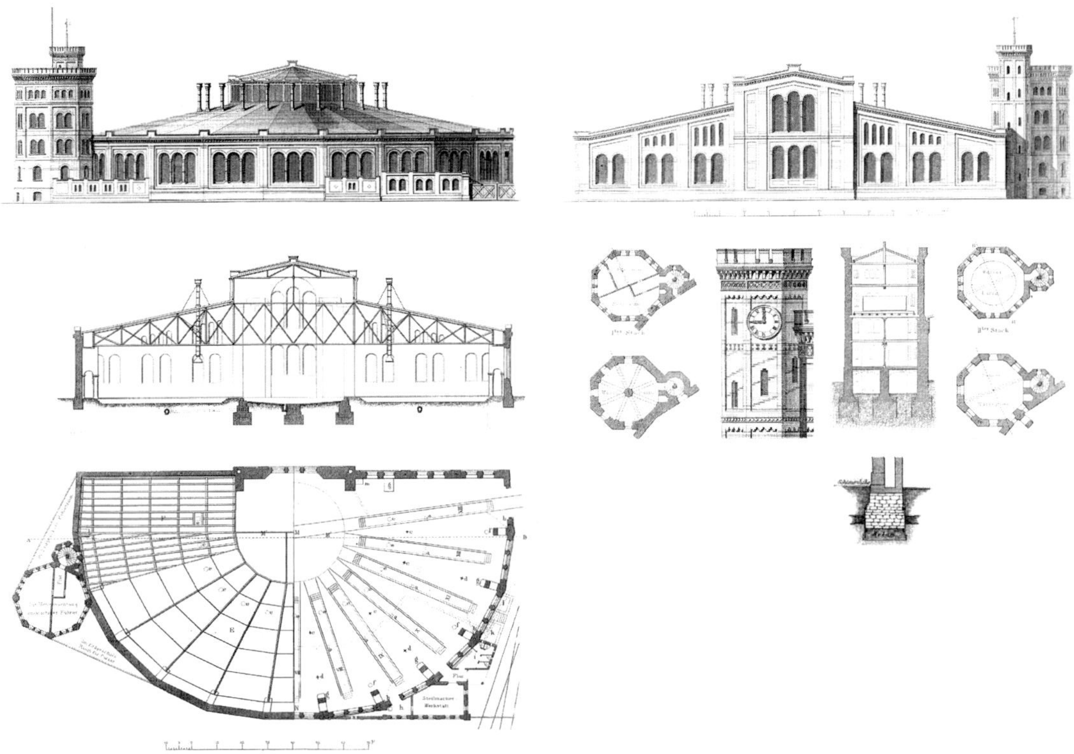

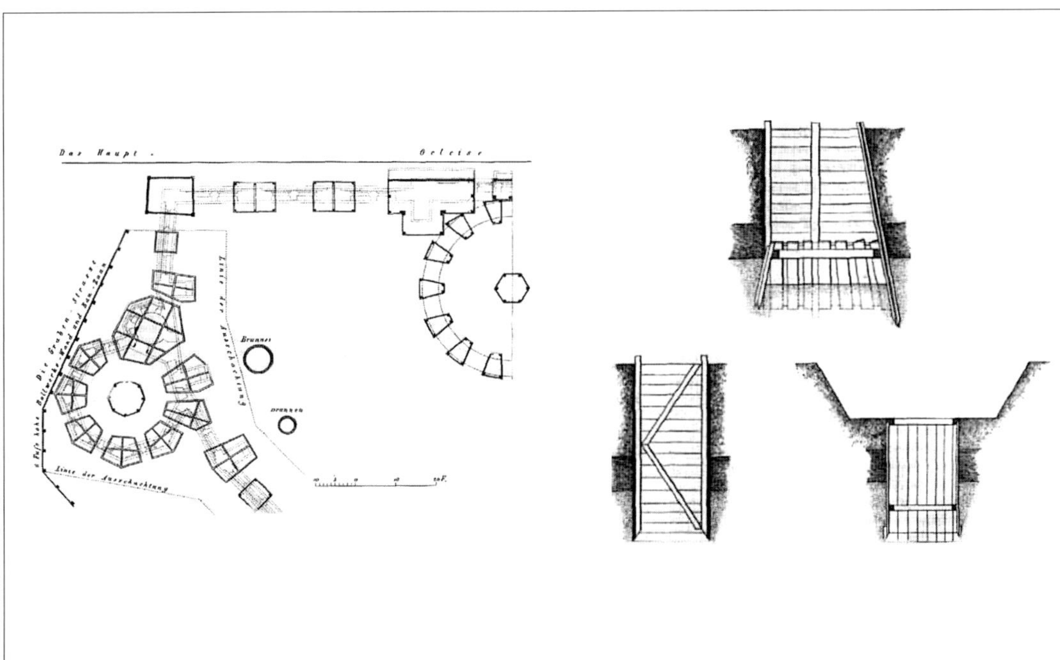
Das Haupt -
Geleise
Brunnen
Linie der Ausschachtung
Die Graben Strasse

opened in 1866, as a traditional timber structure on piles. Small concrete bases are built under the footing of the masonry walls that stand inside and on the back of the sheds, whereas the front walls, along the waterside, are made of timber and stand on foundations made of masonry footings and timber piles.[128] The southern quay, which is 1870 meters long and supports ten sheds, several rails and thirty-seven cranes powered by steam engines, is made entirely of brick masonry standing on the already mentioned sunken well-foundations. The walls of the wells are made of bricks bound in hydraulic mortar, and are connected at the top by brick masonry vaults. According to Hagen, the building of the wells for the *Sandthor* quay involves numerous difficulties, since several wells often slope on one side while sinking into the ground and demand considerable efforts to be straightened out. Despite that, none of them is damaged.[129]

An important corollary of the techniques to building deep underwater foundations concerns the use of different kinds of pressurized chambers that sink into the soil to build wells of considerable depth, which are then filled with concrete and brick masonry. The use of this building technique actually spreads in Germany around the mid-1860s, mainly to build the foundations of bridge piers, but the original concept essentially derives from the idea of the hollow pressurized iron pile that the British surgeon Laurence Holker Potts patents in 1843, and that the engineers William Cubitt and John Wright modify in 1851 to build the foundations of the piers for the Rochester Bridge.[130] Pott's hollow iron pile is opened at the lower end and closed at the top by a cap. It is connected to a pump that sucks the air and provokes a depressurization inside the pile, which descends into the soil by gravity, assisted

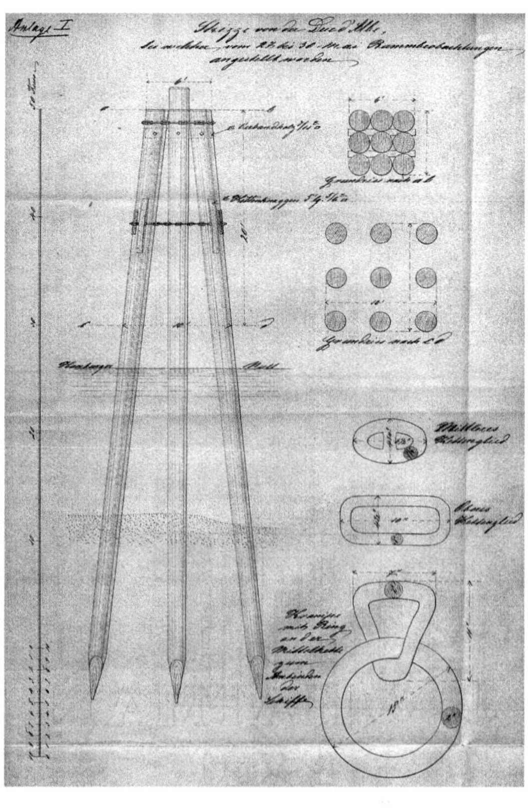

193. A typical dolphin of the Port of Hamburg (SAH, 326-2 I Strom und Hafenbau I, Bau und Unterhaltung der Landungsanlagen auf dem Grasbrook 1838–1863, n° 713, n. fol.).

[128] See Johannes Christian Wilhelm Dalmann, *Der Hafen von Hamburg-Altona, insbesondere die Anlage des Sandthor-Quais in Hamburg*, Berlin, Ernst & Korn, 1868, p. 6.
[129] See Hagen, *Brunnen, Wasserleitung und Fundirungen*...cit., p. 367.
[130] *Ibid.*, pp. 374–413.
[131] See J.-B. Bordot, *Travaux du Pont de Kehl*, in, "Nouvelles annales de la construction," V, 1859, coll. 18–20; Charles Alfred Oppermann, *Travaux du pont du Rhin à Kehl*, in, "Nouvelles annales de la construction," V, 1859, coll. 155–160; J. G. Schwedler, Hipp, *Der Rheinbrückenbau bei Kehl*, in, "Zeitschrift für Bauwesen," X, 1860, coll. 7–46; Emile Vuigner, Emile Fleur de Saint-Denis, *Pont sur le Rhin à Kehl. Détails pratiques sur les dispositions générales et d'exécution de cet ouvrage d'art*, Paris, Dunod, 1861.
[132] About the pneumatic tube by Pott and its relation with pneumatic system developed in Kehl, see Auguste Perdonnet, *Notions générales sur les chemins de fer*, Paris, Lacroix et Baudry, 1859, pp. 222–232.

by atmospheric pressure, while the loosened soil that is collected at the same time inside the pile is pumped out. Once this is achieved, the cap is removed, a new pile is set at the top and the procedure is repeated. Cubit and Wright invert the operating principle of Pott's pile by inflating air, instead of extracting it. The pressure inside the pile prevents groundwater from rising, while workers dig the soil from inside and bring the debris out in buckets, passing through a decompression chamber. A similar technique is used in France to dig wells for mining activities, and the French engineers of the *Compagnie des chemins de fer de l'Est* (East Railway Company), lead by Émile Vuigner, develop a further variation when they build the foundations of the bridge over the Rhine in the town of Kehl, at the boundary between France and the Grand Duchy of Baden, between 1858 and 1861. The bridge is intended to connect the French eastern railways with those in Baden, and the construction of it is therefore based on an international agreement, establishing that the French railway company takes charge of building the piers with the relative foundations, whereas the Baden railway company should construct the bridge deck.[131] Following this accord, the building site for the piers of the bridge in Kehl becomes a privileged observatory for German master builders to learn how to build foundations by means of pressurized chambers. The French engineers combine the concept of the pneumatic pile with that of the diving bell and thereby develop the so-called *Luftkasten* (pressurized caissons).[132] These are rectangular-shaped cast iron caissons, which are open at the bottom, like large-sized diving bells. They are provided with three different pipes on the top, two for the inflation of air and for workers' access through decompres-

194. Anon., Hamburg, view of the port at the Ladungsbrücken. A ship moored at a dolphin is visible in the foreground (Anon., *Hamburgisches Album: eine Sammlung der interessantesten Ansichten von Hamburg und dessen Umgegend; treu nach der Natur gezeichnet und in Stahl gestochen von den berühmtesten deutschen und englischen Künstlern*, Hamburg, Berendsohn, [1860]).

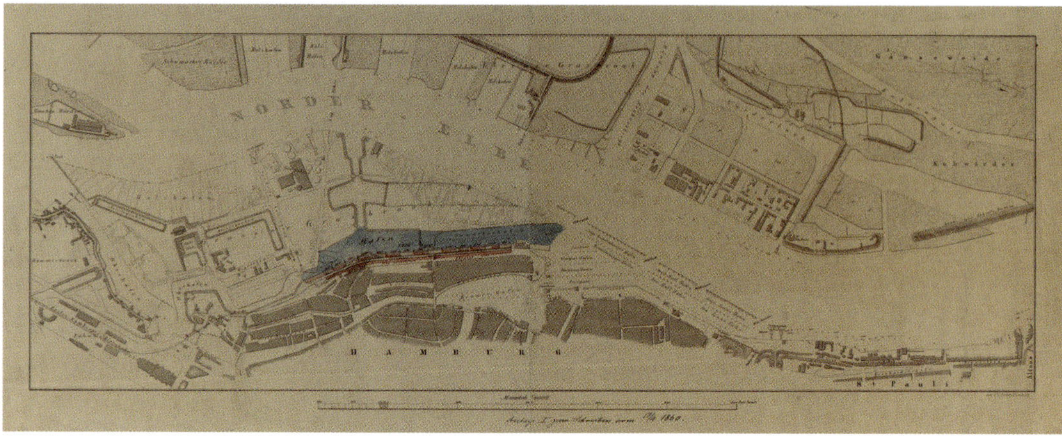

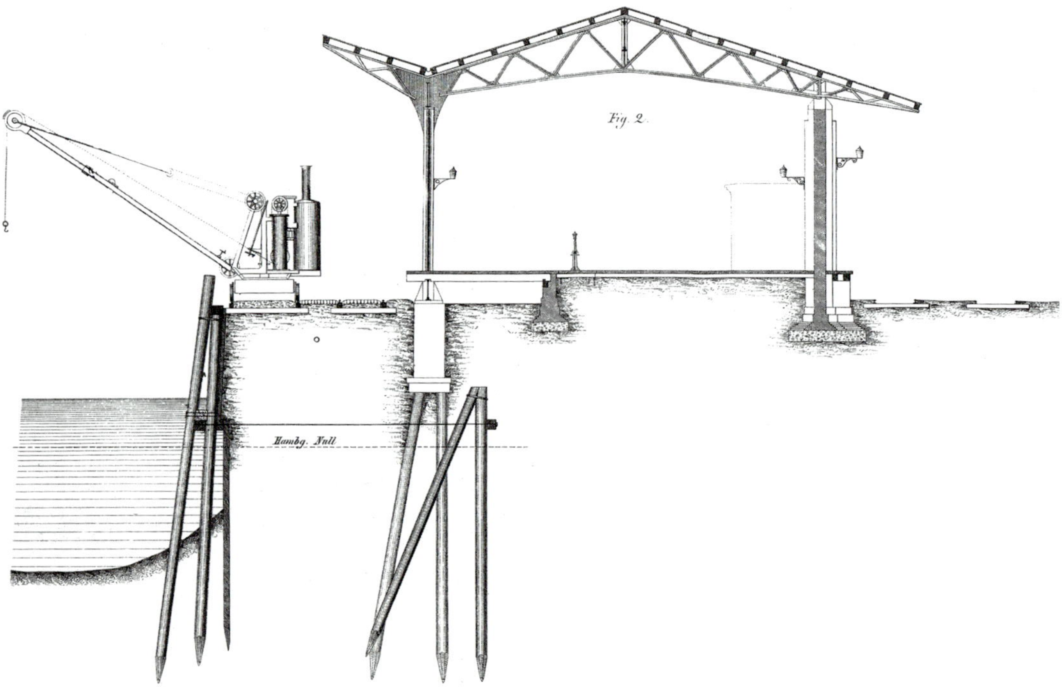

195-196. J. C. W. Dalmann, the *Sandthor* Dock, northern quay, Hamburg, 1860-1866.

195. Plan of the site (SAH – Kartensammlung: 720–1, 126–1, Häfen und Flüsse, Hafenpläne 5, pl. 186–01).

196. Section of the quay and of a shed (J. C. W. Dalmann, *Der Hafen von Hamburg-Altona, insbesondere die Anlage des Sandthor-Quais in Hamburg*, Berlin, Ernst & Korn, 1868).

197-198. J. C. W. Dalmann, the *Sandthor* Dock, Hamburg, southern quay, 1866–1872.

197. Constrution drawings. A stretch of the quay, elevation, plan, sections (SAH – Kartensammlung: 720–1, 126–1, Häfen und Flüsse, Hafenpläne 5, pl. 186–75 III).

198. The base of a crane (SAH – Kartensammlung: 720–1, 126–1, Häfen und Flüsse, Hafenpläne 5, pl. 187– 12).

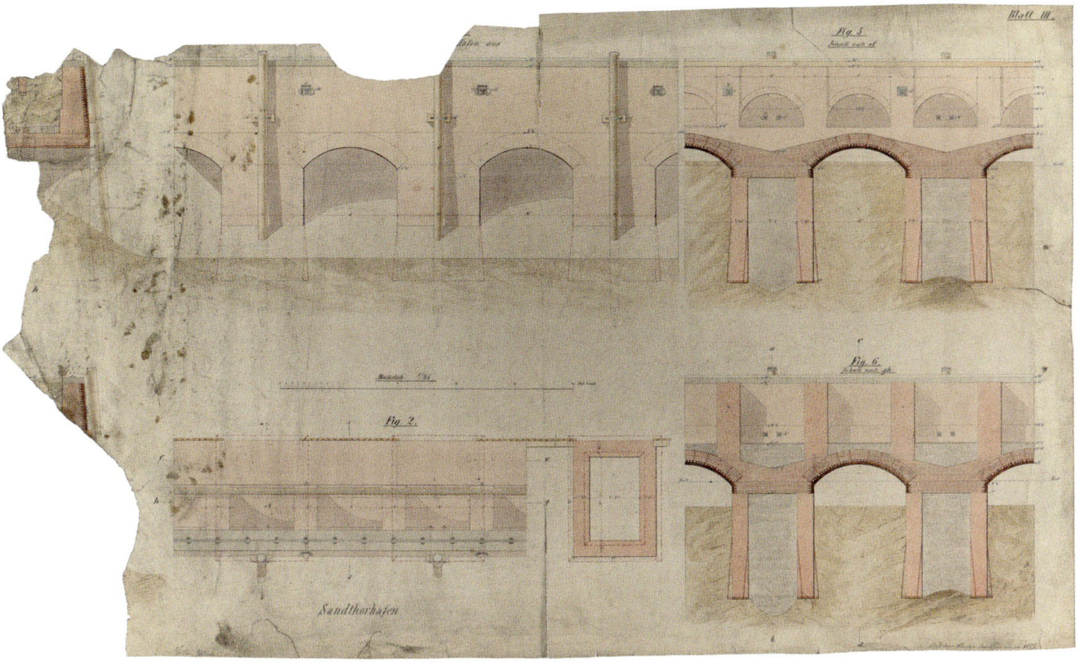

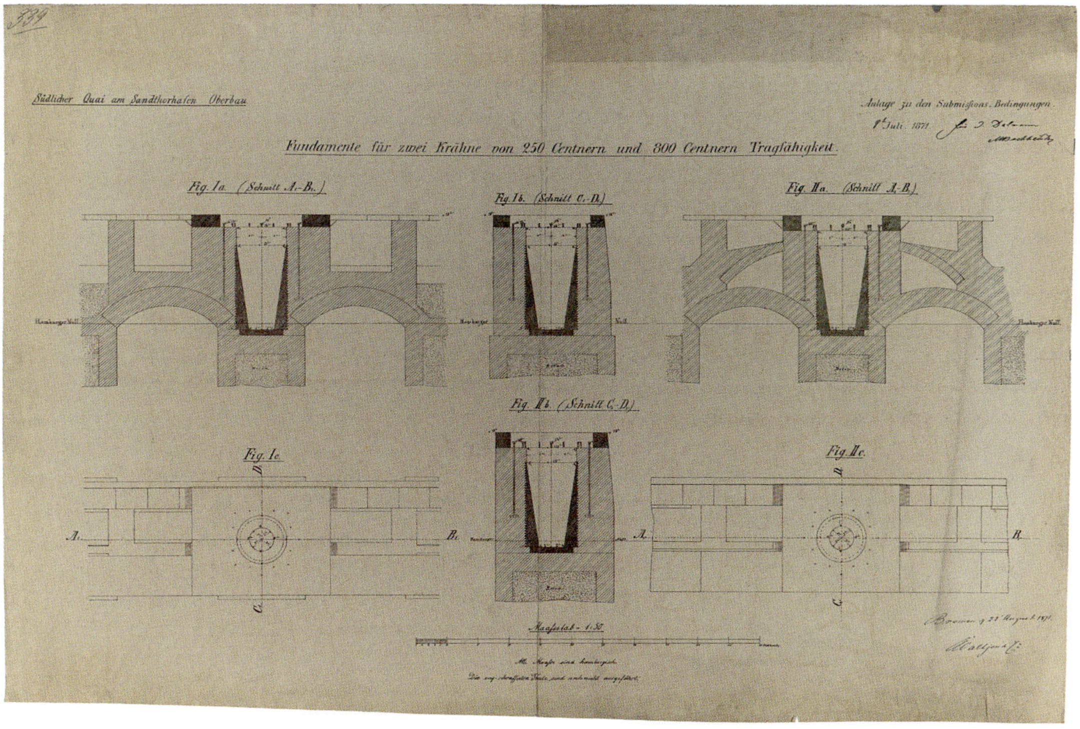
Südlicher Quai am Sandthorhafen Oberbau
Anlage zu den Submissions-Bedingungen.
Fundamente für zwei Krähne von 250 Centnern und 800 Centnern Tragfähigkeit.
Fig. Ia. (Schnitt A-B.)
Fig. Ib. (Schnitt C-D.)
Fig. IIa. (Schnitt A-B.)
Fig. IIb. (Schnitt C-D.)
Fig. Ic.
Fig. IIc.

199-202. E. Vuigner, E. Fleur de Saint-Denis, F. Keller, bridge over the River Rhine, Kehl, 1858–1861.

199. View of the bridge, photograph, early 20th century (personal collection).

200-202. Plans of the site; half elevation and half plan of the bridge; elevation and plan of the piers under construction (J. G. Schwedler, Hipp, *Der Rheinbrückenbau bei Kehl*, in, "Zeitschrift für Bauwesen," X, 1860).

sion chambers, whereas the third pipe stands in water to regulate pressure, and it is provided with a rotating digging machine to extract the soil dug by the workers. Four contiguous caissons, each one measuring 7 meters in length, 5.8 meters in width and 3.6 meters in height, are built up and placed in a square to build the foundations of the abutments, whereas the caissons for the freestanding piers measure 5.5 square meters at the base, and are 3.6 m. high. Three caissons are placed in line to build the foundation of each central pier, whereas four caissons in line are necessary to build the foundations of the two other piers, close to the abutments. The building site of each pier is surrounded by roof-covered scaffoldings; they house all the necessary installations, namely the steam-engine, the pumps, a bridge crane and the system made of jacks, screw and metal bars to suspend the caissons so that they sink both gradually and regularly. Workers in each caisson dig the soil and push the debris toward the central pipe, into a dip filled with water, where the digging machine collects them in buckets that go up to the surface. Contiguous caissons sink simultaneously, while closed shoring walls, made of timber and enveloped in thin metal sheets to reduce the friction with earth, are regularly built on the caissons as they sink, forming enclosures that are then filled with stone-faced walls and concrete. The concrete is made of artificial cement from Colmar, sand and pebbles from the Rhine. It is produced on the scaffolding, poured into the highest timber enclosures and levelled in layers of about ½ – ¾ feet thickness.[133] When the caissons reach a depth of about 20 meters under the riverbed, workers gradually fill the inside spaces with brick-faced walls and concrete, before leaving the caissons in the pipes. These are later pulled out and the remaining hollow cylinders are also filled with concrete. The foundation of each pier finally consists of cast iron caissons filled with brick masonry and concrete, at a depth of about 20 meters, and an enormous stone-faced well filled with concrete.

In 1864, the only freestanding pier of the bridge over the Pregel in Königsberg is built using a single pneumatic caisson, which is similar to the caissons used in Kehl, and is therefore provided with a central pipe for the

[133] See Schwedler, Hipp, *Der Rheinbrückenbau bei Kehl*...cit., col. 26.

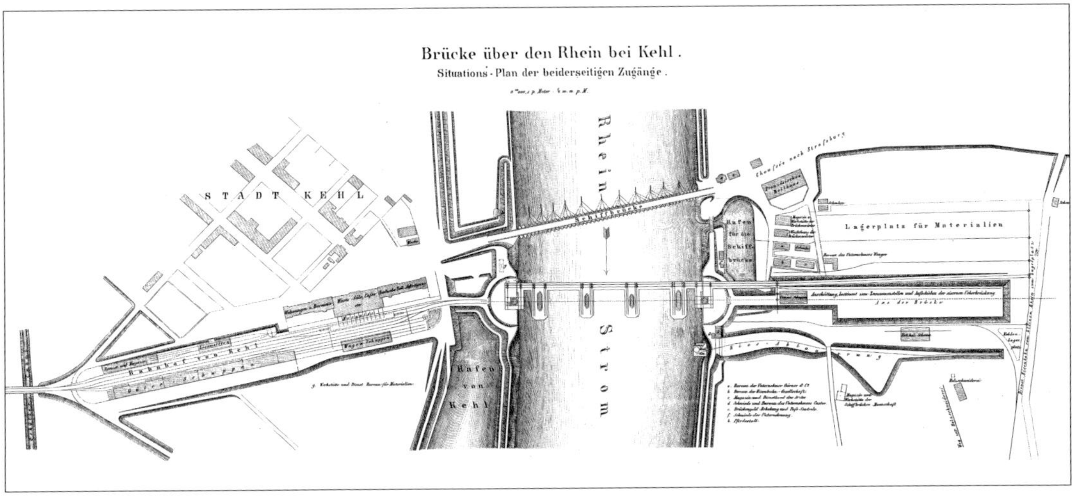
Brücke über den Rhein bei Kehl.
Situations-Plan der beiderseitigen Zugänge.
STADT KEHL
Rhein
Strom
Hafen von Kehl
Lagerplatz für Materialien

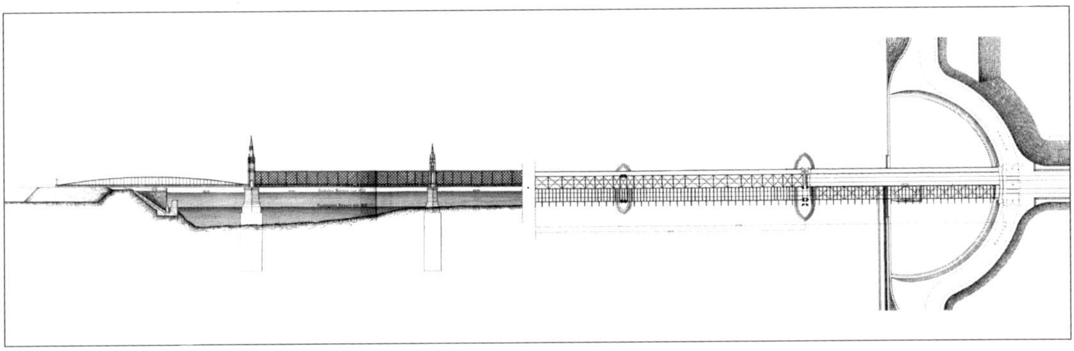

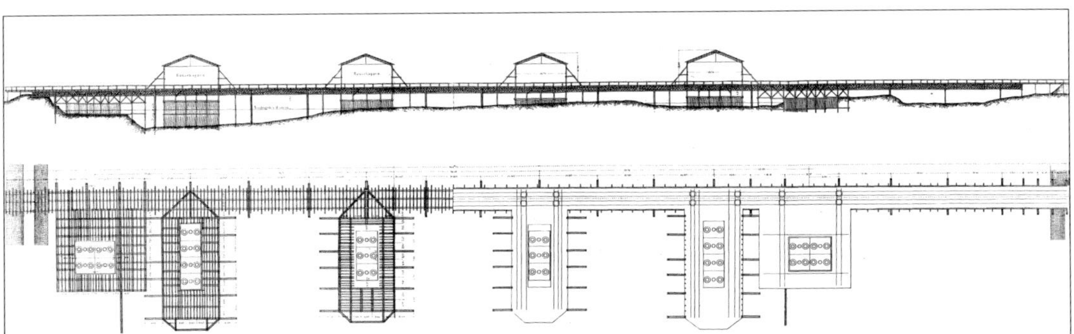

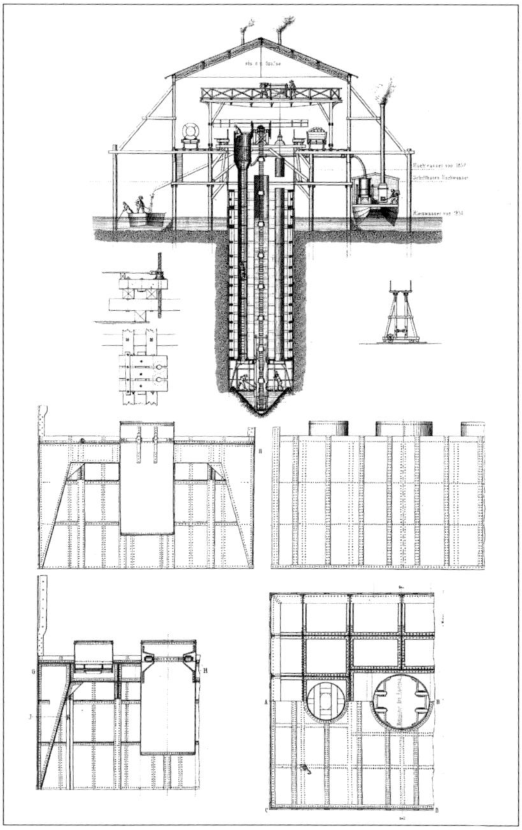

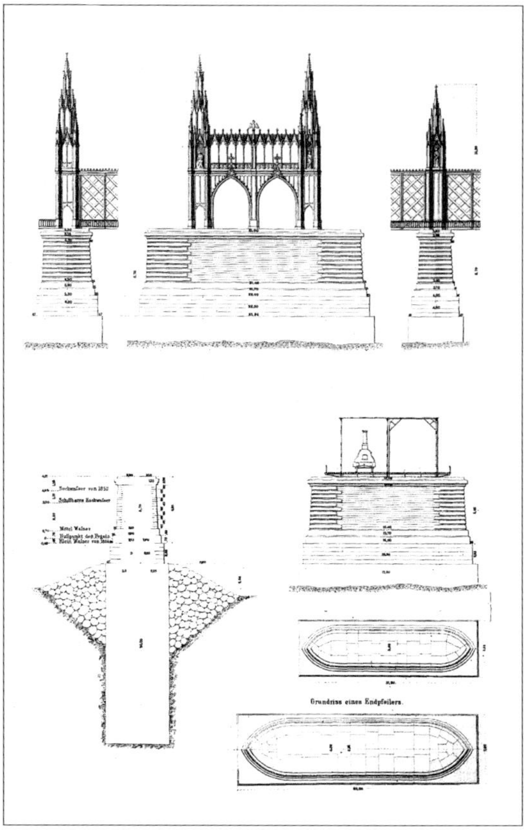

203-204. E. Vuigner, E. Fleur de Saint-Denis, F. Keller, bridge over the River Rhine, Kehl, 1858–1861. Construction of a pier on the French side, the digging process, drawings of the caisson; drawings of a pier, elevations, section, plans (J. G. Schwedler, Hipp, *Der Rheinbrückenbau bei Kehl*, in, "Zeitschrift für Bauwesen," X, 1860).

digging machine, and two lateral pipes for workers' access and air insufflation.[134] The building site is surrounded with scaffoldings that house the cranes, the jacks and the engine that powers the digging machine, while the pump is in a shed on dry land. It is powered by a steam engine, which, at the same time, moves the drums for the production of mortar and concrete inside the same shed. Unlike the case of Kehl, brick masonry is used here to fill the space inside and over the caisson, while concrete is just used to fill the hollow spaces that remain once the pipes are extracted.

Pressurized circular bells, instead of caissons, are used to build the foundations for the freestanding piers of the bridges over the Rivers Oder and Parnitz near Stettin that are constructed between 1866 and 1868, in the context of the already mentioned project for the improvement of the railway from Berlin to Stargard.[135] The two bridges are almost identical, and each consists of a central circular pier supporting a swing bridge, two freestanding piers standing on the riverbed and two abutments. The latter are built on ordinary sunken well foundations, each one composed of three wells, except for the left abutment of the bridge over the Oder, which is built on a stretch of the concrete foundation of the dock that had been built in 1844. The central freestanding piers of both

[134] See Löffler, *Die Fundirung der Eisenbahnbrücke über den Pregel in Königsberg*, in, "Zeitschrift für Bauwesen," XVI, 1866, coll. 517–536; Hagen, *Brunnen, Wasserleitung und Fundirungen*...cit., pp. 391–398.

[135] See Stein, *Erweiterungsbauten der Berlin-Stettiner Eisenbahn...* cit., pp. 6, 28–31.

[136] See Stahl, *Die Construction der Taucherglocken zur Fundirung der Rheinbrücke bei Hamm*, in, "Zeitschrift des Vereins deutscher Ingenieure," vol. XIII, 1869, coll. 185–188; Anon.,

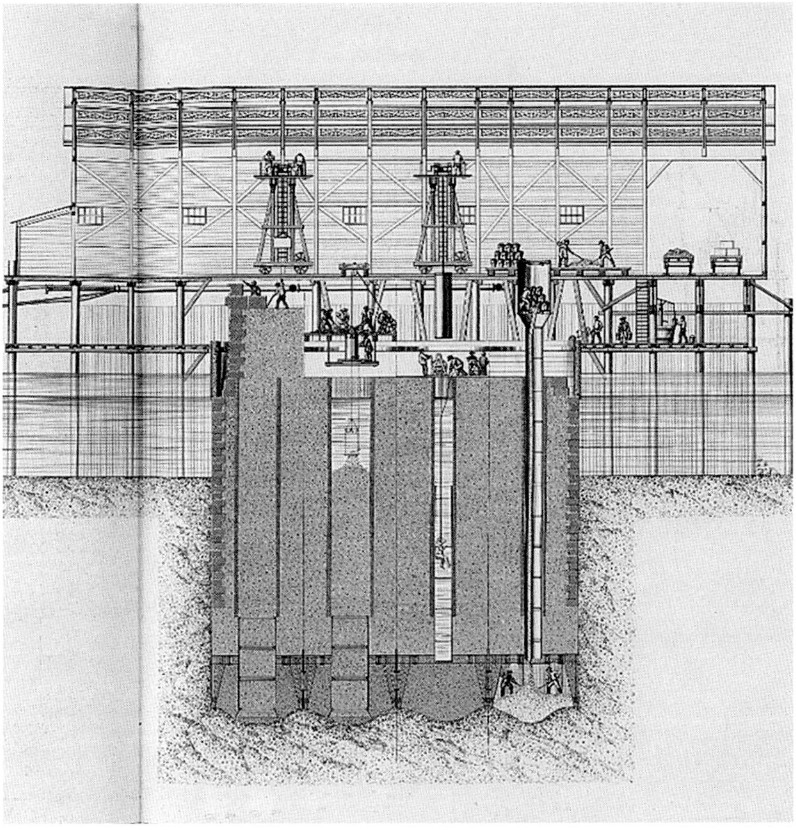

205. E. Vuigner, E. Fleur de Saint-Denis, F. Keller, bridge over the River Rhine, Kehl, 1858–1861. Construction of a pier on the Baden side, the pouring process (E. Vuigner, E Fleur de Saint-Denis, *Pont sur le Rhin à Kehl. Détails pratiques sur les dispositions générales et d'exécution de cet ouvrage d'art*, Paris, Dunod, Atlas, 1861).

bridges are built following the same procedure. The bell is suspended on scaffoldings by means of sixteen couples of chains made of metal bars connected one to each other, and hanging from the screws fixed in jacks. The pressurized bell is provided with two pipes and no digging machine is installed. The well for the central pier of the bridge over the Oder reaches a depth of 42 feet from the medium level of the gauge, whereas the one for the central pier of the bridge over the Parnitz reaches a depth of 25 feet. The wells over the bells are built in brick masonry, and they taper from the bottom to the top. Concrete is only poured onto the bottom of the bell after the well has reached the desired depth, forming roughly 4 ½ feet high bases, while all other hollow spaces are filled with brick masonry. A similar procedure is used to build the freestanding pillars on both sides of the circular ones. These, however, each stand on two wells, which are connected at the top by a masonry vault.

Pressurized bells are also used to build the foundations of two freestanding piers (namely the first two next to the right abutment) of the bridge over the Rhine that is built between 1868 and 1870 in Hamm near Düsseldorf, along the railway connecting Düsseldorf and Neuss.[136] A first plan suggests the use of two single, 9 feet high wrought-iron caissons, which are intended

Mittheilungen aus Vereinen. Verein für Eisenbahnkunde in Berlin, in, "Deutsche Bauzeitung," III, 1869, p. 135; Hermann Zwick, ed., *Jahrbuch über die Leistungen und Fortschritte auf dem Gebiet der practischen Baugewerbe*, vol. I, Leipzig, Carl Scholtze, pp. 324–27; Pichier, *Die Verbindungseisenbahn zwischen Düsseldorf und Neuss mit Ueberbrückung des Rheinstromes oberhalb Düsseldorf*, in, "Zeitschrift für Bauwesen," XXII, 1872, (coll. 237–250, 369–384), coll. 239–245.

206-210. Jackobi, Micks, foundation of the pier for the bridge over the River Pregel, Königsberg, 1864. Plan of the site; plan of the pier under construction; longitudinal and cross section of the pier under construction; the building site of the pier, photograph; longitudinal, cross sections and plan of the mortar and concrete plant (Löffler, *Die Fundirung der Eisenbahnbrücke über den Pregel in Königsberg*, in, "Zeitschrift für Bauwesen," XVI, 1866).

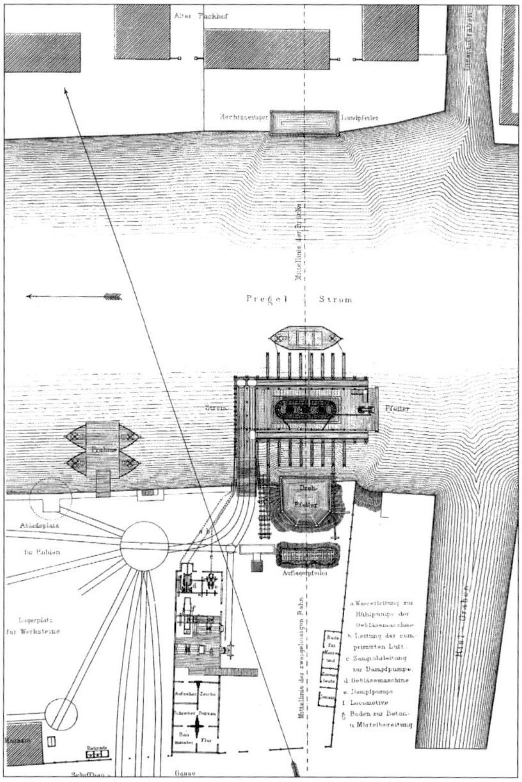

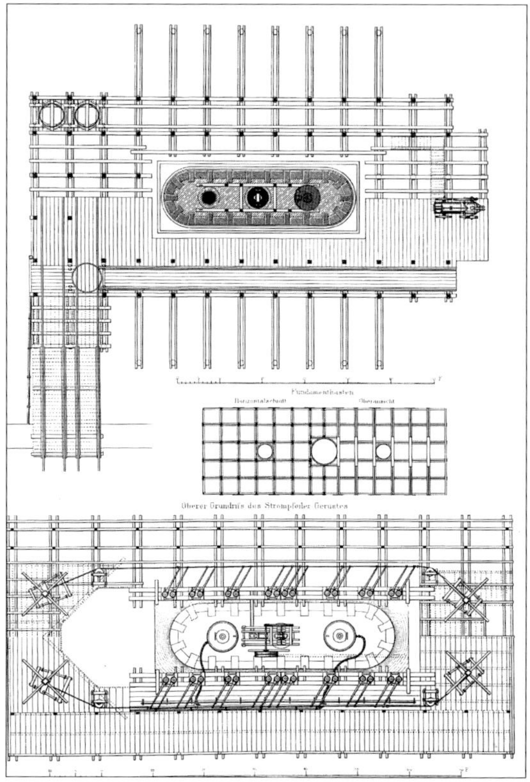

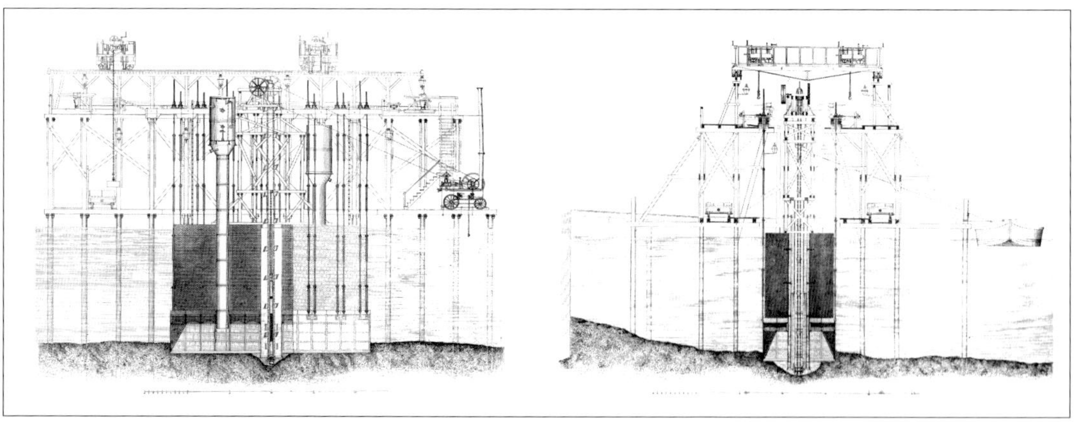

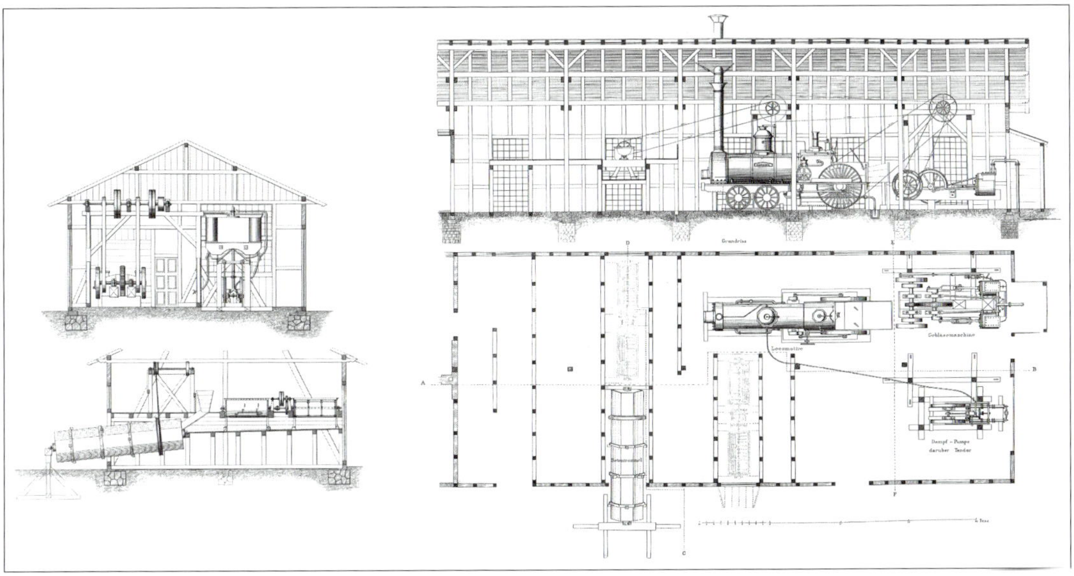
Grundriss
Locomotive
Gebläsemaschine
Dampf – Pumpe
darüber Tender

211-214. Th. Stein, the bridges over the Rivers Oder and Parnitz, Oder Valley, Stettin, 1866–1868. Front elevations and plans; drawings of the pressurized bells; construction of the central pier, section and plan; construction of a side pier section and plan (Th. Stein, *Erweiterungsbauten der Berlin-Stettiner Eisenbahn ausgeführt 1864–1869*, Berlin, Ernst & Korn, 1870).

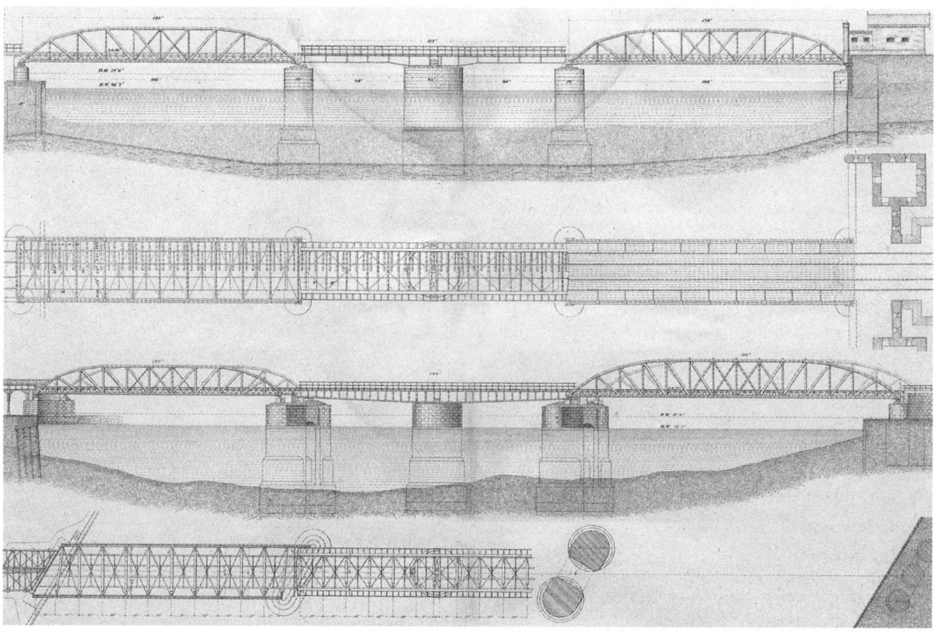

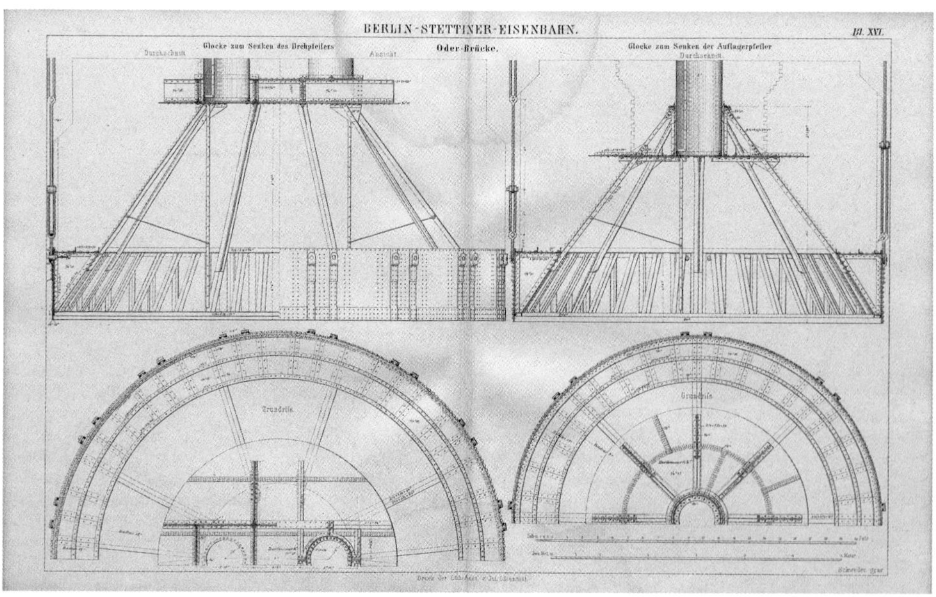

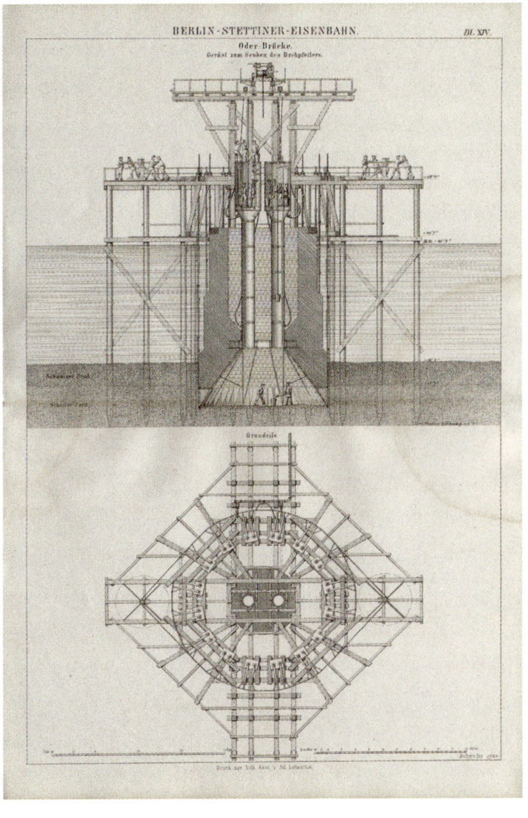
BERLIN-STETTINER-EISENBAHN.
Bl. XIV.
Oder-Brücke.
Gerüst zum Senken des Drehpfeilers.
Grundriss

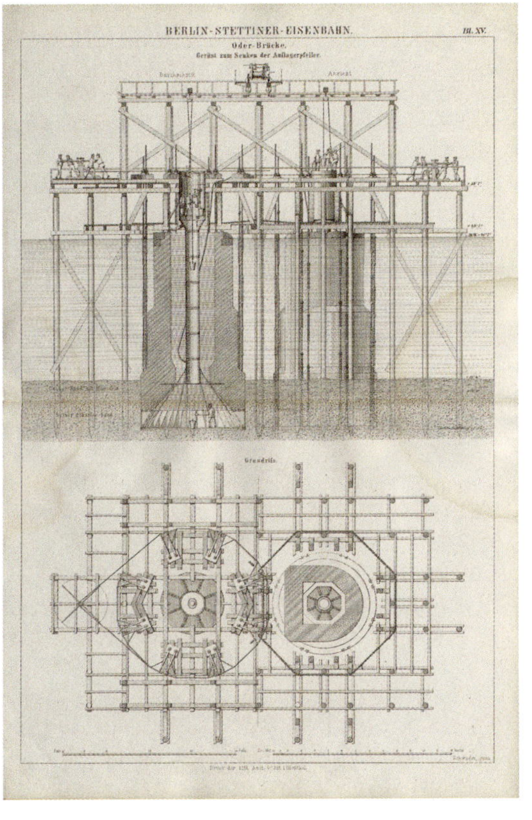
BERLIN-STETTINER-EISENBAHN.
Bl. XV.
Oder-Brücke.
Gerüst zum Senken der Auflagerpfeiler.
Grundriss

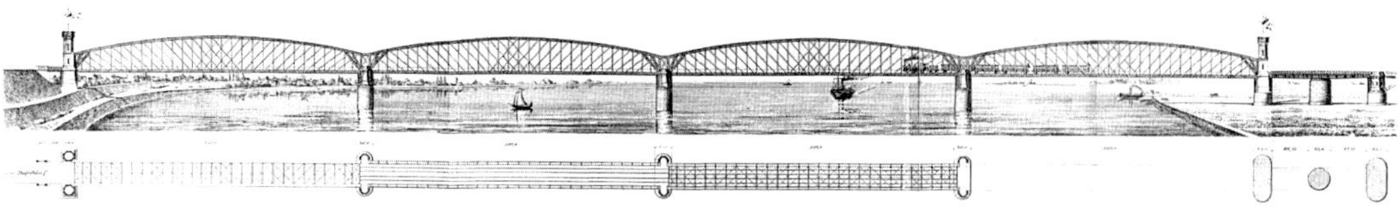

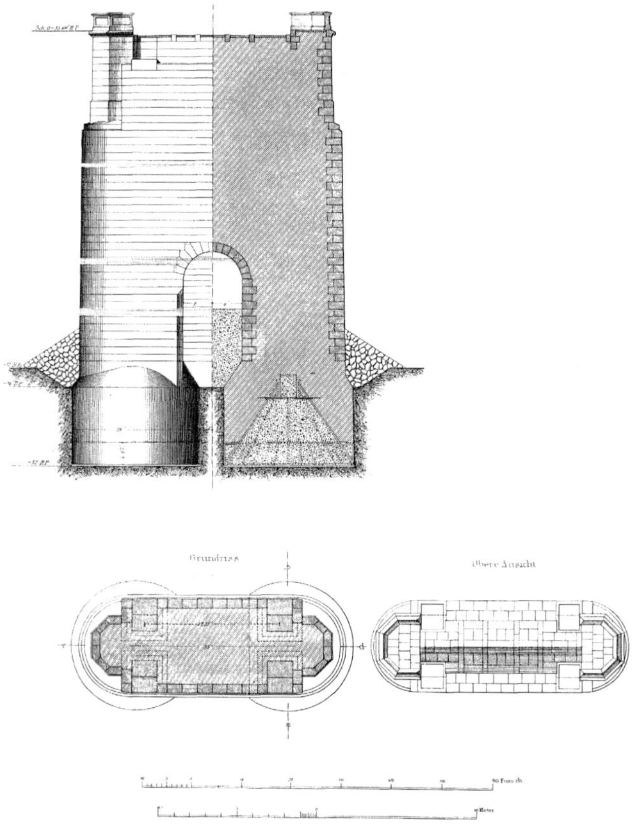
Grundriss
Obere Ansicht

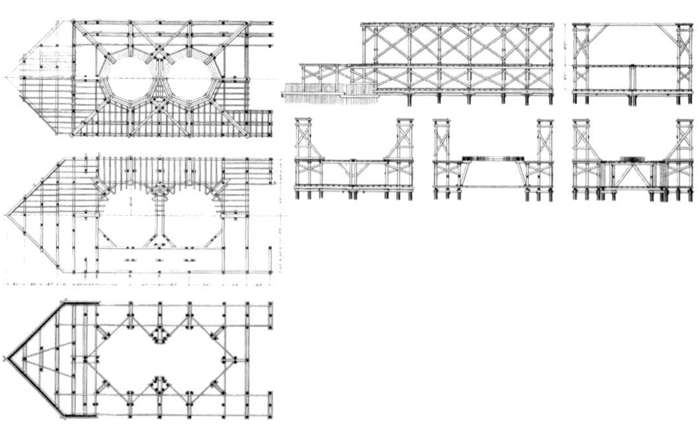

215-217. Bergisch-Märkische and Aachen-Düsseldorfer Railway Societies, bridge over the River Rhine, close to Düsseldorf, 1868-1879. Elevation and plan; half elevation and half, longitudinal section of a pier, plans at different levels; drawings of the scaffoldings (Pichier, *Die Verbindungseisenbahn zwischen Düsseldorf und Neuss mit Ueberbrückung des Rheinstromes oberhalb Düsseldorf*, in, "Zeitschrift für Bauwesen," XXII, 1872).

to gradually be covered with poured concrete inside enclosures. Since it is considered too demanding to build up and steer such caissons, this plan is set aside and two 26 feet-wide diving bells for each pier are used instead of caissons. They are sunk down from usual scaffoldings housing materials and installations.[137] The wells over the bells are filled with brick masonry, while concrete is poured inside the bells once they have reached the planned depth. Mortar and concrete are produced using cement from the *Bonner-Portland-Cement-Fabrik*, mixed with trass and sand. Concrete is also poured underwater into the space between the two well foundations of each pier, which is closed along the major sides by means of sheet piles; each pier essentially consists of two cylindrical pillars, each rising from a well, which are connected at the top by an arch. All other piers of the bridge in Düsseldorf are built on the far more common foundations made of concrete poured by means of a pipe inside enclosures of sheet piles, apart from the foundations of the right pier standing on land, which consist of masonry.[138]

At the turn of the 1870s, besides the development of pressurized caissons and bells, a number of foundations built adopt the technique of the ordinary sunken well. Again in the Oder valley near Stettin, the railway line Breslau-Schweidnitz-Freiburg is improved as of 1869, demanding the construction of five further bridges, among which, the one that crosses the Berlin-Stettin-Stargard railway line over the watercourse called *Kleine Reglitz* is partially built on sunken well foundations.[139] The bridge actually stretches over five spans, two of which are covered with masonry vaults, whereas the remaining three spans are covered with metal constructions. The abutments stand on foundations made of concrete poured inside enclosures of sheeting piles, whereas the two piers in the central part of the bridge each stand on two sunken wells.[140] Several bridges along the railway connecting Berlin, Potsdam and Magdeburg are also built on sunken well foundations, namely the viaduct over the *Schifffahrt* Canal in Berlin and the bridges over the Elbe near Magdeburg, near Biederitz and near Gommern. The first has piers standing along the embankments of the canal, on foundations composed of three different wells in line, which are built in a shored up and partially artificially buried site. Concrete is poured onto the bottom of the well, in spaces that are dug by means of the previously described Indian-spade. Similar constructions made of pillars standing on three sunken wells in line are also built for the other bridges (mentioned above) near Magdeburg, Biederitz and Gommern.[141]

The huge railway bridge that is built between 1868 and 1872 over the River Vistula and its tidal land, near the town of Thorn (today Torun), offers a further example of work partially realized on sunken well foundations. The bridge has two abutments and sixteen piers, ten of which are built, each on three wells in line. One stands in the riverbed of a minor branch of the Vistula, while all the others are built on the tidal land that stretches between the main course of the Vistula and the aforementioned minor branch. The wells are partially filled with concrete and partially with brick masonry. The piers standing

[137] *Ibid.*, col. 241.
[138] *Ibid.*, coll. 244–245.
[139] They are the bridge over the *Große Reglitz*, the bridge ove the *Kleine Reglitz*, a tidal bridge close to the goods station, a bridge passing under the road Stettin-Altdamm, and the bridge over the Parnitz; see Friedrich Eduard Salomon Wiebe, *Die Bauten der Breslau-Schweidnitz-Freiburger Eisenbahnim Oderthal bei Stettin*, in, "Deutsche Bauzeitung," IX, 1875 (n° 71, pp. 353–355, n° 73, pp. 363–365, n° 75, pp. 373–375), n° 73, pp. 363–365; Teichgräber, *Die Brücken im Oderthale*…cit.
[140] The piers of all other bridges in the Stettin Oder valley along the same railway line stand on timber piling foundations.
[141] See Julius Ludwig Quassowsky, *Ueber Fundirungen mit Senkbrunnen nebst Beschreibung einiger Fälle aus der Praxis*, in, "Zeitschrift für Bauwesen," XXIV, 1874 (coll. 297–312), coll. 304–312.

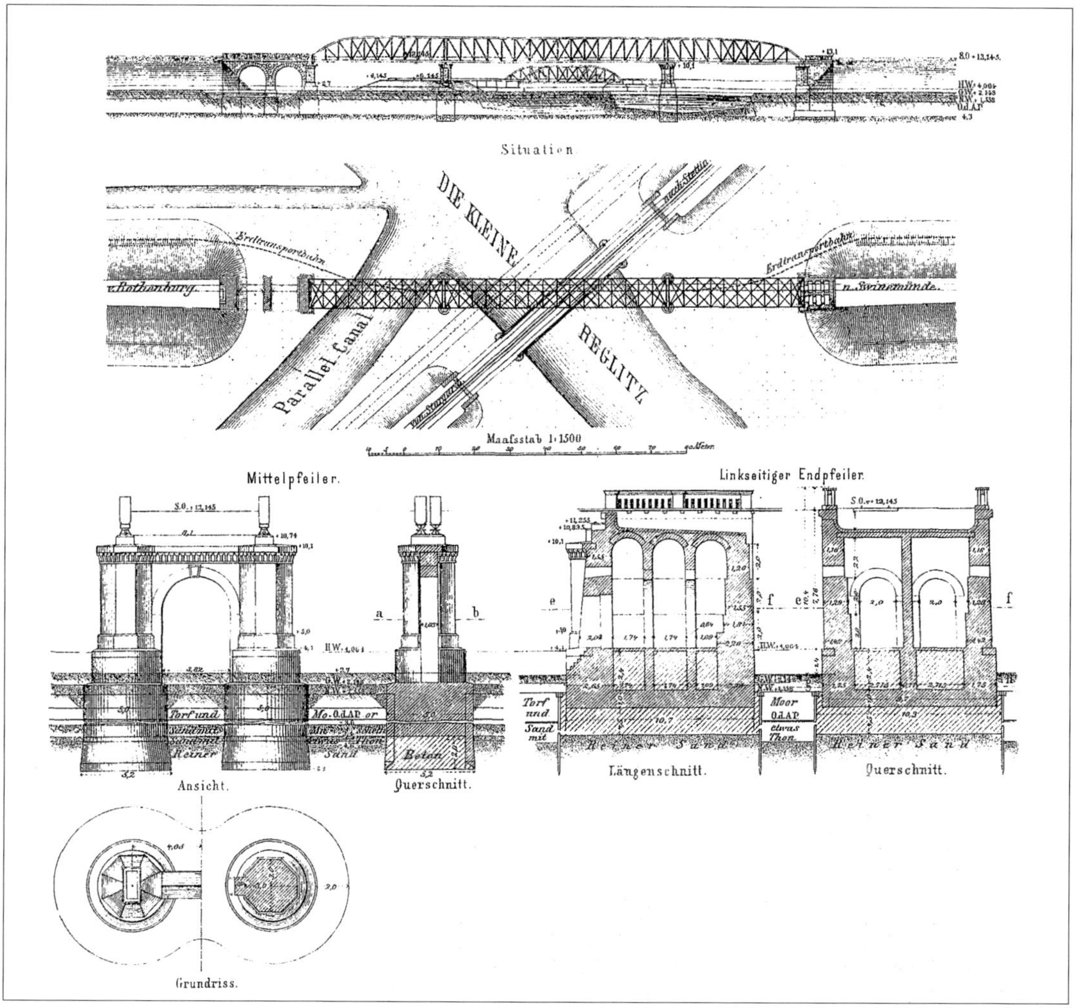

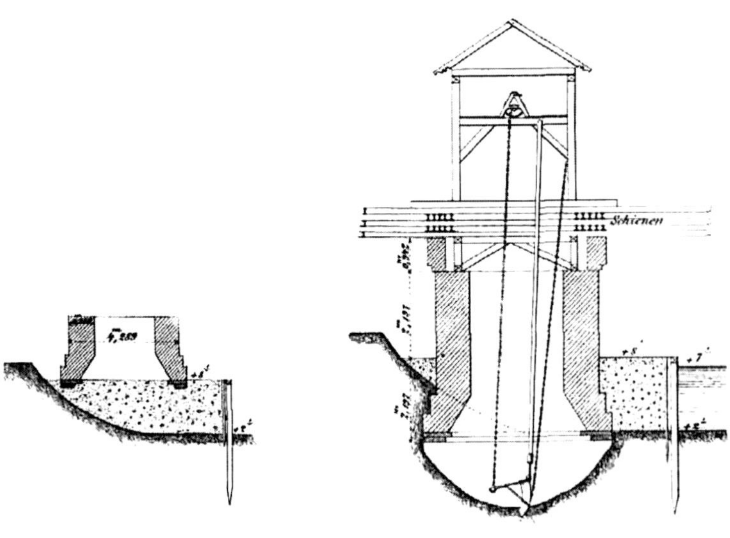

218. Breslau-Schweidnitz-Freiburger Railway Society, bridges over the Watercourse *Kleine Reglitz*, Oder Valley, Stettin, 1869–1874. Elevation of the bridges, plan of the site; elevation, plan and cross section of a pier; cross and longitudinal sections of an abutement (F. E. S. Wiebe, *Die Bauten der Breslau-Schweidnitz-Freiburger Eisenbahnim Oderthal bei Stettin*, in, "Deutsche Bauzeitung," IX, 1875).

219. Berlin-Potsdam-Magdeburger Railway Society, bridge over the River Elbe, close to Magdeburg, 1869–1870. Construction of well foundations (J. L. Quassowsky, *Ueber Fundirungen mit Senkbrunnen nebst Beschreibung einiger Fälle aus der Praxis*, in, "Zeitschrift für Bauwesen," XXIV, 1874).

220-221. Märkisch–Posener Eisenbahngesellschaft, bridge over the River Vistula in Tourun, 1868–1872.

220. Plan of the site, half elevation and half section of a pier on a well foundation; section of a well under construction; half view from above and half horizontal section of a pier; spade to dig (Anon., *Eisenbahnbrücke über die Weichsel bei Thorn*, in, "Zeitschrift für Bauwesen," XXVI, 1876).

221. Construction of a pier on a concrete foundation. The pouring process, cross and longitudinal section; construction of masonry on the concrete fondation, cross and longitudinal section (Anon., *Eisenbahnbrücke über die Weichsel bei Thorn*, in, "Zeitschrift für Bauwesen," XXVI, 1876).

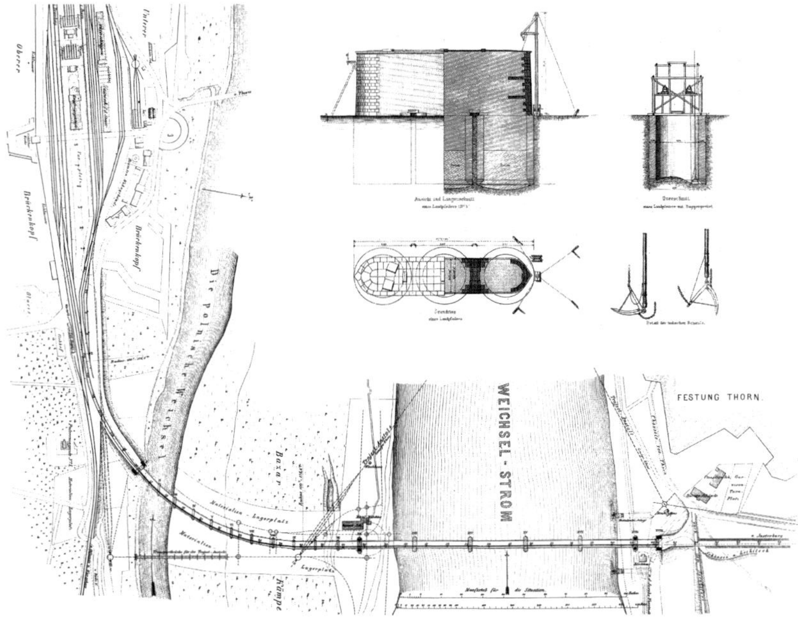

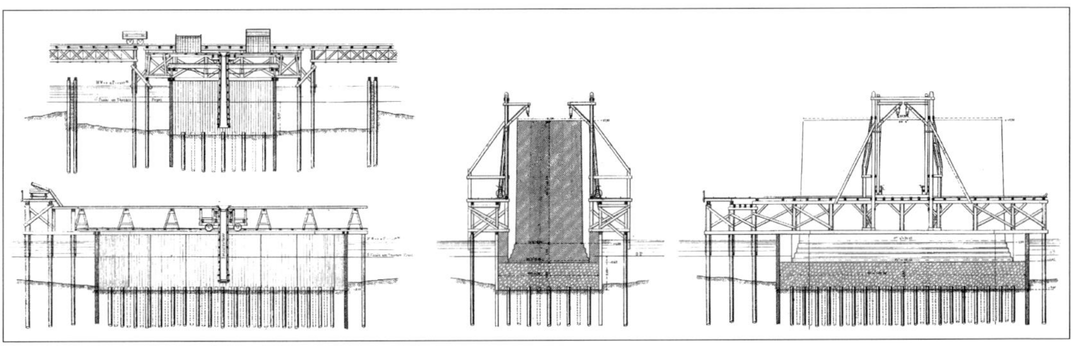

straight in the Vistula are built on foundations made of timber piles rammed inside enclosures of sheet piles. They are covered with a bed of concrete, whereas the abutments are built straight on the soil. Concrete is produced using cement taken from the *Stettiner-Portland-Cement-Fabrik*.[142]

Dalmann also makes further use of sunken well foundations to build the quays of the *Grasbrook* dock, which is completed in 1876. The railway turntables at the junction between the southern quay of the Sandthor dock and the northern quay of the *Grasbrook* dock is, instead, built on a circular foundation made of timber piles, the heads of which are enveloped in a ring of concrete. Dalmann returns in 1875 to this kind of now well-explored building technique, based on the use of timber piles covered by a concrete basement, when he designs the construction of the Strandhafen, which stretches straight along the Elbe.

[142] See Anon., *Eisenbahnbrücke über die Weichsel bei Thorn*, in, "Zeitschrift für Bauwesen," XXVI, 1876, (coll. 35–54, 197–218), coll. 38–49.

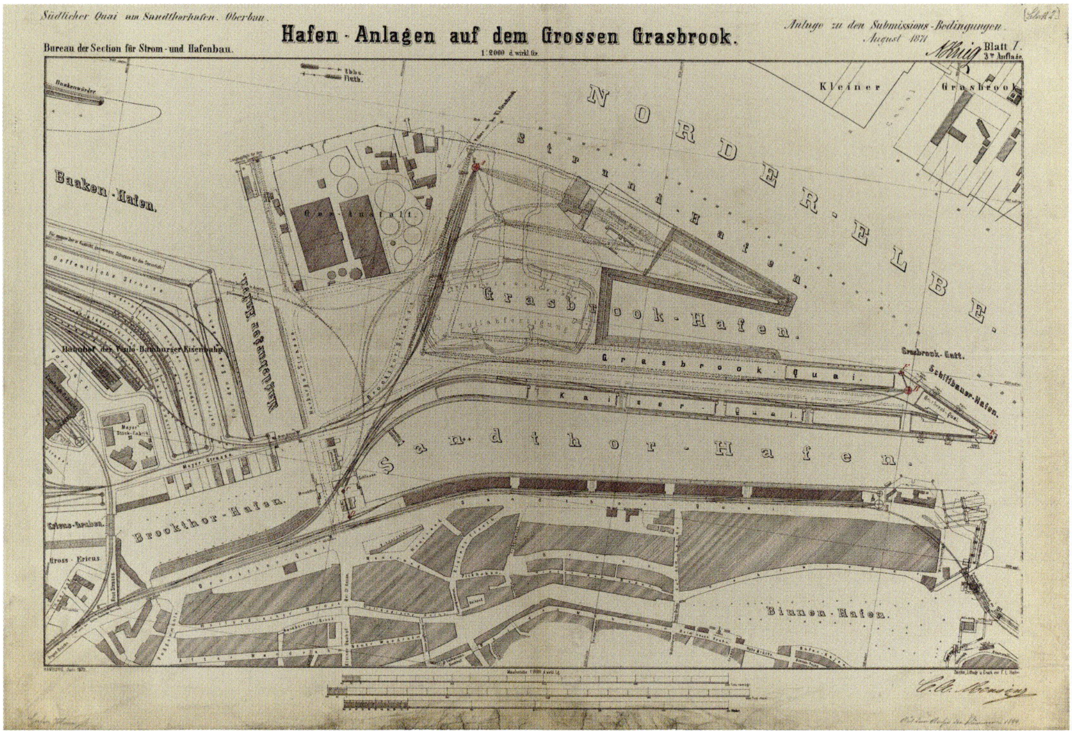
Südlicher Quai am Sandthorhafen. Oberbau.
Hafen-Anlagen auf dem Grossen Grasbrook.
Bureau der Section für Strom- und Hafenbau.
1:2000 d. wirkl. Gr.
Anlage zu den Submissions-Bedingungen.
August 1871
Blatt I.
3te Auflage.
Kleiner Grasbrook
NORDER-ELBE.
Strand-Hafen.
Baaken-Hafen.
Grasbrook-Hafen.
Grasbrook-Quai.
Kaiser-Quai.
Sandthor-Hafen.
Brookthor-Hafen.
Binnen-Hafen.
Schiffbauer-Hafen.

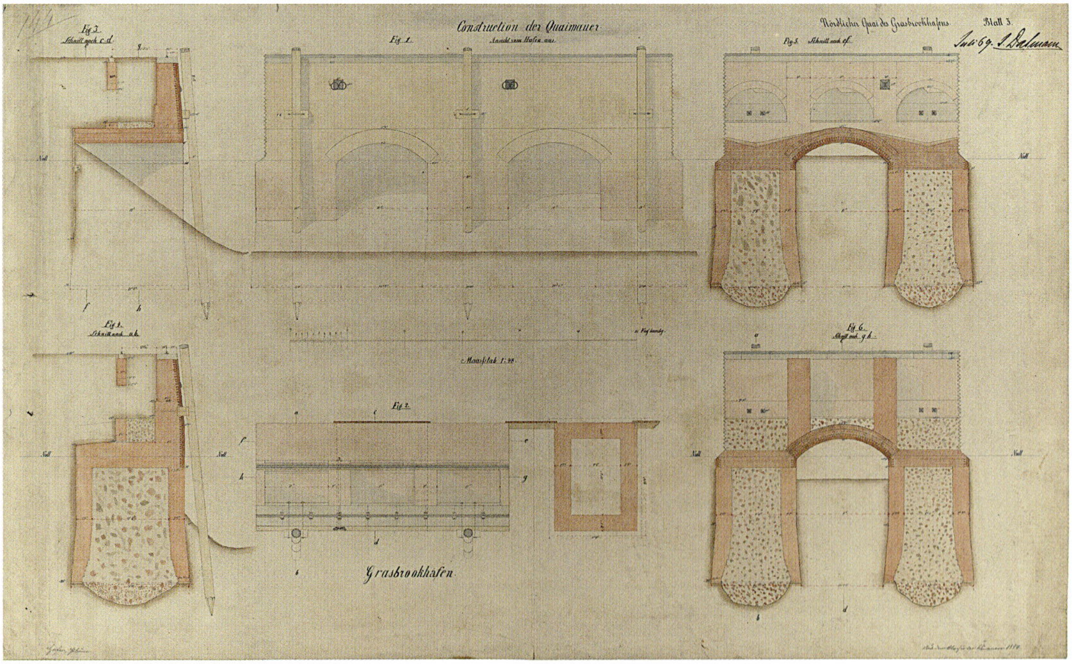
Construction der Quaimauer
Nördlicher Quai des Grasbrookhafens
Blatt 3.
Fig 3. Schnitt nach c.d.
Fig 1. Ansicht vom Hafen aus
Fig 5. Schnitt nach e.f.
Fig 4. Schnitt nach a.b.
Fig 6. Schnitt nach g.h.
Fig 2.
Maasstab 1:48
Grasbrookhafen

222-224. J. C. W. Dalmann, the *Grasbrook* Dock, Hamburg, 1871–1876.

222. Plan of the site (SAH – Kartensammlung: 720–1, 126–1, Häfen und Flüsse, Hafenpläne 5, pl. 187–13a).

223. Cross sections of the quay, elevation, plan and longitudinal sections of a stretch of the quay (SAH – Kartensammlung: 720–1, 126–1, Häfen und Flüsse, Hafenpläne 6, pl. 186–92–3).

224. The well foundations. Construction drawings, 1873 (SAH – Kartensammlung: 720–1, 126–1, Häfen und Flüsse, Hafenpläne 6, pl. 187–31IV).

225. J. C. W. Dalmann, the *Sandthor* Dock, southern quay, Hamburg, 1866–1872. Plan and section of a railway turntable (SAH – Kartensammlung: 720–1, 126–1, Häfen und Flüsse, Hafenpläne 5, pl. 187–14).

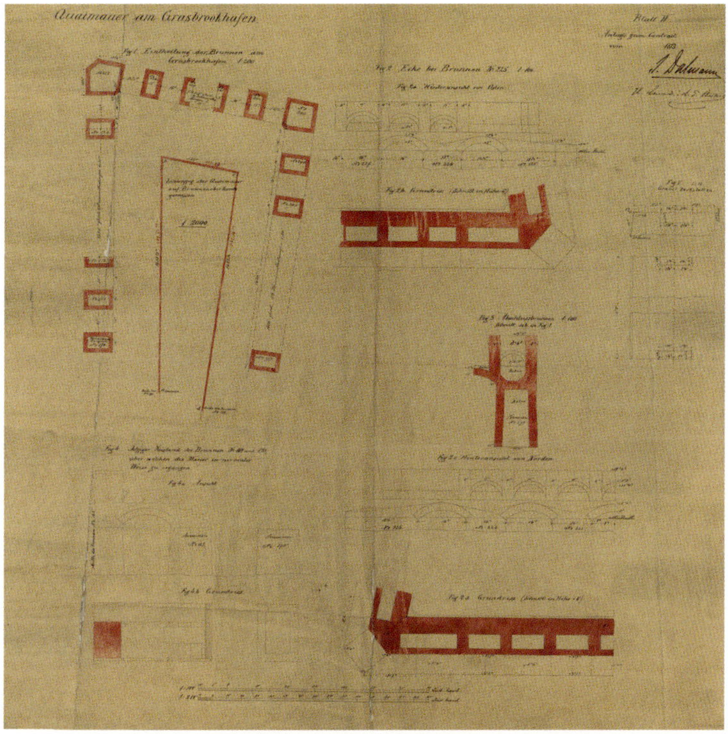

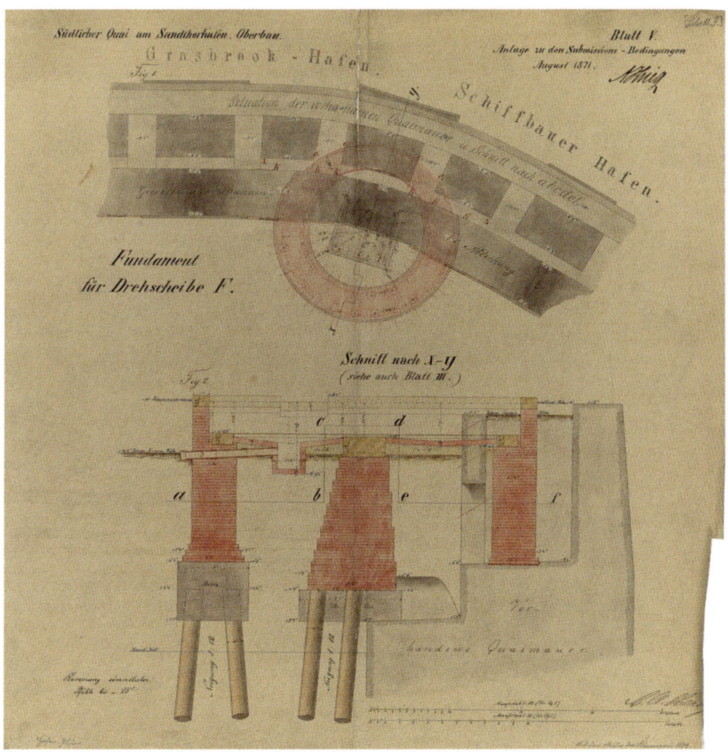

Concrete for above-ground constructions

The acknowledgement of the similarity between lean conglomerate and concrete, around the mid-1850s, triggers the prevalence of concrete over lean conglomerate in above-ground construction. This process takes about ten years and two conditions favour its development: the increasing use of hydraulic binders, which are also recognized as appropriate for lean conglomerate mixtures, especially in the case of foundations and basements; and the awareness that Coignet, in France, as well as British master builders, tends to reduce the amount of cement in concrete as much as possible when they build above ground, thus producing mixtures that are similar to the ones of the German lean conglomerate.

Some literary works capture this change during its development. Seen in this light, the historical excursus about concrete construction that Becker traces in his book about fireproof staircases, from 1857, assumes particular relevance.[143] Becker tries to provide a unitary and synthetic vision of the history of constructions made of lime and cement-base conglomerate, from the ancient Roman ones to those of his time, including constructions belonging to the domain of hydraulic engineering, along with the buildings by Lebrun, Rydin, Prochnow, Leuchs, and the houses that the British army constructs on the Isle of Wight in the early 1850s. Becker's sources concerning the ancient Roman art of building are Rode's translations of Vitruvius' *De Architectura*, and Hirt's *Baukunst nach den Grundsätzen der Alten*. Thus, Becker describes the ancient way of building walls made of masonry faces and concrete filling, with a special regard for the *opus reticulatum*. He also mentions the Romans' skills in building concrete vaults and freestanding concrete walls by means of provisional centrings and formworks, besides the construction of hydraulic works using the indispensable pozzolana.[144] Beyond his description of the different ways Romans used concrete, it is also noteworthy that Becker considers ancient and modern concrete as essentially equivalent. "Both ancient and recent concrete," he writes in this regard, "consist in mixtures of mortar and all kinds of rubble, and sometimes in mixtures of gravel and lime-based mortar." He continues, "The ancients used it to build principally concrete walls, vaults, floors and hydraulic works," hinting, in this way, at the possibility of using concrete to build such kind of works anew.[145] Becker, however, wavers when he writes about the relation between lean conglomerate and concrete. He defines lean conglomerate as a kind of concrete lacking in rubble, and he shares the current conviction that lean conglomerate has prevented the spread of concrete use in above-ground construction in Germany.[146] As an example of lean conglomerate construction, he mentions the walls of a factory along the *Schifffahrt* Canal in Berlin, close to the site of the *Zoologischer Garten*, which he describes as made of mortar (one measure of lime from Rüdersdorf and six measures of sand and gravel) poured and tamped in formworks. He further-

[143] See Becker, *Der feuerfeste Treppenbau*...cit., pp. 52–59.
[144] About the use of concrete in modern hydraulic engineering, Becker's main reference is Bélidor's *Architecture hydraulique;* see Becker, *Der feuerfeste Treppenbau*... cit., pp. 53.
[145] "Der Béton der Alten wie der der Neuern aus einem Gemisch aller Art klein geschlagener Bruch- oder Mauersteine mit Mörtel, zuweilen auch aus einer Mischung von Kies und Kalkmörtel. Hauptsächlich zur Aufführung von Mauern, zu Gußgewölben, zu Estrich, wie zu Wasserbauten wurde der Béton von den Alten vielfach verwendet" (*ibid.*, p. 52).
[146] "Although the use of concrete over ground is known since fourteen-years, lean conglomerate has prevailed" ("In Deutschland ist die Anwendung des Betons zu Hochbauten seit etwa 14 Jahren im Gebrauch, doch im Allgemeinen seit dieser Zeit der Kalk-Sand-Pisé-Bau mehr in Ausübung gekommen") (*ibid.*, p. 59).
[147] *Ibid.*, p. 56.
[148] "Welche große Festigkeit schon Mörtel gewöhnlicher Art, von Kalk ohne wesentliche hydraulische Eigenschaft, mit großem Sandzusatz, wie derselbe zum Kalk-Sand-Pisé-Bau verwendet wird" (*ibid.*, p. 57).
[149] Anon., *Der Werth des*

more reports that, on the same site, four columns have been built by pouring and tamping, in wooden formworks, a kind of concrete made of one measure of Portland cement and six measures of sand, gravel and stone rubble. These same columns are covered with plaster and with a layer of oil painting, to give them the appearance of sandstone.[147] Although he is a major supporter of Portland cement, Becker praises "the great solidity of ordinary mortar, made with lime lacking any hydraulic property, and with much sand, like the one that it is used for buildings made of lean conglomerate."[148] Becker's historical excursus becomes, a few years later, the main source for an article dealing with concrete, lean conglomerate and artificial stones, which is published in 1864 in the "Zeitschrift für praktische Baukunst" under the title *Der Werth des künstlichen Baumaterials* (The Importance of Artificial Building Materials).[149]

The book about concrete building by the Austro-Hungarian hydraulic engineer Johann von Mihálik (1818–1892) also contributes to strengthening, in Germany, the increasingly widespread belief that concrete is appropriate for extended uses, both underwater and underground as well as above ground; the second edition of Von Mihálik's book is published in Berlin in 1860, shortly after having been issued in Vienna.[150] "Concrete," Mihálik clearly states, "is suitable to building with advantages to all kinds of masonry, both underwater and above ground."[151]

The most evident literary expression of how the concepts of lean conglomerate and concrete overlap is evident in the third edition of Engel's book about lean conglomerate, which is published in 1865. Already in the preface, making reference to the use of lean conglomerate, Engle asserts: "This building method, which is closely related to the centuries-old use of lime-based mortar, may have been already known in antiquity, according to Vitruvius' reports, which relate the Roman way of building walls by pouring lime and stone rubble (concrete) into box-shaped contraptions."[152] By then, hydraulic lime, Roman and Portland cements are all described as possible components of lean conglomerate, and some kinds of concrete mixtures taken from Coignet are listed among the ones considered to be appropriate for lean conglomerate,

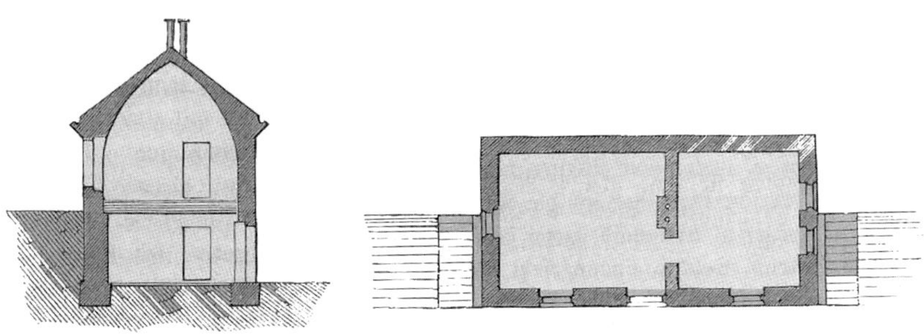

226. C. Ruland, signalman's house along the River Isar, close to Munich, 1858–1859. Cross section and plan (J. D. F. Engel, *Der Kalk-Sand-Pisébau und die Kalksand-Ziegelfabrikation*, Leipzig, C. H. Seemann, 1865).

künstlichen Baumaterials...cit.
[150] Johann von Mihálik, *Praktische Anleitung zum Béton-Bau für alle Zweige des Bauwesens*, Wien, Joseph Stöckholzer v. Hirschfeld, 1859,[2] Berlin, Theobald Grieben, 1860.[2] Full name of Mihálik is János Mihálik von Madunycz; he is born in Arad, which is today in Romania.
[151] "Mit Béton kann Mauerwerk jeder Art, und zwar sowohl ober als unter Wasser" (Von Mihálik, *Praktische Anleitung zum Béton-Bau*...cit., 1860, p. 7).
[152] "Daß diese Baumethode, welche im innigem Zusammenhange mit der seit Jarhtausenden bestehenden Anwendung des Kalkmörtels ist, schon im Alterthume bekannt gewesen sein dürften, beweisen die Berichte des Vitruv, nach welchen die Römer schon kastenartige Vorrichtung zur Herstellung von Mauern aus Kalk und Sreinstücken (Béton) benutzt haben" (Engel, *Der Kalk-Sand-Pisébau und die Kalksand-Ziegelfabrikation*... cit., p. 3). Similarly to what already observed for Becker, Vitruvius' *De Architectura* and Hirt's *Baukunst nach den Grundsätzen der Alten* are Engel's literary references for the art of building in the antiquity; see *ibid.*, pp. 3–4.

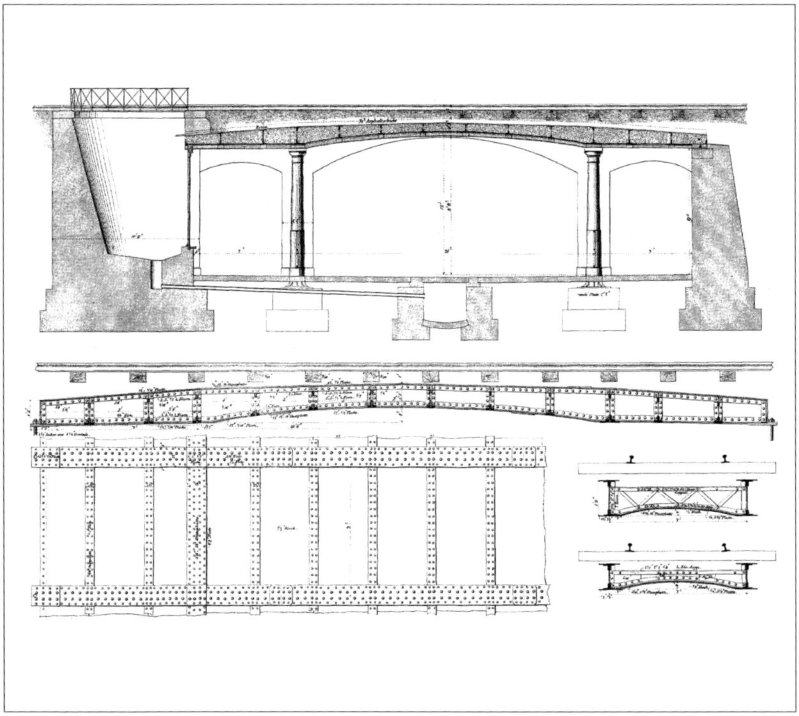

227-228. Niederschlesisch–Märkische Railway, passenger station and entrance tunnel, Görtlitz, 1867–1868.

227. Entrance tunnel. Cross section, girders (Anon., *Der Umbau des Bahnhofes zu Görlitz*, in, "Zeitschrift für Bauwesen," XX, 1870).

228. Passenger station. Cross sections through the buildings and through the entrance tunnel, plan of the groung floor, plan of the basement (Anon., *Der Umbau des Bahnhofes zu Görlitz*, in, "Zeitschrift für Bauwesen," XX, 1870).

229-230. J. v. Schlierholz, signalman's house for the Swabia railways, 1867–1868.

229. Perspective view, front elevation and cross section (J. v. Schlierholz, *Ueber Beton-Verwendung zu Hochbauzwecken, hauptsächlich zu ganzen Gebäuden*, in, "Allgemeine Bauzeitung," XXXV, 1870).

230. Sketch of the signalman's house built by Leube (E. Breuss, W. Günther, K. Kaindl, R. Zrost, *175 Jahre Leube. Chronik eines Familienunternehmens 1838 bis 2013*, Leube, St. Leonhard, 2013).

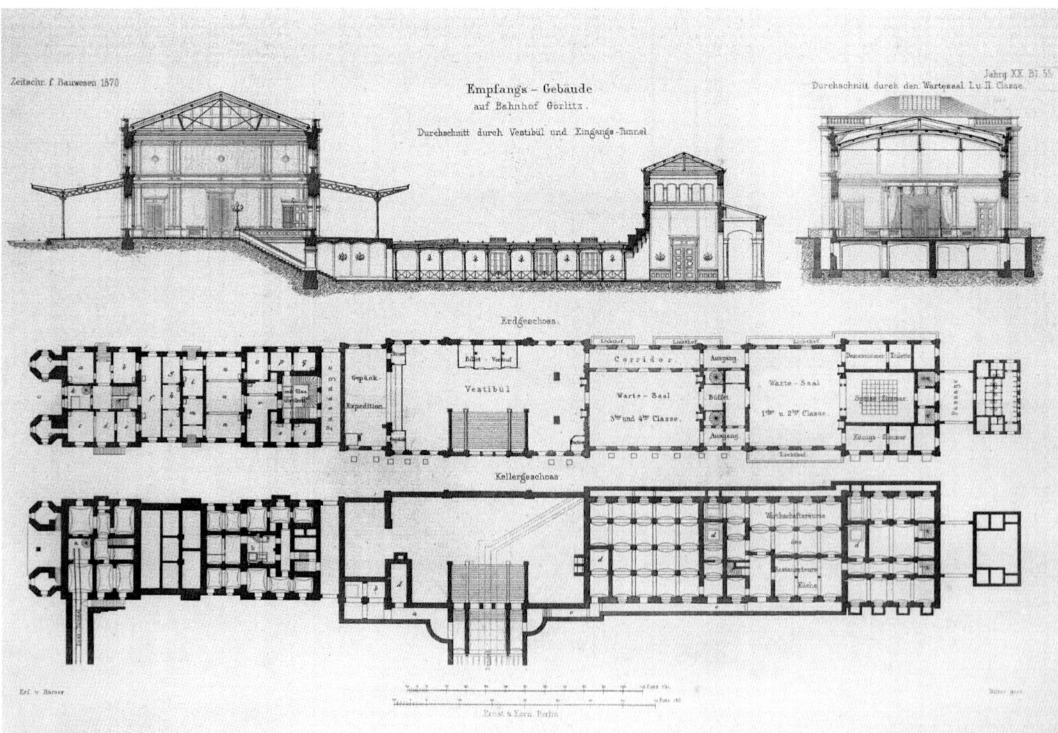

up to the point that Engel resorts to the expression "rammed, hydraulic, and lime-based mass" to briefly describe concrete.[153] Engel considers concrete as an appropriate material for building vaults, and mentions, as pertinent evidence, the ancient Roman concrete vaults, along with the far more recent examples of vaults built by Hoffmann, some others described by Mihálik and the pointed vault of a signalman's house that the master builder Carl Ruland (1806–1871) builds between 1858 and 1859, along the River Isar, close to Munich.[154] Apart from the in-between floor, Ruland's house is entirely made of a quite lean concrete (one measure of hydraulic lime and ten measures of sand and stone rubble), from the foundation to the vault forming the roof. This house is to be seen in relation to the previously described signalman's houses made of lean conglomerate in the region of Hamburg, and it can also be considered one of the earliest examples of a concrete house built in Germany.

As of about the mid-1860s, a crucial period begins for the definitive acknowledgement of concrete as a building material for above-ground construction. The building of railways is, once again, a driving force. In 1865, the Dyckerhoffs' cement company, which has been entrusted with the construction of concrete floors for some stations of the Swabian railways, builds a 4 feet span concrete railway bridge. In the same year, the Dyckerhoffs experi-

[153] See Engel, *Der Kalk-Sand-Pisébau und die Kalksand-Ziegelfabrikation*...cit., p. 67.
[154] *Ibid.*, p. 90; Joseph von Schlierholz, *Ueber Beton-Verwendung zu Hochbauzwecken, hauptsächlich zu ganzen Gebäuden*, in, "Allgemeine Bauzeitung," XXXV, 1870, pp. 260–265. About Carl Ruland, see Anon., *Zur Erinnerung an Carl Ruland, kgl. Regierungs- und Kreisbaurath a. D., Ehrenmitglied des Münchener Architekten- und Ingenieur-Vereins*, in, "Zeitschrift des bayerischen Architekten und Ingenieur-Vereins," IV, 1872, pp. 1–4.

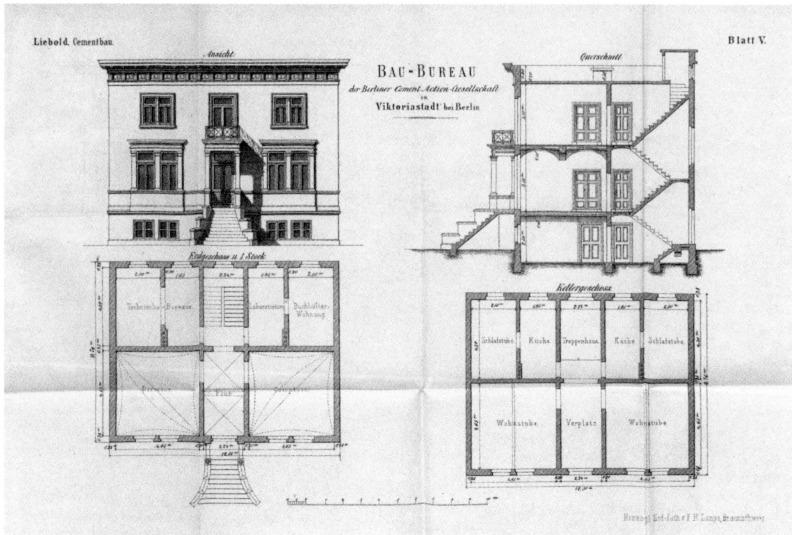

231. Berlin Cement-Bau AG, administration building, *Victoriastadt*, Berlin, approx. 1872. Front elevation, ground and first floor plans, cross section (B. Liebold, *Zement*, Halle, Hendel, 1874).

ment in their own factory with the construction of floors made of iron beams, placed 18 inches from each other, and concrete flat vaults as fillings. The kind of concrete used in both cases is made of one measure of cement, three of sand and four of stone rubble.[155]

Between 1867 and 1868, concrete is used to fill the fields of the extraordinary iron floor that is built to cover a tunnel under the railway station of Görtliz. This tunnel is 100 feet long and 35 feet wide.[156] It is divided into three aisles by two rows of eleven iron columns; the central aisle is 21 feet wide and the lateral ones are each 7 feet wide. The columns bear eleven transversal girders, which are light-arched over the main aisle, and twelve minor girders in each of the ten fields formed by the major girders. This iron structure forms a horizontal frame containing one hundred and ten hollow fields, which are filled with concrete. The entirety is finally covered with asphalt and gravel, forming the base for the rails.

In the wake of the example shown by Ruland, around 1867, the master builder Joseph von Schlierholz (1817–?), who is engaged in the construction of the Swabia railways, considers the possibility of building signalman's houses in conglomerate, seeing that, as he himself asserts, neither natural stones, nor large amounts of good quality bricks are easily available in the region, whereas hydraulic lime and cement are being produced in several local kilns.[157] Schlierholz rejects the use of aerial lean conglomerate (*Kalksand*), which he judges insufficient to produce rapidly solidifying masses, and chooses, instead, to experiment with three different Roman and Portland cement-based concrete mixtures, being aware of Von Pettenkofer's analyses and of Von Mihálik's and Becker's writings. Therefore, the construction of three test-houses, between Ulm and Blaubeuren, is planned.

[155] See Stegmann, *Das Bauunternehmen Dyckerhoff Et Widmann*...cit., pp. 27–31.
[156] See Anon., *Der Umbau des Bahnhofes zu Görlitz*, in, "Zeitschrift für Bauwesen," XX, 1870, (coll. 471–482), col. 474. The subway becomes known as *Jakobstunnel*, since it stretches along the *Jakobsstrasse*.
[157] See Schierholz, *Ueber Beton-Verwendung*...cit.; Dollinger, *Wärterhaus aus Béton*, in, "Deutsche Bauzeitung," IV, 1870, pp. 44–45.
[158] See Anon., *Bahnwärterhäuser aus*

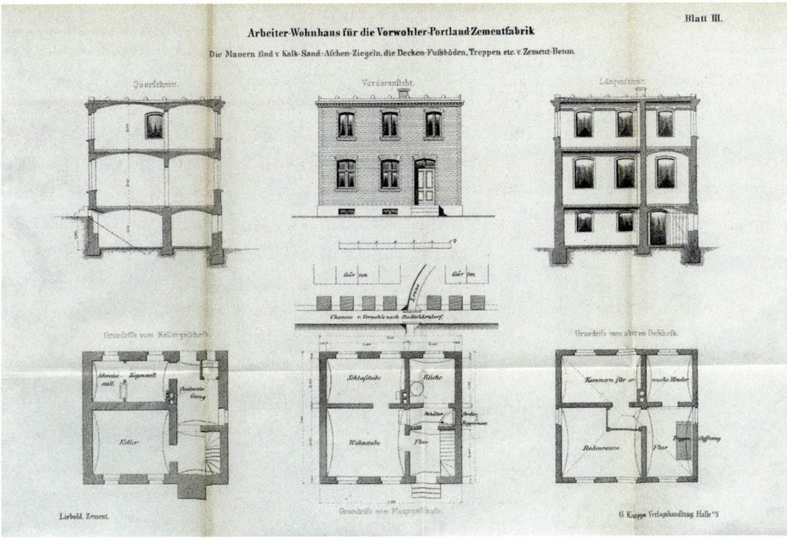

232-233. B. Liebold, workers' dwelling, Vorwohle, approx. 1873.

232. Cross sections, front elevation, longitudinal section, basement, ground and first floor plans, site plan of workers' neighbourhood (B. Liebold, *Zement*, Halle, Hendel, 1874).

233. The junction between a concrete vault and a wall made of sand-lime bricks, first and second solution. Sections (B. Liebold, *Zement*, Halle, Hendel, 1874).

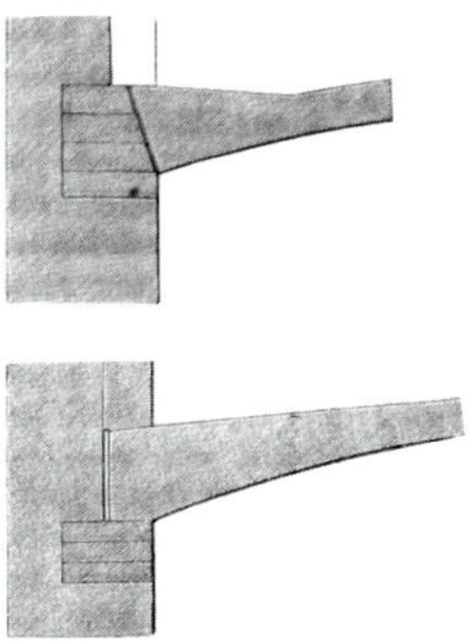

Each house needs a basement covered with two barrel vaults, and a main living space above covered with a pointed vault and divided into two levels by means of an in-between timber floor. Apart from the in-between timber floor, all other bearing walls and vaults are intended to be built in concrete, to pour and tamp in formworks and on centrings. The construction begins in the summer of 1867, and it is achieved in the spring of 1868. Two houses are constructed entirely according to the plans, whereas the third one is covered with an ordinary roof and cement tiles, due to a lack of time. At least one of these houses is built by Leube, specifically the one in Gerhausen.[158] In the following years, the Swabia railway administration orders the construction of further concrete houses, and, according to Schlierholz, about ten are completed before 1870.

The increasing need for housing that is caused by industrial and urban development also provides the occasion to experiment with above-ground concrete construction. The housing question is particularly pressing in Berlin, where the effects of reforms promoted early in the century by Heinrich Friedrich Karl vom und zum Stein (1757–1831) and Karl August von Hardenberg (1750–1822) make the phenomenon of urbanization even more critical. Between 1859 and 1867, the population increases from 418,000 to 702,437 inhabitants, and the housing question gains such an importance that the master builder Franz Gustav Assmann, in a talk given at an assembly of the Berlin *Architektenverein* in 1872, states: "The housing question has now turned into housing crisis."[159] It is in such a context that the entrepreneurs Anton and Alfred Lehmann, based in the manufacturing area of *Rummelsburg*, plan to build a working-class neighbourhood, entirely composed of concrete buildings, according to British models they have heard about from the master builder

Gußmauerwerk, in, "Polytechnisches Journal," XLIX, vol. CXXVII, p. 517; Breuss, Günther, Kaindl, Zrost, *175 Jahre Leube*...cit., p. 35.

[159] "Aus der Wohnungsfrage ist aber inzwischen eine wirkliche Wohnungsnoth geworden" (Franz Gustav Assmann, *Die Wohnungnoth in Berlin*, in, "Zeitschrift für Bauwesen," XXIII, 1873, (coll. 111–130), col. 111).

Albrecht Konstantin Türrschmiedt (1817–1871).[160] In 1870, on behalf of the Lehmanns, the engineer Alexis Riese goes on a journey to England, where he learns about how to produce and apply concrete, while the Lehmanns buy a roughly 20 hectare wide land plot, on the eastern side of *Rummelsburg*, where the new neighbourhood is intended to be built. According to Türrschmidt, it is through a contact with the British master builder "Wonnacott" that Riese learns about concrete. The master builder that Türrschmiedt mentions could be Thomas Robjohns Wonnacott (1834–1918), who builds a concrete house in the first half of the 1870s in Grays, in the county of Sussex.[161] Once back in Berlin, Riese builds an experimental concrete house composed of two levels and a basement. He uses a concrete mixture composed of Portland cement from Stettin, sand, pebbles from Rüdersdorf, rubble taken from the demolition of other buildings and slags from a nearby gas manufacturing installation, which he pours into wrought-iron metal formworks of the type that Charles Drake patented in the United Kingdom in 1867.[162] As with the metal formworks, the concrete mixture must also to be seen in terms of Riese's journey to England, since – as reported in an article from 1873 focusing on the construction of concrete houses – British builders mix "Portland cement and the most different materials that are affordably available close to the building sites, like sea sand, coal ash, slags, gravel, rubble," and the ratio between cement and other aggregate varies between 1:7 and 1:13.[163] Such a kind of concrete is quite lean and therefore reminiscent of Engel's hydraulic, lime-base mass. Türrschmiedt welcomes this construction with enthusiasm, claiming that concrete will make it possible to do without masons, who are on strike in Berlin during that period.

The reliability of Riese's concrete test-house is verified and recognized by the Berlin *Baupolizei* and by the Berlin magistrate on a visit that takes place in November of 1871.[164] Meanwhile, the Lehmanns found a building company, the *Cement-Bau A.G.*, and once the Berlin *Baupolizei* approves the use of concrete, the construction of the new dwellings begins.[165] The early houses that are built in 1872 (about 12) only have the walls made of con-

[160] See Lehmann, *Bericht über einen Versuch mit Konkretbau*, in, "Verhandlungen des Vereins zur Beförderung des Gewerbefleisses in Preussen," LI, 1872, pp. 37–38. About concrete houses in England, see Friedrich Wiebe, *Bauwerke aus Concret*, in, "Deutsche Industriezeitung," 1870, pp. 7–8, and in, "Zeitschrift für Bauhandwerker," XIV, 1870, pp. 61–62. Friedrich Türrschmiedt founds the *Verein für die Fabrikation von Ziegeln, Kalk und Cement* (Society for the manufacturing of bricks, lime and cement) in 1865, together with the master builder Friedrich Eduard Hoffmann (1818–1900). About Türrschmiedt, see Lämmerhirt, *Albrecht Türrschmiedt*, in, "Deutsche Bauzeitung," V, 1871, pp. 330–331. About the development of Rummelsburg and the early steps to the building of a worker settlement, see Armin Niemeyer, *Ein Vorläufer des Betonbaues am Rande Berlins*, in, Ulrich Großmann, Klaus Freckmann, Fred Kaspar, Ulrich Klein, ed., *Aus den Forschungen des Arbeitskreises für Haus- und Siedlungsforschung*, Marburg, Jonas, 1991, pp. 97–108; Christina Czymay, *"Colonie Victoriastadt" in Berlin-Lichtenberg*, in, Berlin Landesdenkmalamt, ed., *Beiträge zur Denkmalpflege in Berlin*, n° 12, Berlin, Schelzky und Jeep, 1997, pp. 60–63; Id., *Die Colonie Victoriastadt – Betonhäuser in Berlin Rummelsberg*, in, Hartwig Schmidt, ed., *Zur Geschichte des Stahlbetonbaus. Die Anfänge in Deutschland 1850 bis 1910*, Berlin, Ernst & Sohn, 1999, pp. 11–15.

[161] See Albrecht Konstantin Türrschmiedt, *Berlin erstes Concretehaus*, in, "Notizblatt des Vereins für die Fabrikation von Ziegeln, Kalk und Cement," 1871, n° 3, n. p. About Thomas Robjohns Wonnacott and the concrete house he builds in Grays, see Peter Raby, *Alfred Russel Wallace*, Princeton, New Jersey, Princeton University Press, 2001.

[162] Heike Hinz, "Berlin-Lichtenberg, Türrschmidtstrasse 17. Frühe Betonhäuser in der Victoriastadt ab 1871. Zur Entstehung und Sanierung des Wohnhauses in der Türrschmidtstrasse 17," final report for "Aufbaustudiengang Denkmalpflege," Technische Universität Berlin, 2002; Anon., *Ueber die Anwendung des Beton zur Herstellung von Wohnhäusers*...cit., p. 28.

[163] "Zur Erzeugung des concretes sind in England in Mischung mit Portlancement die verschiedensten Materialien in Anwendung gekommen, wie solche in der nähe von Baustellen eben am wohlfeilsten zu haben waren, Z.B. Meersand, Kohlenasche, Schlacken, Kies, Steinbrocken" (Anon., *Ueber die Anwendung des Beton zur Herstellung von Wohnhäusers*, in, "Zeitschrift für Bauhandwerker," XVII, 1873, (pp. 15, 28–31), p. 15); about concrete mixtures used in England, see also Bernhard Liebold, *Zement*, Halle, Hendel, 1874, pp. 41–42, n. 1.

[164] See Hinz, "Berlin-Lichtenberg..." cit.

[165] See Niemeyer, *Ein Vorläufer des Betonbaues*...cit., p. 100.

234. O. Tafel, Villa Merkel under construction, Esslingen, approx. 1873, photograph (courtesy of the *Galerien der Stadt Esslingen am Neckar*).

235. B. Liebold, Villa Planck, Holzminden, 1877. Front elevation, cross section, groundfloor, roof and first floor plans (B. Liebold, *Wohnhaus des Hrn. F. Planck in Holzminden*, in, "Deutsche Bauzeitung," XI, 1877).

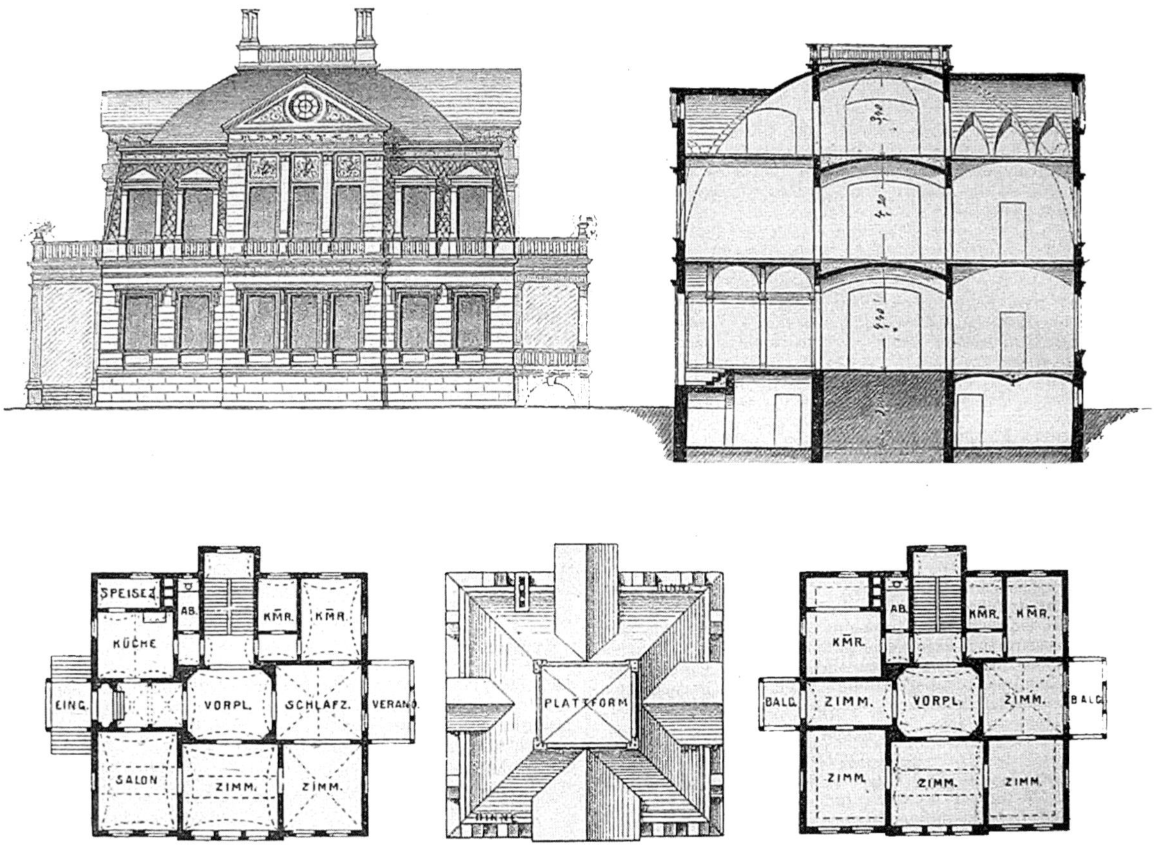

236. General Inspection des Ingenierurs-Corps und der Festungen, experimental construction of a bombproof iron-concrete floor in Berlin-Tegel. Cross and longitudinal sections, 1875 (GStA IV. HA Preußische Armee, Rep. 16 Militärvorschriften, n° 667, fol. 1v).

crete, whereas the floors and the staircases are made of wood.[166] The following ones are rather almost entirely made of concrete, including staircases and floors. These consist of single flat vaults covering small rooms, whereas each larger room is covered with two flat vaults resting on two bearing walls and on a central iron double T beam. Chimneys are built as hollow pipes inside the walls, by putting provisional round poles in the formworks at appropriate places, before the concrete is poured.[167] The concrete mixture used to build the walls is quite lean, being composed of 10% Portland cement, 15% sand and 75% coal ash. The use of slags is thought to improve the health of the houses, keeping walls dry. The mixture used to build the vaults contains a larger percentage of cement, about 14%, probably as a precaution.[168] The strength tests that are carried out before work starts prove that this kind of vaults can withstand 3750 kg/m^2 against an expected strain of 750 kg/m^2. The neighbourhood is only partially constructed (about sixty buildings out of two hundred), and it is given the name of *Victoriastadt*, referring to the queen Victoria, a detail that reinforces the connection with British culture.

In 1872, just as the construction of *Victoriastadt* begins, the entrepreneur Godhard Friedrich Julius Prüssing and the businessman Friedrich Planck found a cement factory in the town of Vorwohle, in the Duchy of Brunswick, which operates under the name of *Vorwohler Portland-Cement-Fabrik*. Shortly after the foundation of the cement factory, they contact the master builder Berhard Liebold (1843–1916), who is based in the nearby town of Holzminden, where he runs a private office and is a teacher at the Holzminder Baugewerkschule.[169] Liebold already has some experience with cement pipes, since he built a canalisation in a factory in Karlsruhe, and he is probably also aware of the concrete houses under construction in Berlin, through periodicals that are available in the library of the *Baugewerkschule*.[170] Together with Liebold, Prüssing and Planck found a building company, the *Vorwohler Cement-Baugesellschaft Prüssing, Planck & Co.*, in 1873, with the evident aim of exploiting the Portland cement they produce.[171] In the same period, Liebold designs a group of dwellings for Prüssing and Planck to build in Vorwohle, for the workers of the local cement factory. They have foundations made of stone, walls made of sand-lime bricks and vaults of concrete. The sand-lime bricks for the walls are produced using a mixture of seven measures of ash and one measure of slaked lime, according to instructions that Liebold himself develops, and they are moulded in a Bernhardi press.[172] Concrete for the flat vaults and for the staircases is more prudently composed of one measure of cement, two of sand and four of rubble.[173] The construction of the early vaults goes through a singular accident, which is probably due, once again, to the excessive belief in the adhesion between cement mortar and bricks. Liebold, in fact, at first draws concrete vaults that do not rest perpendicularly on the sand-lime brick walls, but on banquettes of hard-fired bricks and fat lime-based mortar

[166] See Ernst Kanow, *"Colonie Victoriastadt." Eine Berliner Wohnsiedlung mit mehr als hundert Jahre alten Wohnhäusern aus Beton*, in, "Architektur der DDR," XXX, 1981, (pp. 50–53), p. 51.
[167] *Ibid.*, p. 52.
[168] See Anon., *Das Cementbau*, in, "Zeitschrift für Bauhanwerker," XVI, 1872, (pp. 187–188), p. 188; *Mittheilungen aus Vereinen*, in, "Deutsche Bauzeitung," VI, 1872, p. 385; Anon., *Versuch mit Concretbau*, in, "Zeitschrift für Bauhandwerker," XVII, 1873, p. 143; Liebold, *Zement*...cit., pp. 76, 106.
[169] See Matthias Seeliger, *Betonbau in der Provinz - Die Vorwohler Zementbaugesellschaft*, in, Uta Hassler, ed., *Häuser aus Beton. Vom Stampfbeton zum Grosstafelbau*, Tübingen, Berlin, Ernst Wasmuth, 2004, pp. 47–57.
[170] See Liebold, *Zement*...cit., pp. 40–47, 85; Seeliger, *Betonbau in der Provinz*...cit., p. 48.
[171] See *ibid.*, p. 49.
[172] See Liebold, *Zement*...cit., p. 54.
[173] *Ibid.*, p. 106.

3, Die Erddecke ist über dem Scheitel der Betonübermauerung 0,90 m. stark zu machen. Wo ausnahmsweise die Gesammtstärke der Decke (von 1,50 m. über den Bombenbalken) etwas reducirt werden muß, ist dies durch Verminderung der Erddecke, unter entsprechender Verstärkung der Betondecke zu erzielen, wobei etwa für je 3 cm. Reduction der Erddecke 1 cm. Beton zuzulegen ist. Eine erhebliche Verschwächung der Deckenstärke erheischt jedoch die Anwendung stärkerer Balkenprofile.

4, Die inneren Mauerlatten sind nicht, wie bisher, bündig mit der Mauerfläche zu verlegen, sondern um etwa 5 cm. von der letzteren abzurücken. Eine weitere Sicherung gegen das Abbrechen der Mauerkante ist dadurch zu erzielen, daß die obersten Ziegelschichten mit etwas Auskragung nach dem Innern des Hohlraumes verlegt, oder – wo ein fester Haustein in größeren lagerhaften Platten billig zu beschaffen ist – die Kronen der tragenden Mauern mit solchen abgedeckt werden, wodurch zugleich der Vortheil einer besseren Vertheilung der Druck- und Stoßwirkungen auf das tragende Mauerwerk erreicht wird.

5, Wenn der einzudeckende Hohlraum, wie in der Regel bei Hohltraversen, mit einer Stirnmauer abschließt, welche nur von einer Erddecke von gleicher Dicke, wie über der Bombendecke, gegen das Mörserfeuer geschützt ist, so darf die Eisendecke nicht an der inneren Fläche der Stirnmauer aufhören, sondern muß noch mit ein paar Balken, resp. in einer Breite von etwa 50 bis 70 cm. auf die Krone der Stirnmauer übergreifen, um das Durchschlagen von hier auftreffenden Mörsergranaten zu verhindern. Diese auf die Stirnmauer zu verlegenden Balken können jedoch von erheblich schwächerem Profil sein und mit etwas weiterem Zwischenraume verlegt werden.

Querschnitt

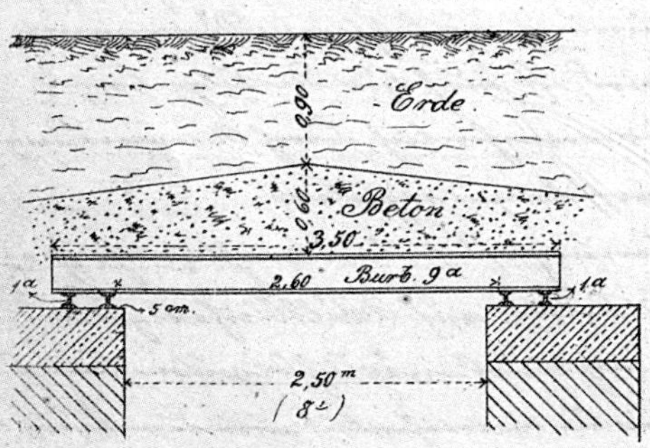

Längenschnitt

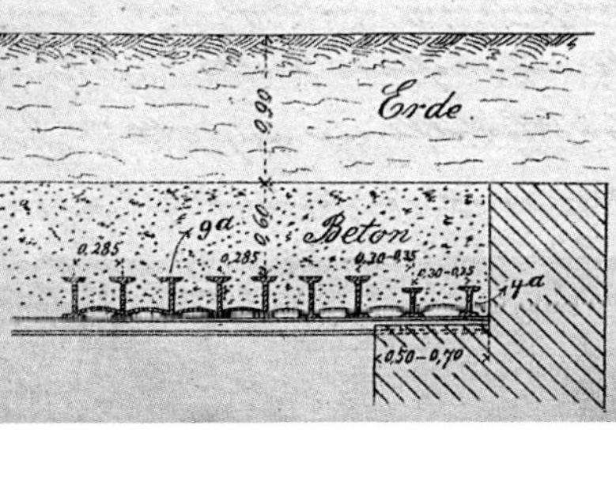

6, Die bis auf Weiteres für bombensichere Eisendecken bei den verschiedenen Spannungen zu wählenden I Profile sind aus der nachfolgenden Tabelle zu entnehmen. Die nebenstehenden Skizzen veranschaulichen als Beispiel die hiernach sich ergebende Construction der Bombendecke für einen 8' (2,50 m.) weiten Hohlraum.

Ta-

forming angles of around 18 degrees with the imposts of the concrete vaults. Liebold believes that the brick masonry banquettes and the concrete vaults bind so firmly together that they act as monolithic elements. "In order to avoid a weakening of the walls," Liebold later relates in this regard, "I made building banquettes of hard-fired bricks […] since I thought that the concrete mass of the vaults could weld with the bricks of the banquettes."[174] Such construction, of course, fails and, only in a second attempt, does Liebold conceive vaults that rest perpendicularly on the banquettes inside the walls.

In the 1870s, along with dwellings for the working-class, some concrete houses for entrepreneurs are also constructed. The Stuttgart architect Otto Tafel (1838–1914) builds a villa for the train manufacturer Oskar Merkel in Esslingen am Necker, around 1873, while Liebold builds a number of additional concrete buildings in the area between the River Weser and the Harz Mountains. Between 1876 and 1877, he builds the local inn and a dwelling for the local factory of asphalt in Vorwohle; in 1877, in Holzminden, he builds Planck's residence and, nearby in Kreiensen, he builds the Maigatter Hotel.[175] Being the residence of an entrepreneur involved in the production of cement and concrete, Planck's house is considered an exemplary building, conceived to test and demonstrate the properties of cement and concrete. The large number of various kinds of vaults employed to cover the rooms and halls of the house are evidence of such a purpose. Particularly impressive is the groin vault forming the roof, which is originally even provided with a flat terrace in the central part of it. The concrete for the walls and for the upper vaults is composed of cement, sand and gravel, whereas for the inner vaults and for the staircase, it is made of cement, sand and coal slags.[176]

Despite these experimental works, the general construction of buildings entirely made of concrete remains quite modest. In 1874, the *Verband deutscher Architekten- und Ingenieurvereine* (Associated Society of the German Architects and Engineers Societies) debates about the appropriate application of concrete in architecture, realizing that concrete is suitable for functional buildings without aesthetic requirements. Nevertheless, when the matter is again discussed ten years later, the number of concrete buildings in Germany remains extremely modest, even in the domain of the functional buildings.[177] The supposed unsuitability of concrete to provide aesthetic qualities, therefore, does not justify the limited number of concrete buildings completed in Germany in the third quarter of the century. The reason is rather to be found in the importance that the German building culture traditionally devotes to ordinary masonry. Concrete only very gradually becomes a part of construction practices, mainly by replacing the masonry in single parts of the buildings, namely where its extraordinary monolithic behaviour is considered particularly advantageous. The reference principally goes to vaults and flat

[174] "Die Widerlager der Gewölbe liess ich, um eine Schwächung der Umfassungsmauern zu verhindern […] von gut gebrannten Mauerziegel herstellen, weil ich glaubte, dass sich mit diesen die Bétonmasse des Gewölbes gut und innig verbinden würde" (Liebold, *Zement*…cit., p. 100).

[175] See Anon., *Ein Beitrag zum Betonieren von Hochbauten*, in, "Oeberösterreiche Bauzeitung," II, 1897, (pp. 69–70), p. 69; Seeliger, *Betonbau in der Provinz*...cit., p. 56.

[176] See Bernhard Liebold, *Wohnhaus des Hrn. Kaufmann F. Planck in Holzminden*, in, "Deutsche Bauzeitung," XI, 1877, pp. 458–459.

[177] In 1884, Schlierholz carries out statistical research about the number and the type of concrete constructions in Germany, on behalf of the *Verband deutscher Architekten- und Ingenieurvereine* and he only finds 17 houses, 2 stables, factories and some basins; see Verein Deutscher Portland-Cement-Fabrikanten, ed., *Der Portland-Cement und seine Anwendungen im Bauwesen*, Berlin, Ernst Toeche, 1892, p. 185.

[178] "Für die Aus- und Anbemauerung ist fortan nicht mehr Ziegelmauerwerk, sondern Beton auf Wölbblechen zu verwenden, und ist der Herstellung eines Möglichst guten Betons besondere Sorgfalt zuzuwenden, da vornehmlich der Festigkeit der Betonmasse, insbesondere auch derjenigen

vaults used as filling panels in iron floors. Noteworthy are the floors of the corridors of the Frankfurt *Alte Oper*, built between 1873 and 1880. They consist of iron beams and concrete slabs. Concrete is poured on falseworks, which are provided with provisional boxes on the surface, in order to endow the ceilings with the aspect of old panelled ones. And it is also worth mentioning the early use of concrete for the construction of bombproof floors in military shelters. These had traditionally consisted of iron beams, laid close to each other and supporting a high floor made of brick masonry. In the summer of 1874, the *General Inspection des Ingenieur- Corps und der Festungen* tests, on the firing range of Tegel in Berlin, a new kind of bombproof floor. It is made of a network of iron double T-beams, the minors of which are placed extremely close to each other, with the space between them covered by means of little metal vaults; this armour of iron beams is covered with a bed of concrete which is 60 cm high in the middle and slopes toward the extremities. "Henceforth, brick masonry," asserts the report following the tests, "will no longer be used to build over the beams; concrete is instead to be poured over little metal vaults and it should be as good as possible, produced with care, since the distribution of thrust produced by projectiles over the iron beams depends on the solidity of concrete, especially on the part that lies between the beams."[178]

Further studies about Portland cement and the regulation of its main physical features

Along with the previously observed development of building skills with concrete, during the third quarter of the century, the study of cement goes through further evolutions, via chemical analyses and strength tests. Around the mid-1850s, and following the model of Von Pettenkofer's research, the *Württemberg Centralstelle für Gewerbe und Handel* (Centre for Manufacturing and Trade)[179] studies three different kinds of English Roman cements and one kind of Portland cement.[180] The raw materials, the cement powders, as well as a hardened cubic specimen for each kind of cement are analysed, with the aim of comparing the different chemical compositions and finding the modifications that occur during the different steps of the cement's production and use.[181] The outcomes are then compared with the ones of similar analyses on two kinds of local hydraulic lime taken from the town of Horb am Neckar, southwest of Stuttgart. In accordance with Von Pettenkofer's explanations, this comparisons shows that local marls contain a lower amount of clay than the mixtures of lime and clay that are usually burnt to produce Portland cement, and that, moreover, local hydraulic lime is usually not burnt enough.[182]

Approximately during the same period that these cement studies take place in Stuttgart, the chemist Georg Feichtinger from Munich carries out similar investigations.[183] He studies a kind of Portland cement from England and

zwischen den Balken, das Maaß der Verteilung der Stoß und Sprengwirkung einschlagender Geschosse auf mehrere Balken abhängt" (General Inspection des Ingenierus- Corps und der Festungen, "Vorläufige Bestimmungen über die mit Rücksichtigkeit auf die Wirkung des gezogenen 28 cm Mörsers den Bombensicheren Eisendecken fortan zu gebende Construction," Berlin, March 11, 1875 (GStA, IV, HA Preußische Armee, Rep. 16 Militärvorschriften, n° 667).

[179] A public institution founded in 1848 in Stuttgart, with aims similar to those of the Prussian *Technische Deputation*.

[180] Carl Knauß, *Chemische Untersuchung einiger englischen hydraulischen Kalke*, in, "Gewerbeblatt aus Württemberg," 1855, n° 4, pp. 25–32, and in, "Polytechnische Journal," CXXXV, 1855, pp. 361–369.

[181] The raw materials from which Portland cement was burnt are, however, not available (Knauß, *Chemische Untersuchung*...cit., p. 362).

[182] *Ibid.*, p. 369.

[183] See Georg Feichtinger, *Ueber die chemischen und physikalischen Eigenschaften mehrerer bayerischen hydraulischen Kalke im Vergleich mit Portland-Cement; zugleich ein Beitrag zur Theorie des Erhärtens der hydraulischen Kalke*, in, "Polytechnisches Journal," CLII, 1859, pp. 40–61, 108–118.

237-238. G. Schwartz, levers to test tensile and compression strength of cement-based mortar specimens, 1867 (W. Michaelis, *Die hydraulischen Mörtel insbesondere der Portland-Cement in chemisch-technischer Beziehung für Fabrikanten, Bautechniker, Ingenieure und Chemiker*, Leipzig, Quant & Händel, 1869).

five different kinds of Bavarian hydraulic lime. His conclusions align with Von Pettenkofer about the origin of the different physical properties of Portland cement and Bavarian hydraulic limes. In particular, he emphasises the quality of the English Medway clay, which is usually used to produce Portland cement, and, on this topic, he emphatically states: "Bavarians need to search for a similar kind clay, if they aim at getting a Portland cement-equivalent product."[184]

Apart from such comparative investigations, which essentially involve consequences on a local scale, some essential chemical studies are carried out during the 1860s, focusing the question of cement hardening, namely how water comes to combine with substances composing the cement. Such studies contribute to improving the general scientific knowledge about hydraulic limes and cements. The question of cement hydration, which is already recognizable in previous studies about hydraulic lime and mortar (John, Fuchs, Von Pettenkofer), crystallises during this period, giving rise to two different approaches to the matter. The first develops in the wake of Fuchs' theory about the hardening of hydraulic lime, which Von Pettenkofer essentially believes to explain the hardening of Portland cement. This is thought to occur by a combination that develops between lime and silica by means of water, once silica has been set free from the other silicates' components during the burning of lime and clay. The second approach, alternately, is based on the belief in the essential difference between the hardening process of Portland cement versus that of hydraulic lime. The German promoter of this second approach is the chemist August Winkler, a scholar of the better-known Karl Jacob Löwig (1803–1890), who teaches chemistry at the university of Breslau.[185] Winkler divides hydraulic limes into two groups. One includes the Portland cements, the other the so-called Roman cements, namely all hydraulic limes that are producible by burning marls with a moderate fire, and besides, pozzolana, trass, brick powder and other similar materials. The latter are thought to contain unslaked lime once burnt, and to harden according to the principles of Fuchs' theory; whereas Portland cement is thought to contain no slaked lime at all once burnt, but rather a complex silicate made of alkali, lime, silica, silicic acid, clay and iron oxide, which is supposed to arise during the firing of lime and clay. Such a silicate is believed to undergo a decomposition process when in contact with water, while some lime combines with silica and clay, and the remaining part of it, with carbonic acid.[186] Feichtinger does not agree at all with Winkler. He strongly maintains the applicability of Fuchs' theory to explain the hardening process of Portland cement, which he further describes as a three step-process consisting of a preliminary reaction of water with all the components of cement, followed by the combination of silica and lime through water, and by the eventual combination between the residual hydrated lime and carbonic acid. Feichtinger and Winkler engage in a genuine quarrel over the hardening process of Portland cement, by way of articles

[184] "Wollen wir demnach ein dem Portland-Cemente ähnliches Product erhalten, so muß vor allem ein Thon gefunden werden, der in der chemischen Zusammensetzung ähnlich ist dem Thone vom Medwayflusse" (Feichtinger, *Ueber die chemischen und physikalischen Eigenschaften...* cit., p. 61).

[185] See August Winkler, *Ueber hydraulische Mörtel*, in, "Journal für praktische Chemie," vol. LVII, 1856, pp. 444–462, and in, "Polytechnische Journal," vol. CXLII, 1856, pp. 106–121; about the relation between Winkler and Löwig see, in particular, p. 108, n. 18.

[186] See Winkler, *Ueber hydraulische Mörtel*...cit.; Id., *Ueber Portland-Cement*, in, "Chemisches Centralblatt," III, 1858, pp. 481–483, and in, "Polytechnische Journal," vol. CXLIX, 1858, pp. 262–265; Friedrich Knapp, *Ueber Cemente*, in, "Polytechnisches Journal," CCXXII, 1876, (pp. 147–153), p. 148.

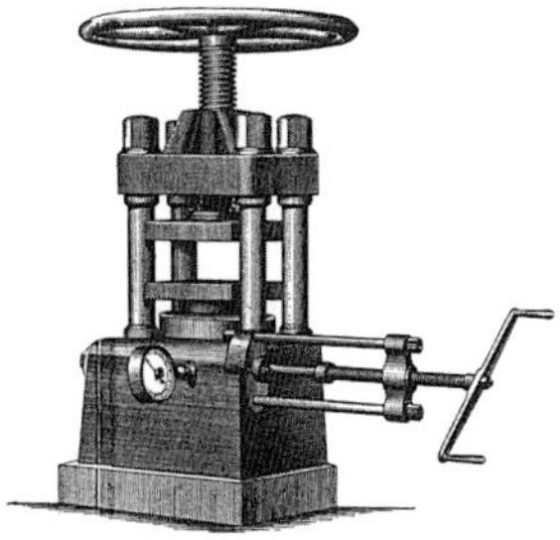

published in the "Polytechnisches Journal" between 1864 and 1865, and they never come to an agreement.

In the wake of Winkler's explanation, the chemist Wilhelm Heldt (1823–1865), a Liebig scholar, develops a further contribution to the comprehension of the hardening process of Portland cement, through research that he publishes in 1865.[187] Heldt asserts that, during the burning of Portland cement, lime combines with the main components of clay, silica and alumina, forming two compounds that he calls lime-aluminate and lime-silicate. These two compounds are then believed, once again, to decompose to their essential components by means of water, so that part of lime combines with silica and clay, while the remaining part of it combines with carbonic acid.

In such a framework, the Berlin chemist Wilhelm Michaelis (1840–1911) plays the critical role of connecting the theoretical achievement from the domain of chemistry to that of practical building needs. Michaelis studies chemistry at the University of Berlin and attends practical training at the cement factory in Wildau. Once he completes the training, he goes back to Berlin, where he continues to study at the local *Bergakademie* (school of mines) and graduates in 1867 having written a dissertation about Portland cement, which is then published in the "Journal für praktische Chemie."[188] Two years later Michaelis publishes a comprehensive book about the essential components of hydraulic mortar, with special regard to Portland cement. He relates about its chemical features, focusing especially on the hardening process, but, unlike most chemists, he also describes how to manufacture and test cement, besides giving a synthetic account of the main physical features that a good kind of Portland

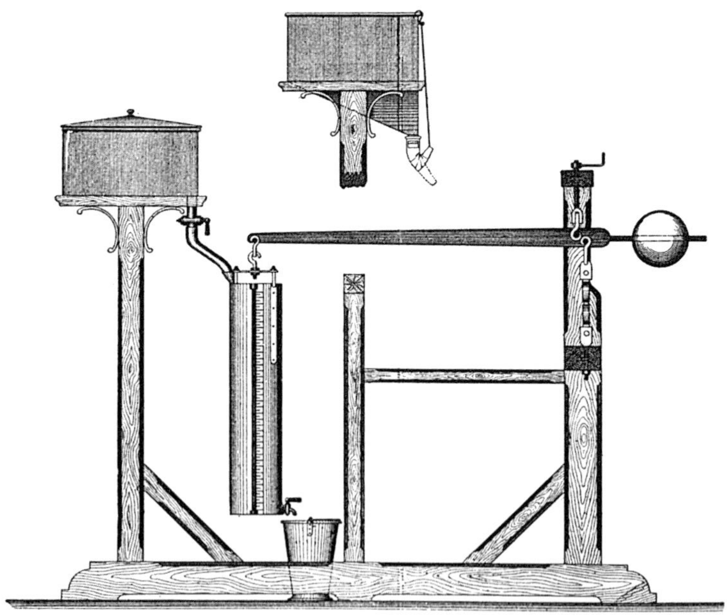

[187] See Wilhelm Heldt, *Studien über die Cemente*, in, "Journal für praktische Chemie," XCIV, 1865, pp. 129–161, 202–237; Knapp, *Ueber Cemente*...cit., pp. 148–149.

[188] See Wilhelm Michaelis, *Ueber den Portland-Cement*, in, "Journal für praktische Chemie," vol. C, 1867, pp. 257–303.

[189] See Michaelis, *Die Hydraulischen Mörtel*...cit., pp. 186–188, 192–194. The exact proportions given by Michaelis are 80 "äquivalent" of silic acid, between 15 and 25 äquivalent of clay and iron oxide, and between 210 and 230 äquivalent of lime; the term "äquivalent" is used to mean atom at that time; see entry "Aequivalente" in, *Herders Conversations-Lexikon*, vol. I, 1857², p. 54; Knapp, *Ueber*

239-241. W. Michaelis, H. Früling, A. Rudeloff, press for compressive strength test, levers for tensile strength tests, 1876 (W. Michaelis, *Zur Beurtheilung des Cementes*, Berlin, A. Seidel, 1876).

cement should have. Considering the compositions of "a series of recognized good cements," Michaelis establishes that Portland cement should contain approximately 25% silica, 69% lime and a small amount of clay and iron oxide. He essentially believes that these substances form aluminate and lime silicate compounds during the burning process, and he shares Winkler's belief about the lack of unslaked lime in Portland cement once adequately burnt. He justifies the typical and often observed phenomenon of lime slaking as the result of side factors such as the presence of coal slags from the kiln, or the use of mixtures of lime and clay that have not been assembled accurately enough; however, he rules out the idea that such phenomena could be part of the essential properties of cement. Again in accordance with Winkler, he also believes that the chemical principles of Roman and Portland cement hardening are different, and that the latter essentially hardens by simply combining its main components (aluminate and lime silicate compounds) with water. He acknowledges the existence of other complex chemical reactions, but he considers them to be secondary aspects of the hardening process.[189]

Concerning the main physical features of Portland cement, Michaelis explains that it "should consist of powder that is as fine and homogeneous as possible, it should harden strongly but not too fast, except in a few cases (constructions in rough water, sealing of cracks in walls underwater, etc.), it should neither expand nor shrink, it should be homogeneously and agreeably light-grey coloured like a stone, without any brown spots, it should show strong cementation properties, bind with stones, tolerate the addition of sand, and it must be free from any adulteration."[190]

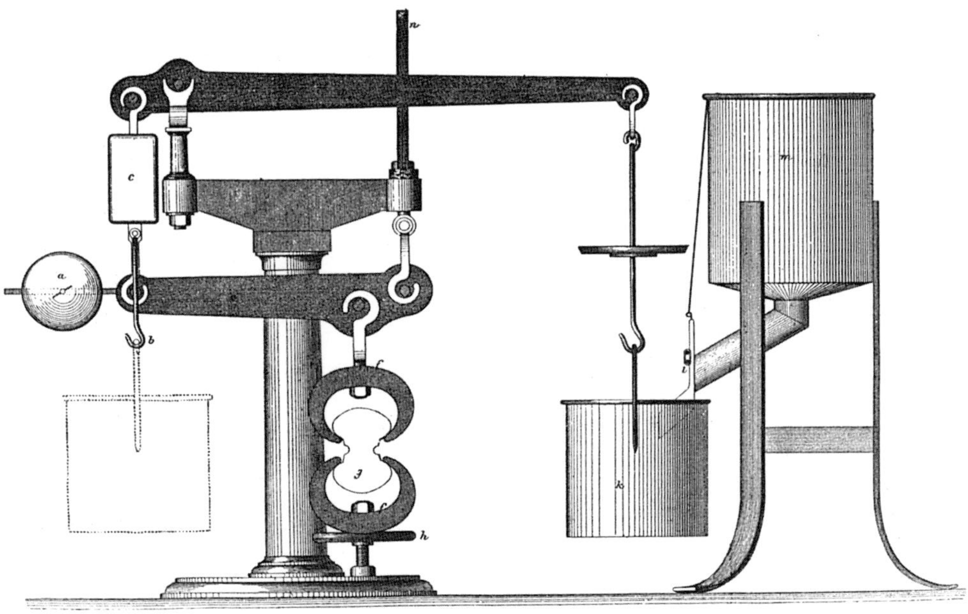

Cemente, pp. 149–150.
[190] "Portland-Cement soll ein möglichst feines und gleichmässiges Pulver darstellen; er soll stark, aber mit Ausnahme einiger weniger Fälle (Bauten in bewegtem Wasser, Dichten von Sprüngen in Wassermauern etc.) nicht zu schnell erhärten; er soll dabei sein Volumen nicht verändern, weder Contraction noch Volumvermehrung zeigen; er soll eine gleichmässige, angenehme, hellgraue Steinfarbe annehmen und durchaus keine braune Flecke bekommen; er soll stark verkittende Eigenschaften haben, stark am Stein haften und einen hohen Sandzusatz vertragen; er soll endlich frei von Verfälschungen sein" (Michaelis, *Die Hydraulische Mörtel*...cit., p. 223).

To verify the quality of Portland cement, Michaelis attributes great importance to strength tests on specimens made of pure hardened cement and specimens made of cement-based mortar. In his book from 1869, he describes a long series of strength tests carried out on specimens made of different kinds of mortar since the early decades of the century in France, England, Denmark and Germany. Among the ones in Germany, he mentions the flexural strength test that the *Oberbergamt* in Bonn carries out on specimens made of trass-based mortar; the compression strength tests that Lentze performs to verify the quality of the cement-based mortar he intends to use to build the piers of the bridges in Dirschau and Marienburg; the flexural strength tests by professor Julius Manger from the Berlin *Gewerbe-Institute*, who studies, inter alias, cement from the *Stettiner Portland-Cement Fabrik*; some compressive tests that are undertaken in 1862 in Siegburg in Rhineland; the tensile strength tests that the *Bauinspektor* Dircksen carries out when he builds a further stretch of the previously mentioned *Verbindungsbahn* in Berlin; and the tensile and flexural tests that the master builder G. Schwartz executes using an appropriate machine that he conceives expressly for that purpose in 1867.[191] This machine, which is essentially a second-class lever, can be set for tensile and for compressive strength tests. Among the different kinds of possible strength tests, Michaelis considers the tensile one to be the most appropriate for verifying the quality of Portland cement.

In 1872 Michaelis founds, together with H. Frühling and A. Rudeloff, a chemical-technical bureau equipped with a workshop to test building materials, and expressly designs two machines to test their compressive and tensile strengths. The first is a manual press, the second is a first-class lever, equipped with two stirrups to charge a specimen along with a special container, which can be filled either with sand or with water. Michaelis' workshop produces and sells these machines.[192]

It is likely on the basis of such knowledge and expertise that the *Deutsche Verein für Fabrikation von Ziegeln, Thonwaaren, Kalk und Cement* (Society of Bricks, Earthenware, Lime and Cement Makers – mentioned shortly below as *Verein*), which was founded in 1865 on the initiative of the previously mentioned master builder Türrschmiedt, entrusts Michaelis, in January 1875, with the writing of instructions for verifying and testing the quality of Portland cement.[193] Such instructions are to be submitted to the *Verein*, and, if approved, they should be adopted as rules, in order to provide order in the domain of cement manufacturing and trade. Michaelis compares the outcomes of several strength tests – especially referring to the experiments on the strength of cement that the British engineer of the London Metropolitan Boards of Works, John Grant, carries out as of 1858 to build a drainage system in South London – and draws seventeen conclusions. The first seven can be considered real prescriptions; they concern the weight, the minimal strength and the grain texture of cement. The other thirteen can be, instead, considered as advice for builders rather than instructions for proving the quality of cement.[194] Such theses are harshly criticized at the twelfth general assembly of the *Verein*, which takes place in January 1876, and are, consequently, rejected.

[191] See Michaelis, *Die Hydraulische Mörtel*...cit., pp. 229–259. For the tests by Manger, see Julius Manger, *Der Stettiner Portland-Cement in Versuche und Erfahrungen dargestellt und beleuchtet*, Stettin, F. Heffenland, 1860; Id., *Untersuchungen über die Festigkeit von reinen und gemisehlen Cementen*, in, "Zeitschrift für Bauwesen," IX, 1859, coll. 523–534.

[192] See Wilhelm Michaelis, *Zur Beurtheilung des Cementes*,

The controversy essentially arises over the few strict prescriptions contained in the theses, namely the one imposing that cement should be sold by weight, the one establishing a correlation between the grain and the price of cement, and the one about the minimum required tensile strengths. The first two are considered as pertaining to the domain of business rather than that of technology, while the values of minimal required tensile strength are considered high and unrelated to customers' needs.[195] The opposition against Michaelis' theses is so great that some cement makers even mistrust their scientific reliability. In order to overcome the deadlock, the *Verein* entrusts a committee made of architects and advocates of the cement makers, lead by Hugo Delbrück (1825–1900), who is at that time the director of the *Stettiner Portland-Cement-Fabrik*, with the development of new instructions.[196] The committee produces a document made of six "resolutions," accounting for the weight, the setting time, the grain-size and the volume variation of cement, besides prescribing that the cementation capacity is to be confirmed by means of tensile strength tests on specimens made of cement-based mortar, using an improved machine by Michaelis. The resolutions are approved at the thirteenth general assembly of the *Verein*, which takes place in January of 1877, and are adopted by the Berlin *Architekten Verein* (Berlin Society of Architects) and by the *Verein Berliner Baumarkt* (Society of the Berlin Building Market) as norms, being the first shared and compulsory rules to evaluate cement. Shortly after the approval of the new resolution, a number of cement makers led by Michaelis foment a new opposition, which, however, has, no effective consequence. One year later the norms are even submitted for the approval of the minister of trade and industry, and with that, they rise from the domain of shared rules among professional associations to the wider domain of national regulations and laws.

With the development and approval of a regulation about the features of Portland cement, an essential chapter of German construction history is achieved. It involved the radical transformation of a building culture that used to exalt the quality of aerial lime and mortar, limiting the use of hydraulic mortar made of lime and pozzolanic materials to a few cases, to a building culture in which the use of hydraulic artificial cement became prevalent. At the same time, essential skills to build with concrete develop, progressing from the observations of foreign, and more advanced, expertise. The scientific approach to the study of lime and cement, and the effort to establish rules in the domain of construction, even for the empirical practice of strength tests, are crucial features of this evolution and affect the subsequent development of reinforced concrete. As a matter of fact, Joseph Monier's empirical method to produce objects and construction elements made of concrete and metal nets is studied on scientific bases specifically in Germany, in the first half of the 1880s, leading to a clear formulation and definition of reinforced concrete, and providing an early mathematical model to calculate a reinforced concrete slab. This can be considered the result of the evolution of knowledge and expertise about cement and concrete that occurs approximately during the first three quarters of the 19th century.[197]

Berlin, A. Seidel, 1876, p. 5.
[193] See Anon., *Die Cementfrage*, in, "Thonindustrie-Zeitung," I, 1876, pp. 2–3.
[194] See Michaelis, *Zur Beurtheilung des Cementes*, pp. 45–46.
[195] See Anon., *Die Cementfrage*, in, "Thonindustrie-Zeitung," I, 1876, pp. 37–38.
[196] *Ibid.*, p. 3.
[197] On this matter, see Gustav Adolf Wayss, *Das System Monier (Eisengerippe mit Cementumhüllung) in seiner Anwendung auf das gesammte Bauwesen*, Berlin, Seidel & Cie, 1887.

Bibliography and archival sources

This bibliography only gathers the primary sources; the secondary sources are mentioned in the notes to the text. It is compiled in chronologic order, in accordance with the writing style and with the aim of providing readers with an overview of the topics of interest that developed in the technical literature during the analysed period.
The chapters' sequence – periodicals, books and archival sources – results from the research methodology. First printed sources were analysed, in order to trace a sequence of important facts and topics, which were then studied in detail through archival sources.

Articles from periodicals

Christian Ludwig Ziegler, *Von dem Terras, der Puzzolana, und deren Sobstitutionsarten*, in, "Hannoversches Magazin," XI, 1773, coll. 81–94.

Jordan, *Nachricht*, in, "Hannoversches Magazin," XX, 1782, n° 19, coll. 301–304.

Mitglieder des Königl. Preuss. Ober-Bau-Departments, *Vorrede*, in, "Sammlung nützlicher Aufsätze und Nachrichten, die Baukunst betreffend," I, 1797, pp. III–IX.

David Gilly, *Über die Gründung der Gebäude auf ausgemauerte Brunnen*, in, "Sammlung nützlicher Aufsätze und Nachrichten, die Baukunst betreffend," I, 1799, pp. 137–138, vol. II, 1798.

Paul Ludwig Simon, *Ueber die Natur des Kalksteins*, in, "Sammlung nützlicher Aufsätze und Nachrichten die Baukunst betreffend," I, 1799, pp. 96–112, II, 1800, pp. 53–78, 84–119.

Anon., *Veflag aan Zijne Excellentie den Minister van Eeredienst an Binnenlandsche Zaken, door de Eerste Klasse van het Koninklijk Instituut van Wetenschappen, Letterkunde en Schoone Kunsten, betrekkelijk den haar opgedragen last, nopens het fluk der buitenlandse, en de sedert 1789 uitgevondene binnenlandse Tras of Cement (Bericht an Se. Excell. Den Minister der Gottesverehrung und der innern Angelegenheit, von der ersten Klasse des Königl. Instituts der Wissenschaften, Literatur und schönen Künste, betreffend den an sie gelangten Auftrag in Ansehung des ausländischen, und des seit 1789 erfundenen inländischen Trasses oder Cement)*, in, "Allgemeine Literatur-Zeitung," 1812, n° 212, coll. 1–8.

Johann Wolfgang Döbereiner, *Ueber das Verhalten des Kalks zu dem Kiesel- und Thonkali auf nassem Wege und über andere verwandte Gegenstände*, in, "Journal für Chemie und Physik," vol. X, 1814, pp. 113–123.

Christoph Arzberger, *Römisches Cement*, in, "Kunst- und Gewerbe-Blatt des polytechnischen Vereins für das Königreich Bayern," V, 1819, coll. 598–600.

Louis Joseph Vicat, *Sur la Chuax et le Mortier en général et en particulier; sur la différence entre le mortier de chaux de coquilles de moules et de pierres calcinaires, avec la théorie des mortiers*, in, "Annales de Chimie et Physique," XIX, 1821, pp. 15–25.

W. Z., *"Über Kalk und Mörtel im Allgemeinen und den Unterschied zwischen Muschelschalen und Kalksteinmörtel insbesondere; nebst Theorie des Mörtels". Eine von der Holländischen Gesellschaft der Wissenschaften der gekrönte Preisschrift, von J. F. John. 1819*, in, "Jenaische Allgemeine Literatur-Zeitung," 1821, n° 178, coll. 457–461.

Louis-Joseph Vicat, *Ueber Kalk und Mörtel. Ein Auszug aus Hrn. Vicat's, Brücken- und Strassenbau-Inspektors, Recherches expérimentales sur les chaux de construction, les bétons et les mortiers ordinaires; et du Supplément inédit relatif à la fabrication des pouzzoulanes artificielles*, in, "Polytechnische Journal," 1821, vol. IV, pp. 280–293.

Richard Jakob August von Voit, *Ueber die Bereitung eines guten Mörtels durch Anwendung des richtigen Verhältnisses des Sandes zum Kalk; auch über die Quantität des Mörtels zum Mauerwerk*, in, "Polytechnisches Journal," II, 1821, vol. IV, pp. 293–303.

Anon., *Ueber die Stiftung eines Vereines zur Beförderung des Gewerbfleißes in Preußen, nebst dessen Statut*, in, "Polytechnisches Journal," 1821, vol. IV, pp. 486–494.

Peter Ludwig Maréchaux, *Würdigung sämmtlicher bis jetzt bekannt gewordener Methoden, das Getreide, mehrere Jahre hindurch, ohne Nachtheil für dasselbe, aufzubewahren*, in, "Polytechnisches Journal," 1821, vol. V, pp. 334–376.

Johann Friedrich John, *Sur la Chuax et le Mortier en général et en particulier; sur la différence entre le mortier de chaux de coquilles de moules et de pierres calcinaires, avec la théorie des mortiers*, Louis Joseph Vicat ed., in, "Annales des Mines ou Recueil de mémoires sur l'exploitation des mines, et sur les sciences qui s'y rapportent," vol. VII, 1822, pp. 467–482.

Anon., *Zweckmäßige Getreide-Magazine*, in, "Monatsblatt für Verbesserung des Landbauwesens du für zweckmäßige Verschönerung des Bayerischen Landes," II, 1822, pp. 9–13.

Anon., *Universal- oder Parolik-Kitt (Parolic Cement). Von Th. Gill*, in, "Polytechnisches Journal," 1822, vol. VIII, p. 19.

Anon., *Ueber die Silos oder Erdgruben zur Aufbewahrung des Getreides*, in, "Polytechnisches Journal," 1822, vol. IX, pp. 329–331.

Preisaufgaben des Vereins. Vierte Preisaufgabe, in, "Verhandlungen des Vereins zur Beförderung des Gewerbfleißes in Preußen," I, 1822, p. 22.

Anon., *Ueber den Mörtel und dessen Verbesserung*, in, "Neues Kunst- und Gewerbblatt," IX, 1823, pp. 109–114.

Anon., *Ueber Wasser-Mörtel*, in, "Neues Kunst- und Gewerbblatt," IX, 1823, pp. 285–287.

Coudray, *Interessante Notizen über den Betrieb des Bauwesens im Großherzogthume Sachsen-Weimar*, in, "Monatsblatt für Bauwesens und Landesverschönerung," III, 1823, pp. 17–20.

Anon., *Ueber die Silos oder Erdgruben zur Aufbewahrung des Getreides*, in, "Polytechnisches Journal," 1823, vol. X, p. 123.

Anon., *Neuer Kitt oder künstlicher Stein, worauf Jak. Frost, Baumeister zu Finchley in Middlesex, dd. 11. Junius 1822 sich ein Patent geben ließ*, in, "Polytechnisches Journal," 1823, vol. XI, pp. 117–118.

Anon., *Ueber die Vortheile von Tickell's römischen Kitt (Roman Cement)*, in, "Polytechnisches Journal," 1823, vol. XI, pp. 249–250.

Pierre Berthier, *Analyse verschiedener Kalksteine in Bezug auf den besten Mörtel*, in, "Polytechnisches Journal," 1823, vol. XI, pp. 350–363.

Louis-Joseph Vicat, *Bemerkungen, als Antwort auf den (vorstehenden) Aufsaz des Hrn. Berthier, Ingénieur en chef des Mines, über die Theorie des Mörtels*, in, "Polytechnisches Journal," 1823, vol. XI, pp. 363–373.

Louis-Joseph Vicat, *Bemerkung über einige Erscheinungen bei dem Brennen des gemeinen und künstlichen Kalkes*, in, "Polytechnisches Journal," 1823, vol. XII, pp. 429–433.

Anon., *Ueber die Korngruben oder sogenannten Silos*, in, "Polytechnisches Journal," 1824, vol. XIII, pp. 255–259.

Louis-Joseph Vicat, *Untersuchungen über die Anwendung des hydraulischen Mörtels statt des Gipses zum Modelliren*, in, "Polytechnisches Journal," 1824, vol. XV, pp. 186–207.

Clement-Louis Treussart, *Ueber den hydraulischen Mörtel*, in, "Polytechnisches Journal," 1824, vol. XV, pp. 341–345.

Anon., *Auszug aus den Protokollen der monatlichen Versammlungen in dem Jahre 1823*, in, "Verhandlungen des Vereins zur Beförderung des Gewerbfleißes in Preußen," III, 1824, (pp. 13–18), p. 16.

Johann Nepomuk Fuchs, *Ueber ein neues nutzbares Produkt aus Kieselerde und Kali*, in, Karl Wilhelm Gottlob Kastner, ed., *Archiv für die gesammte Naturlehre*, 1825, vol. V, pp. 385–412.

Anon., *Neue Verbesserung in Verfertigung künstlicher Steine, worauf Joh. Aspdin, Maurermeister zu Leeds in Yorkshire, am 21. Oktober 1824. sich ein Patent geben ließ*, in, "Polytechnisches Journal," 1825, vol. XVI, pp. 304–305.

Samuel Bayshan, *Methode, Vasen, Urnen, Beken und andere Zierrathe, die man bisher aus Stein und Marmor arbeitete, aus einer Verbindung von Artikeln zu verfertigen, die ehevor niemahls zu diesem Zweke gebraucht wurden, worauf Samuel Bayshan Gentleman zu New-castle-underline, in Staffordshire, am 26. Juli 1821. sich ein Patent geben ließ*, in, "Polytechnisches Journal," 1825, vol. XVI, pp. 305–308.

Louis-Joseph Vicat, *Thatsachen zur Aufhellung der Theorie über Kalkmörtel*, in, "Polytechnisches Journal," 1825, vol. XVII, pp. 481–484.

Louis-Joseph Vicat, *Neuer hydraulischer Mörtel*, in, "Polytechnisches Journal," 1825, vol. XVIII, p. 480.

Anon., *Auszug aus den Protokollen der monatlichen Versammlungen in den Monaten Juli und August des laufenden Jahres*, in, "Verhandlungen des Vereins zur Beförderung des Gewerbfleißes in Preußen," 1825, vol. IV, (pp. 129–133), p. 133.

Anon., *Hrn. Aspdin's künstliche Steine (Portland-Cement)*, in, "Polytechnisches Journal," 1826, vol. XIX, pp. 588–589.

Anon., *Ueber Theorie der Mörtel von Hrn. Vicat*, in, "Polytechnisches Journal," vol. XX, 1826, pp. 106–107.

Clement-Louis Treussart, *Notiz des General Treussart über die Bereitung der künstlichen Puzzolanen oder Trasse*, in, "Polytechnisches Journal," 1826, vol. XXI, pp. 40–47.

Louis-Joseph Vicat, *Neue Thatsachen zur Theorie der Kalkmörtel*, in, "Polytechnisches Journal," 1826, vol. XXI, pp. 432–437.

Louis-Joseph Vicat, *Nachträgliche Bemerkung über die Mörtel*, in, "Polytechnisches Journal," 1826, vol. XXII, pp. 166–167.

Verzeichniß der im Königreiche Preußen im Jahre 1825 ertheilten Patente, in, "Verhandlungen des Vereins zur Beförderung des Gewerbfleißes in Preußen," 1826, col. V, p. 51.

Anon., *Technologie*, in, "Allgemeine Literatur-Zeitung," 1827, vol. II, col. 480.

Pierre-Simon Girard, *Ueber einige natürliche und künstliche Puzzolanen*, in, "Polytechnisches Journal," 1827, vol. XXIII, pp. 60–65, 1827, vol. XXV, pp. 404–409.

J. J. Hawkins, *Ueber römischen Mörtel (roman Cement); aus einem Schreiben des Hrn. J. J. Hawkins an die Pennsylvania Society for the promotion of the internal improvement*, in, "Polytechnisches Journal," 1827, vol. XXIII, pp. 65–69.

John Phillips Beavan, *Mörtel zum Bauen und zu anderen Zweken, worauf Joh. Phil. Beavan, Gentleman in Clifford Street, Middlesex, in Folge einer von einem im Auslande wohnenden Fremden ihm gemachten Mittheilung sich am 7. December 1825 ein Patent ertheilen ließ*, in, "Polytechnisches Journal," 1827, vol. XXIII, pp. 558–559.

Louis-Joseph Vicat, *Bemerkung über künstliche Puzzolanen*, in, "Polytechnisches Journal," 1827, vol. XXIV, pp. 175–177.

Pierre-Simon Girard, *Notiz über die natürlichen und künstlichen Puzzolanen*, in, "Polytechnisches Journal," 1827, vol. XXV, pp. 409–420.

Peter Christian Wilhelm Beuth, *Der Tunnel unter der Themse*, in, "Verhandlungen des Vereins zur Beförderung des Gewerbfleißes in Preußen," VI, 1827, pp. 286–294.

Anon., *Ueber Straßenbau*, in, "Polytechnisches Journal," 1828, vol. XXX, pp. 318–319.

Friedrich Wilhelm Elsner, *Über die Anwendung des Béton-Mörtels zum Fundamentiren unter Wasser*, in, "Journal für die Baukunst," 1829, vol. I, pp. 236–245.

Anon., *Auszug aus den Protokollen der Sitzungen in den Monaten April, Mai und Juni 1829*, in, "Kunst- und Gewerbe-Blatt des polytechnischen Vereins für das Königreich Bayern," XV, 1829, coll. 389–392.

Johann Nepomuk Fuchs, *Ueber Kalk und Mörtel*, in, "Beilage zum Kunst- und Gewerbe-Blatt des polytechnischen Vereins für das Königreich Bayern," XV, 1829, coll. 1–56, and in, "Die neuesten Forschungen im gebiete der technischen und ökonomischen Chemie," vol. III, 1829, pp. 1–26, 132–162.

Anon., *Ueber die Bereitung der hydraulischen Cemente*, in, "Polytechnisches Journal," 1829, vol. XXXI, pp. 433–435.

Antoine Raucourt de Charleville, *Bemerkungen über den anfänglichen und zukünftigen Widerstand der Mörtel*, in, "Polytechnisches Journal," 1829, vol. XXXI, pp. 435–440.

Preisaufgaben für die Jahre 1829 und 1830. Dritte Preisaufgabe, in, "Verhandlungen des Vereins zur Beförderung des Gewerbfleißes in Preußen," VIII, 1829, p. 31.

Anon., *Ueber die mit Roman Cement, aus der Fabrik von Francis, White und Francis, in London, angestellten Versuche*, in, "Verhandlungen des Vereins zur Beförderung des Gewerbfleißes in Preußen," VIII, 1829, pp. 123–126.

Oynhausen, Von Decken, *Beschreibung der Sammlung einiger nutzbaren Mineralien und Hüttenproducte aus England, welche der Verein der Güte der Herren von Oeynhausen und von Decken verdankt*, in, "Verhandlungen des Vereins zur Beförderung des Gewerbfleißes in Preußen," VIII, 1829, (pp. 140–167), pp. 163–165.

Butzke, *Beschreibung der Ziegeldeckung der Dächer nach böhmischer Art*, in, "Journal für die Baukunst," 1830, vol. II, (pp. 217–236), pp. 217–223.

Anon., *Die Schwung-Ramme, ein Beitrag zum Grundbau*, in, "Journal für die Baukunst," 1830, vol. II, pp. 341–350.

Zimmermann, *Einige Bemerkungen über die Festigkeit, Mischungsverhältnisse und Zubereitung des Bétons, oder des Mauerwerkes aus klein geschlagenen, mit Mörtel untermengten Steinen, dessen man sich zuweilen, um Fangdämme und Wasserschöpfen zu sparen, zur Fundamentierung von Bauwerken unter Wasser bedient*, in, "Journal für die Baukunst," 1830, vol. III, pp. 1–31.

Braun, *Über Anwendung des Trass-Bétons zur Fundamentierung der Gebäude*, in, "Journal für die Baukunst," 1830, vol. III, pp. 112–117.

[Johann Heinrich] Krahmer, *Bemerkungen über die Dauer und Festigkeit des in England erfundenen Roman-Cement. Nebst Nachrichten von einigen andern Cementen*, in, "Journal für die Baukunst," 1830, vol. III, pp. 476–483.

Johann Friedrich Wilhelm Dietlein, *Grundzüge der Vorlesungen in der Königl. Bau-Akademie zu Berlin über Strassen- Brücken- Schleusen- Canal- Strom- Deich und Hafen-Bau*, in, "Journal für die Baukunst," 1830, vol. III, pp. 33–87, 327–358, 420–464, 1831, vol. IV, pp. 90–120, 302–341, 1832, vol. V, pp. 36–72, 155–203, 245–279, 315–385; esp. vol. III, pp. 330, 341, 344–346, vol. IV, pp. 303, 311.

Charles-François Mallet, *Über den Cement von Pouilly und andere Zemente*, in, "Journal für die Baukunst," 1831, vol. IV, pp. 52–80.

Otto Christian Friedrich Reinhold, *Beiträge zur Bestimmung der Höhe der Geweolbsteine und Stärke der Widerlager und Mittelpfleiler grosser massiver Brücken; durch viele Beispiele erläutert und insbesondere für Practiker gesammelt und zusammengestellt*, in, "Journal für die Baukunst," 1831, vol. IV, (pp. 363–407), p. 399.

Anon., *Steinmörtel mit Milch bereitet*, in, "Polytechnisches Journal," 1831, vol. XL, pp. 236–237.

Preisaufgaben des polytech. Vereins in Bayern, in, "Polytechnisches Journal," 1831, vol. XL, (pp. 387–389), p. 388.

Anon., *Ueber Aufbewahrung des Getreides in Silo's und die Benuzung des hydraulischen Mörtels zur Erbauung derselben*, in, "Polytechnisches Journal," 1831, vol. XXXIX, pp. 427–433.

Louis-Joseph Vicat, *Bemerkungen beim Bau der Brücke über die Dordogne bei Souillac*, ed. Friedrich Wilhelm Dietlein, in, "Journal für die Baukunst," 1832, vol. V, pp. 73–86.

Friedrich Panzer, *An die Redaktion des Kunst- und Gewerbe-Blattes*, in, "Kunst- und Gewerbe-Blatt des polytechnischen Vereins für das Königreich Bayern," vol. X, 1832, coll. 111–144.

Preisvertheilung des polytechnischen Vereins, in, "Kunst- und Gewerbe-Blatt des polytechnischen Vereins für das Königreich Bayern," XVIII, 1832, (coll. 643–649), col. 649.

Anon., *Würdigung der Arbeit des Professors und Akademikers Fuchs über Kalk und Mörtel*, in, "Kunst- und Gewerbe-Blatt des polytechnischen Vereins für das Königreich Bayern," XVIII, 1832, coll. 756–761.

Charles-Pierre Gourlier, *Bericht des Hrn. Gourlier über eine Abhandlung des Hrn. Lebrun d. jüng., Maumeister zu Alby, Dept. du Tarn, die Anwendung des Grund oder Steinmörtel (béton) zum Baue ganzer Wohnhäuser betreffend*, in, "Polytechnisches Journal," 1832, vol. XLIV, pp. 114–119.

Anon., *Ein neues Baumaterial*, in, "Polytechnisches Journal," 1832, vol. XLVII, pp. 73–74.

Anon., *Verzeichniß der vom 24. Nov. 1832 bis 22. Jan. 1833 in England ertheilten Patente*, in, "Polytechnisches Journal," 1832, vol. XLVII, (pp. 308–310), p. 308.

[Johann Christian] Wutzke, *Bemerkungen über den Bromberger Kanal*, in, "Preussische Provinzial-Blätter," vol. VII, 1832, pp. 337–348.

Moritz Karl Ernst von Prittwitz, *Beschreibung einer mechanische Vorrichtung zur Bereitung des Mörtels*, in, "Verhandlungen des Vereins zur Beförderung des Gewerbfleisses in Preußen," XI, 1832, pp. 283–284.

Anon., *Ueber den hydraulischen Kalk*, in, "Annalen der Pharmacie," 1833, vol. V, pp. 241–246.

Walchner, *Ueber die Brauchbarkeit vaterländischer Thon- und Mergelarten zu Wassermörtel und über die Anwendung derselben zu ökonomischen Bauten*, in, "Landwithschaftliches Wochenblatt für das Grossherzogthum Baden," 1833, pp. 153–154, 321–324.

Preisaufgaben für die Jahren 1833–34. Erste Preisaufgabe betreffend die Anfertigung eines hydraulischen Cements aus inländischen Materialien von gleicher Güte als der Englische Roman Cement, in, "Verhandlungen des Vereins zur Beförderung des Gewerbfleißes in Preußen," XII, 1833, pp. 30–31.

Hampel, *Beschreibung des in den Jahren 1830 und 1831 für die Cadetten-Anstalt zu Potsdam erbauten Lazareth-Gebäudes*, in, "Journal für die Baukunst," 1834, vol. VII, pp. 131–139.

Karl Zelger, *Hydraulischer Kalk*, in, "Korrespondenten von und für Deutschland," n° 52, 1834; in, "Königlich Bayerisches Intelligenzblatt für den Ober-Main-Kreis," 1834, n° 27, p. 257; and in, "Tagblatt der Stadt Bamberg," 1835, n° 117, p. 468.

Statuten des Gewerbe-Vereins für das Königreich Hannover, in, "Mittheilungen des Gewerbe-Vereins Hannover," 1834,coll. 3–10.

Feller, *Bekanntmachingen vermischten Inhalts. Baumaterialien-Lieferung*, in, "Oeffentlicher Anzeiger als Beilage des Amtsblatts der königlchen Regierung zu Oppeln," 1834, pp. 618–619.

Anon., *Ueber eine neue Benutzung des hydraulischen Zements oder römischen Kittes von Pouilly*, in, "Polytechnisches Journal," 1834, vol. LII, pp. 220–224.

Anon., *Ueber die Anwendung von Schieferplatten als Unterlagen für Eisenbahnen*, in, "Polytechnisches Journal," 1834, vol. LIII, p. 457.

Anon., *Die im November 1834 zu München gehaltene Industrieausstellung*, in, "Polytechnisches Journal," 1834, vol. LIV, (pp. 393–427), p. 410.

Friedrich Wilhelm Otto Ludwig Freiherr von Reden, *Bericht über die von dem Gewerbe-Vereine für das Königsreich Hannover in den Monaten Mai und Juni 1835 veranstaltete erste Ausstellung inländlicher gewerblicher Erzeugnisse*, in, "Mittheilungen des Gewerbe-Vereins für das Königsreich Hannover," 1835, coll. 527–529.

Anon., *Ein Beispiel der Anwendung des Steinmörtels zur Ausbesserung von Bauten*, in, "Polytechnisches Journal," 1835, vol. LVI, pp. 152–153.

Charles-Chrétien-Constant Vauvilliers, *Bericht des Hrn. Vauvilliers über eine Abhandlung des Hrn. Thomassin, in welcher mit Steinmörtel gebaute Bahnen als vorzüglicher als die Eisenbahnen gepriesen werden*, in, "Polytechnisches Journal," 1835, vol. LXI, 1835, pp. 225–228.

Anon., *Auszug aus den Protokollen der Versammlung des Vereins in den Monaten Novemebr und Dezember d. I. J.*, in, "Verhandlungen des Vereins zur Beförderung des Gewerbfleißes in Preußen," XIV, 1835, (pp. 277–280), pp. 279–280.

Adolph Schwarzenberg, Carl Anton Henschel, Johann Werner Henschel, Johann Heinrich Buff, *Gutachten die Brauchbarkeit des Koch'schen Cements betreffend. Protokoll 18ten Juni 1836*, in, "Gewerbeblätter für Kurhessen," I, 1836, pp. 137–139.

Joseph Jopling, *Ueber Platformen für Eisenbahnen*, in, "Polytechnisches Journal," XVII, 1836, vol. LX, pp. 2–5.

Anon., *Ueber die Mörtelbereitung*, in, "Polytechnisches Journal," XVII, 1836, vol. LX, pp. 406–407.

Anon., *Ueber die Benutzung des Steinmörtels zu verschiedenen Bauten*, in, "Polytechnisches Journal," 1836, vol. LXII, p. 155.

Anon., *Über Traß und Kunst-Cement (Aus "Übersicht der Rheinischen und Eisleber erloschenen Vulkane u. von Freiherr von der Weck.")*, in, "Wöchentliches Beiblatt zum Allgemeinen Organ, enthaltend die Verhandlungen und Mittheilungen des Gewerb-Vereins zu Köln," 1836, n° 48, pp. 191–192, n° 50, pp. 195–196.

Anon., *Technische Notizen auf einer Reise nach dem Unter-Rhein, im Herbst 1835*, in, "Zeitschrift über das gesammte Bauwesen," I, 1836, pp. 101–107.

Karl August Menzel, *Ueber die Dorn'schen flachen Lehmdächer, mit Rücksicht auf das zu Eldena im Jahre 1836 ausgeführte flache Lehmdach am dortigen Gasthofe*, in, "Allgemeine Bauzeitung," II, 1837, (n° 2–3, pp. 9–24), n° 2, pp. 9–10, n. 1.

Anon., *Hilfsmittel bei Bauführungen. Ueber den römischen Zement*, in, "Allgemeine Bauzeitung," II, 1837, pp. 390–392.

Lünzner, *Versuche mit kassler Cement*, in, "Gewerbeblätter für Kurhessen," II, 1837, pp. 143–144; in, "Polytechnisches Centralblatt," III, 1837, pp. 600–601.

Johann Daniel Wilhelm Eduard Engelhard, *Instruction für junge Architekten zu Reisen in Italien*, in, "Journal für die Baukunst," 1837, vol. XI, pp. 1–24, 155–174, 203–238, 303–333, esp. pp. 308–309, 312.

Carl Gustaf Rydin, *Das gegossene Kalkmörtelhaus. Eine neue Bauart von C. H. Rydin* [sic!], ed. Heinrich Friedrich Franz Körte, in, "Möglin'sche Jahrbücher der Landwirthschaft," 1837, vol. II, pp. 151–181.

Charles Blacker Vignoles, *Ueber ein Eisenbahnsystem mit hölzernen Längebalken als Unterlage*, in, "Polytechnisches Journal," XVIII, 1837, vol. LXIV, (pp. 2–12), p. 9.

Anon., *Ueber eine neue Art von hohlen Schienen für Eisen- oder Schienenbahnen*, in, "Polytechnisches Journal," 1837, vol. LXV, pp. 9–11.

Friedrich Panzer, *Mittel gegen das Versickern des Wassers in Sandboden*, in, "Allgemeine Bauzeitung," III, 1838, p. 402.

Carl Wilhelm Dempp, *Ueber die Benutzung des Mergelkalkes zu hydraulischem Kalk im Allgemeinen, und dessen Anwendung in Bayern insbesondere*, in, "Allgemeine Bauzeitung," III, 1838, (n° 49, pp. 442–443, n° 50, pp. 447–450, n° 51, pp. 454–457), n° 49, p. 442.

Anon., *Ueber die Verwendung des natürlichen und nachgeahmten Erdharzes zu Fußpfaden, Fahrbahnen und architektonischen Zweken in Frankreich*, in, "Polytechnisches Journal," XIX, 1838, vol. LXIX, (pp. 426–432), p. 427.

Anon., *Ueber die Anwendung von Steinmörtel zum Straßenbaue*, in, "Polytechnisches Journal," XIX, 1838, vol. LXX, pp. 459–460.

Hoffmann, *Notizen über den Kalk vom Moritzberge bei Oppeln, und dessen Verwendung zum Mörtel bei Wasserwerken zu Breslau*, in, "Notiz-Blatt des Architekten-Verein zu Berlin," 1839, pp. 18–19.

Heinrich Joseph Alois von Pechmann, *Ueber die angewendeten Mittel, um die in Sand gegrabenen Theile des Ludwigskanales wasserdicht zu machen*, in, "Allgemeine Bauzeitung," V, 1840, pp. 375–382.

Friedrich Wilhelm Otto Ludwig Freiherr von Reden, *Amtlicher Bericht über die siebzehnte Versammlung des deutschen Naturforscher und Aertzte zu Pyrmont im September 1839*, in, Lorenz Okenfuß, ed., *Isis oder Encyclopädische Zeitung von Oken*, 1840, (coll. 928–930), col. 929.

Karl von Körber, *Ueber den Gebrauch der Santorinerde*, in, "Allgemeine Bauzeitung," VI, 1841, pp. 266–268.

Otto Christian Friedrich Reinhold, *Sammlung practischer Erfahrungen und Vorschriften, Cemente, Mörtel und Bétons betreffend*, in, "Journal für die Baukunst," 1841, vol. XV, pp. 88–100, 131–159.

Heinrich Joseph Alois von Pechmann, *Über die Mittel, welche angewendet werden, um die in Sandboden gegraben Theile des Ludwigs-Kanals wasserdicht zu machen*, in, "Journal für die Baukunst," 1841, vol. XV, pp. 160–179, 1845, vol. XXII, pp. 1–14.

Heinrich Engelhardt, announcement in, "Nordhäuser Nachrichtenblatt," March 4, 1841.

Carl Gustaf Rydin, *Gegossenes Kalkmörtel-Haus. Eine neue Bauart von Rüdin* [sic], ed. Ström, in, "Kunst- und Gewerbe-Blatt des Polytechnischen Vereins für das Königreich Bayern," XXVIII, 1842, coll. 502–513; and in, "Polytechnisches Journal," 1842, vol. LXXXV, pp. 423–431.

Anon., *Auszug aus den Protokollen der Versammlungen des Vereins in den Monaten September und Oktober d. l. J.*, in, "Verhandlungen des Vereins zur Beförderung des Gewerbfleisses in Preussen," XXI, 1842, p. 139.

[Eduard] Römer, *Die Düsseldorf-Elberfelder Eisenbahn*, in, "Allgemeine Bauzeitung," VIII, 1843, (pp. 60–77), pp. 62–64.

Alfred Lecointe, *Bemerkungen über einige mechanische Verfahrungsarten zur Bereitung des Mörtels und Betons*, in, "Allgemeine Bauzeitung," VIII, 1843, pp. 399–400.

François-Martin Lebrun, *Practische Abhandlung über die Kunst mit Béton zu bauen*, in, "Journal für die Baukunst," 1843, vol. XIX, pp. 135–186, 195–253, 287–346.

Carl Maximilian von Bauernfeind, *Ueber hydraulischen Kalk und Mörtel aus Natur- und Kunstprodukten des Fichtelgebirgs*, in, "Kunst- und Gewerbe-Blatt des polytechnischen Vereins für das Königreich Bayern," XXVIII, 1843, coll. 624–647.

Anon., *Zusammenstellung der Nachrichten über die seit 1838 im Königreiche Hannover neu entstanden oder wesentlich vergrößerten und verbesserten Fabriken*, in, "Mittheilungen des Gewerbe-Vereins für das Königreich Hannover, 1842–1843," n° 32, 1843, (coll. 325–408), coll. 399–400.

W. Winkelmann, *Ueber die Fundirung der Schleuse n° 93 des Ludwigs-Canals an der Ausmündung desselben in die Regnitz bei Bamberg*, in, "Zeitschrift für praktische Baukunst," III, 1843, pp. 231–235.

Johann Daniel Wilhelm Eduard Engelhard, *Über den Casseler Cement*, in, "Journal für die Baukunst," 1844, vol. XX, pp. 165–188, 189–198.

Anon., *Auszug aus den Protokollen der Versammlungen des Vereins in den Monaten Januar und Februar d. l. J.*, in, "Verhandlungen des Vereins zur Beförderung des Gewerbfleisses in Preussen," XXIII, 1844, p. 29.

W. A. Kramer, *Bericht über die Deliberation-Versammlung vom 31. October 1844*, in, "Verhandlungen und Schriften der Hamburgischen Gesellschaft zur Beförderung der Künste und Nützlichen Gewerbe. Verhandlungen vom Jahre 1844," 1845, vol. I, pp. 167–171.

Johann Friedrich Voigt, *Bericht der am 2. Mai 1844 niedergesetzten Commission, zur Beratung über den Bauplan für das Gebäude der Patriotischen Gesellschaft*, in, "Verhandlungen und Schriften der Hamburgischen Gesellschaft zur Beförderung der Künste und Nützlichen Gewerbe. Verhandlungen vom Jahre 1844," 1845, vol. I, pp. 120–137.

Emil Müller, *Vom Portland-Zement*, in, "Journal für die Baukunst," 1846, vol. XXIII, pp. 368–372; and in, "Allgemeine Bauzeitung," XI, 1846, pp. 111–113.

Königl. Regierung Abtheilung des Innern, *Vermischte Nachrichten*, in, "Amtsblatt der Königlichen Regierung zu Potsdam und der Stadt Berlin," May 15, 1846, p. 166.

Anon., *Die Eisenbahn zwischen Berlin und Königsberg*, in, "Illustrierte Zeitung," vol. VI, 1846, n° 144, p. 218.

Franz Stauffert, *Über die Baumaterialien des alten und des neuen Griechenlands, und über die geognostischen Verhältnisse dieses Landes*, in, "Journal für die Baukunst," 1846, vol. XXIII, (pp. 101–130, 268–319), pp. 278–279.

Franz Ludwig von Welden, *Ueber die Eigenschaften des Santorin-Mörtels und seiner Anwendung bei Bauten in und außer Wasser*, in, "Kunst- und Gewerbe-Blatt des polytechnischen Vereins für das Königreich Bayern," XXXII, 1846, coll. 683–693.

Gottfried Menzel, *Verfahren bei der Fabrication der Ziegel und des Mörtels auf der königl. preuß. Ziegelei bei Joachimsthal*, in, "Verhandlungen des Vereins zur Beförderung des Gewerbfleißes in Preußen," XXV, 1846, pp. 52–73; and in, "Polytechnisches Journal," 1846, vol. CII, pp. 194–219.

Johann Hermann Maack, *Einiges über die Bereitung und Anwendung des Concrets oder Betons, mit besonderes Rücksicht auf die Fundirung des Gebäudes der Patriotischen Gesellschaft*, in, "Verhandlungen und Schriften der Hamburgischen Gesellschaft zur Beförderung der Künste und Nützlichen Gewerbe. Verhandlungen vom Jahre 1845," 1846, vol. I, pp. 263–270.

Heinrich Hübbe, *Bericht über die 368ste Deliberations-Versammlung vom 18. Dezember 1845*, in, "Verhandlungen und Schriften der Hamburgischen Gesellschaft zur Beförderung der Künste und Nützlichen Gewerbe. Verhandlungen vom Jahre 1845," vol. I, pp. 593–600.

Carl Hoffmann, *Beton-Fundirungen bei den Wasserbauten an der Ruhr*, in, "Notiz-Blatt des Architekten-Vereins zu Berlin," vol. I, 1847, pp. 5–8.

Anon., *Notiz über den in Moabit angefertigten Cement: nach einer Mittheilung der Herren Haslinger und Schondorff*, in, "Notiz-Blatt des Architekten-Vereins zu Berlin," vol. I, 1847, pp. 12–13.

[Wilhelm Salzenberg], *Santorin-Erde*, in, "Notiz-Blatt des Architekten-Vereins zu Berlin," vol. I, 1847, p. 16.

Krafft, *Ueber eine Art massiv zu bauen mit reinem Sande und sehr wenig Kalk*, in, "Kunst- und Gewerbe-Blatt des polytechnischen Vereins für das Königreich Bayern," XXXIII, 1847, coll. 789–792.

Franz Carl Leonhard Elsner, *Ueber Hohofenschlacken als hydraulische Cämente*, in, "Verhandlungen des Vereins zur Beförderung des Gewerbfleißes in Preußen," XXVI, 1847, pp. 89–90; in, "Polytechnisches Journal," 1847, vol. CVI, pp. 312–322; and in, "Kunst- und Gewerbe-Blatt des polytechnischen Vereins für das Königreich Bayern," XXXIII, 1847, coll. 686–688.

Johann Daniel Friedrich Engel, *Der Bau in Kalk-Sand-Pisée*, in, "Zeitschrift für praktische Baukunst," VII, 1847, coll. 9–14.

Joseph Egle, *Notizen über die Kunkurrenz zur Lieferung eines Bauplanes für die St. Nikolai-Kirche in Hamburg*, in, "Allgemeine Bauzietung," XIII, 1848, pp. 123–227.

P. Schulze, *Anwendung von Santorin-Erde und Puzzolane*, in, "Notiz-Blatt des Architekten-Vereins zu Berlin," 1848, vol. II, pp. 25–26.

L. Schwan, *Beschreibung des Baues einer Kaimauer zu Stettin an der Oder vorlängs des dortigen Eisenbahnhofes, insbesondere des dabei in Anwendung gebrachten Gründungsverfahrens mittelst Beton aus Wolliner Mergelkalk*, in, "Zeitschrift für Praktische Baukunst," VIII, 1848, coll. 145–162.

Heinrich Müller, *Von den Fundirungen*, in, "Zeitschrift für praktische Baukunst," VIII, 1848, pp. 207–237.

Franz Carl Leonhard Elsner, *Ueber die Zusammensetzung des Santorins und dessen technische Verwendung*, in, "Verhandlungen des Vereins zur Beförderung des Gewerbfleißes in Preußen," XXVIII, 1849, pp. 86–87; in, "Polytechnisches Journal," 1849, vol. CXIII, pp. 157–158.

Anton Hopfgartner, *Analyse zweier hydraulischen Kalke, eines englischen (Portland-Cement) und eines deutschen hydraulischen Kalkes*, in, "Polytechnisches Journal," 1849, vol. CXIII, pp. 351–357.

Max Joseph von Pettenkofer, *Bemerkungen zu Hopfgartner's Analyse eines englischen und eines deutschen hydraulischen Kalkes*, in, "Polytechnisches Journal," 1849, vol. CXIII, pp. 357–371.

Karl Emil Franz Schafhäutl, *Das Portland- und Roman-Cement. Ein Beitrag zur Geschichte der Cemente oder hydraulischen Mörtel in England, nebst einem Anhange über die Theorie der Erstarrung der Mörtel und über den glänzenden Stucco der Alten*, in, "Polytechnisches Journal," 1851, vol. CXXII, pp. 186–208, 267–293.

Xaver Johannes Landerer, *Ueber die Santorinerde und deren Gebrauch als Zahnkitt*, in, "Polytechnisches Journal," 1851, vol. CXXII, p. 468.

Anon., *Ueber die mit der Anwendung von Santorin-Erde in Preussen gemachten Versuche*, in, "Zeitschrift für die Bauwesen," I, 1851, coll. 293–302, 347–356.

Johann Carl Leuchs, *Der Bau mit Mauern aus künstlichem Stein in einem Stück*, in, "Allgemeine Polytechnische-Zeitung und Handlings-Zeitung," 1852, n° 6, pp. 41–43.

Anon., *Deutschland. Berlin*, in, "Frankfurter Volkszeitung," December 29, 1852, n° 311, n. p.

Johann Caspar Garthe, *Die neuesten englischen Versuche über Vergleichung des Portland- und Roman-Cements*, in, "Polytechnisches Journal," 1852, vol. CXXIV, pp. 25–48.

Ludwig Benjamin Henz, *Die Restauration des Diemel-Viadukt*, in, "Zeitschrift für den Bauwesen," II, 1852, coll. 15–34.

Anon., *Auszug aus der Mittheilung des Herrn Ministers für Handel etc. an die Kamern über den Stand des Baues der Ostbahn, der Westphälischen und Saarbrücker Eisenbahn in November des Jahres 1851*, in, "Zeitschrift für Bauwesen," II, 1852, p. 126.

C. Studt, *Ueber die auf die Exercier-Platz zu Breslau erbaute Halle für die Schlesische Industrie-Ausstellung im Jahre 1852*, in, "Zeitschrift für Bauwesen," II, 1852, coll. 533–538.

M. Gottgetreu, *Das Gymnasium zu Anclam*, in, "Zeitschrift für Bauwesen," II, 1852, coll. 455–458.

Anon., *Die neuen Lagerhäuser in Leipzig*, in, "Zeitschrift für praktischen Baukunst," XIII, 1853, coll. 97–100.

Königliches Polizei Präsidium, Karl Ludwig Friedrich von Hinckeldey, *Neue Bau-Ordnung für Berlin*, in, "Zeitschrift für praktische Baukunst," XIII, 1853, coll. 133–152.

Hesse, *Das neue erbaute Sommer-Lazareth bei der Charité-Kranken-Heil-Anstalt in Berlin*, in, "Zeitschrift für Bauwesen," III, 1853, coll. 343–350.

Friedrich Adler, *Das Neue Museum in Berlin*, in, "Zeitschrift für Bauwesen," III, 1853, (coll. 23–34, 571–586), col. 576.

Anon., *Verfahren zur Verfertigung von Wasserleitungröhren aus hydraulischem Kalke*, in, "Kunst- und Gewerbe-Blatt des polytechnischen Vereins für das Königreich Bayern," XL, 1854, coll. 99–102.

J. Karlinger, *Verfahren zur Verfertigung von Wasserleitungröhren aus hydraulischem Kalke*, in, "Kunst- und Gewerbe-Blatt des polytechnischen Vereins für das Königreich Bayern," XL, 1854, coll. 99–102; and in, "Polytechnisches Journal," 1854, vol. CXXXII, pp. 202–204.

Gebrüder Born, *Ueber Verfertigung von Wasserleitung-Röhren aus Cement*, in, "Polytechnisches Journal," 1854, vol. CXXXIV, pp. 136–140.

P. R. Brecht, *Kirche zu Brodwin*, in, "Zeitschrift für Bauwesen," IV, 1854, coll. 209–212.

Winterstein, *Bericht über die Reise des Architekten-Vereins und des Vereins für Eisenbahnkunde nach Königsberg*, in, "Zeitschrift für Bauwesen," IV, 1854, coll. 543–571.

Ludwig Benjamin Henz, *Der Bau des Neisse-Viaducts bei Görlitz in der Niederschlesisch-Märkischen Eisenbahn*, in, "Zeitschrift für Bauwesen," V, 1855, coll. 281–344.

Carl Knauß, *Chemische Untersuchung einiger englischen hydraulischen Kalke*, in, "Gewerbeblatt aus Württemberg," 1855, n° 4, pp. 25–32; and in, "Polytechnisches Journal," 1855, vol. CXXXV, pp. 361–369.

Johann Carl Wilhelm Lentze, *Die im Bau begriffenen Brücken über die Weichsel bei Dirschau und über die Nogat bei Marienburg*, in, "Zeitschrift für Bauwesen," V, 1855, coll. 445–458.

Karl Wilhelm Drewitz, *Das neue Wachtgebäude am Unterbaum zu Berlin*, in, "Zeitschrift für Bauwesen," V, 1855, coll. 467–472.

Paul Emanuel Spieker, *Bericht über eine Probe auf Tragvermögen von Backsteinen aus geschlemmter Erde*, in, "Zeitschrift für Bauwesen," V, 1855, coll. 474–479.

Karl Wilhelm Drewitz, *Casernement für das Königliche Garde-Dragoner-Regiment auf dem sogenannten Upstall vor dem Halleschen Tor bei Berlin*, in, "Zeitschrift für Bauwesen," V, 1855, coll. 521–542.

August Winkler, *Ueber hydraulische Mörtel*, in, "Journal für praktische Chemie," 1856, vol. LVII, pp. 444–462; and in, "Polytechnisches Journal," 1856, vol. CXLII, pp. 106–121.

François Coignet, *Benutzung des geformten und zusammengepreten Mörtels*, in, "Polytechnisches Journal," 1856, vol. CXL, pp. 101–104; and in, "Kunst- und Gewerbe-Blatt des polytechnischen Vereins für das Königreich Bayern," XLII, 1856, coll. 404–408.

Anon., *Ueber Häuserbau mit Mörtel*, in, "Kunst- und Gewerbe-Blatt des Polytechnischen Vereins für das Königreich Bayern," XLII, 1856, p. 576.

Cremer, *Mörtelmachine, welche bei idem Bau des Schifffahrtsschleuse des Berlin-Spandauer Kanals bei Plötzensee in Anwendung gekommen ist*, in, "Zeitschrift für Bauwesen," VI, 1856, coll. 192–194.

Johannes Christian Wilhelm Dalmann, *Hafen-Anlagen in Frankreich und Holland*, in, "Zeitschrift für Bauwesen," (VI, 1856, coll. 361–392, VII, 1857, coll. 313–348), VII, 1857, coll. 329–339.

Ad. Braubach, *Main-Brücke bei Frankfurt in der Main-Neckar-Eisenbahn*, in, "Zeitschrift für den Bauwesen," VI, 1856, coll. 479–488.

Friedrich Hoffmann, *Der Bahnhof der Berlin-Hamburger Eisenbahn in Berlin*, in, "Zeitschrift für Bauwesen," VI, 1856, coll. 487–496.

C. Hennicke, *Ueber die Anlage von Abtritten in Wohngebäude und öffentlichen Anstalten*, in, "Zeitschrift für Bauwesen," VII, 1857, coll. 123–144.

Hermann Lohse, *Die Rheinbrücke bei Cöln*, in, "Zeitschrift für Bauwesen," VII, 1857, coll. 307–314.

Theodor Weishaupt, *Die Homberg-Ruhrorter Rheintraject-Anstalt*, in, "Zeitschrift für Bauwesen," VII, 1857, coll. 347–390.

Wilhelm Salzenberg, *Einige Bemerkungen über die Gewölbeconstruction des antiken Rom*, in, "Zeitschrift für Bauwesen," VII, 1857, coll. 424–429.

Anon., *Ueber Portland-Cement*, in, "Chemisches Centralblatt," III, 1858, pp. 481–483; and in, "Polytechnisches Journal," 1858, vol. CXLIX, pp. 262–265.

August von der Heydt, *Circular-Verfügung vom 19. April 1858, die Aufstellung von Baumaterialien-Verzeichnissen in den einzelnen Baukreisen betreffend*, in, "Zeitschrift für Bauwesen," VIII, 1858, (coll. 369–371), col. 369.

E. H. Hoffmann, *Mittheilungen über Gewölbe aus Stampfmörtel*, in, "Zeitschrift für Bauwesen," VIII, 1858, coll. 453–456.

Stein, *Die Bauanlagen der Kreuz-Cüstrin-Frankfurter Eisenbahn*, in, "Zeitschrift für Bauwesen," VIII, 1858, coll. 459–486.

J.-B. Bodot, *Travaux du Pont de Kehl*, in, "Nouvelles annales de la construction," V, 1859, coll. 18–20.

Charles Alfred Oppermann, *Travaux du pont du Rhin à Kehl*, in, "Nouvelles annales de la construction," V, 1859, coll. 155–160.

Georg Feichtinger, *Ueber die chemischen und physikalischen Eigenschaften mehrerer bayerischen hydraulischen Kalke im Vergleich mit Portland-Cement; zugleich ein Beitrag zur Theorie des Erhärtens der hydraulischen Kalke*, in, "Polytechnisches Journal," 1859, vol. CLII, pp. 40–61, 108–118.

Johann Daniel Friedrich Engel, *Ueber Kalkziegelfabrikation und Kalkziegelbau*, in, "Polytechnisches Journal," 1859, vol. CLIII, pp. 100–109.

Sanftleben, *Anfertigung und Verwendung von Portland-Cement-Röhren zu Chaussee-Durchlässen*, in, "Zeitschrift für Bauwesen," IX, 1859, coll. 417–420; and in, "Polytechnisches Journal," 1859, vol. CLIV, pp. 421–423.

Julius Manger, *Untersuchungen über die Festigkeit von reinen und gemischten Cementen*, in, "Zeitschrift für Bauwesen," IX, 1859, coll. 523–534.

Anon., *Die Gitterbrücken der Weichsel bei Dirschau und Marienburg*, in, "Warschauer Zeitung," 1860, n° 43, 45, 47, 49, 50, 51, n. p.

J. G. Schwedler, Hipp, *Der Rheinbrückenbau bei Kehl*, in, "Zeitschrift für Bauwesen," X, 1860, coll. 7–46.

Witzeck, *Die Gründung der Gebäude des Thüringisches Bahnhofes bei Leipzig*, in, "Zeitschrift für Bauwesen," X, 1860, coll. 213–222.

E. H. Hoffmann, *Ueber Anwendung von Stampfmörtel bei kleinen Brückengewölben*, in, "Zeitschrift für Bauwesen," X, 1860, coll. 263–266.

Herrmann, *Hochaltar in der Stephanskirche zu Gartz, aus Portland-Cement erbaut*, in, "Zeitschrift für Bauwesen," X, 1860, coll. 311–316.

Stein, *Ueber den Bau eiserner Thurmspitzen*, in, "Zeitschrift für Bauwesen," X, 1860, coll. 481–490.

Anon., *Steinpresse zur Fabrikation von Ziegel- und Pisé-steinen*, in, "Zeitschrift für Praktischen Baukunst," XXI, 1861, coll. 127–128.

Verein für Eisenbahnkunde zu Berlin, *Protokoll des Vereins für Eisenbahnkunde zu Berlin, 9 October 1860*, in, "Zeitschrift für Bauwesen," XI, 1861, coll. 334–348.

Johann Carl Wilhelm Lentze, *Beschreibung einiger Hülfseinrichtungen von dem Bau der Weichsel- und Nogat-Brücken. II. Die Cementhütte und die Mörtelmühle für den Bau der Weichselbrücke*, in, "Zeitschrift für Bauwesen," XII, 1861, coll. 375–380.

Hermann Bleibtreu, *Ueber Dauerhaftigkeit des Portland-Cement-Mörtels*, in, "Zeitschrift für Bauwesen," XI, 1861, coll. 495–502.

G. Dulk, *Notizen über den Ludwigs- (Main-Donau-) Kanal*, in, "Zeitschrift Für Bauwesen," XII, 1862, coll. 227–238.

Architekten Verein zu Berlin, *Mittheilungen aus Vereinen. Architekten-Verein zu Berlin. Versammlung am 31. Mai 1862*, in, "Zeitschrift für Bauwesen," XII, 1862, col. 563.

Anon., *Die Bauanlagen der Saarbrücken-Trier-Luxemburger Eisenbahn*, in, "Zeitschrift für Bauwesen," XIII, 1863, coll. 43–58.

Assmann, *Verfahren bei Geraderichtung des Thurmes an der evangelischen Kirche in Gleiwitz*, in, "Zeitschrift für Bauwesen," XIII, 1863, coll. 67–70.

Hermann Lohse, *Die Rhein-Brücke bei Cöln*, in, "Zeitschrift für Bauwesen," XIII, 1863, (coll. 175–196, 335–70), coll. 178–192, 189–196.

Emil Hermann Hartwich, *Rheinbrücke bei Coblenz*, in, "Zeitschrift für Bauwesen," XIV, 1864, coll. 385–416, 529–579.

Wilhelm Heldt, *Studien über die Cemente*, in, "Journal für praktische Chemie," XCIV, 1865, pp. 129–161, 202–237.

Anon., *Ueber künstliches Baumaterial*, in, "Landwirthschaftliches Centralblatt für Deutschland," XIII, 1865, pp. 375–380.

H. Tellkampf, *Ueber den Bau von Wärterhäusern mit Umfassungsmauern von Kalkpisé*, in, "Zeitschrift für praktische Baukunst," XXVI, 1865, coll. 113–122.

H. Weise, *Locomotiv-Haus der Berlin-Potsdam-Magdeburh Eisenbahn in Berlin*, in, "Zeitschrift für Bauwesen," XV, 1865, coll. 435–458.

Architekten Verein zu Berlin, *Mittheilungen aus Vereinen, Versammlung am 18. November 1865*, in, "Zeitschrift für Bauwesen," XVI, 1866, col. 320.

Verein für Eisenbahnkunde zu Berlin, *Protokoll vom 14. November 1865*, in, "Zeitschrift für Bauwesen," XVI, 1866, col. 330.

Löffler, *Die Fundirung der Eisenbahnbrücke über den Pregel in Königsberg*, in, "Zeitschrift für Bauwesen," XVI, 1866, coll. 517–536.

Anon., *Die Kunststeingießerei im Dienste der moderne Baukunst*, in, "Zeitschrift für praktische Baukunst," XXVI, 1866, coll. 23–28.

Anon, *Der Werth des künstlichen Baumaterials*, in, "Zeitschrift für praktische Baukunst," XXVI, 1866, coll. 117–122.

Wilhelm Michaelis, *Ueber den Portland-Cement*, in, "Journal für praktische Chemie," vol. C, 1867, pp. 257–303.

Anon, *Ueber Kalkpisébau*, in, "Zeitschrift für praktische Baukunst," XXVII, 1867, coll. 162–163.

Anon. *Bahnwärterhäuser aus Gußmauerwerk*, in, "Polytechnisches Journal," 1868, vol. CLXXXVII, p. 517.

Hermann von der Hude, *Die Kunsthalle in Hamburg*, in, "Zeitschrift für Bauwesen," XVIII, 1868, coll. 3–8.

Anon., *Die Künstlerichen Verzierungen des neuen Stadttheaters zu Leipzig*, in, "Zeitschrift für praktische Baukunst," XXVIII, 1868, coll. 309–314.

Verein für Eisenbahnkunde in Berlin, *Mittheilungen aus Vereinen*, in, "Deutsche Bauzeitung," III, 1869, p. 135.

Anon., *Die X. Versammlung des Vereins mittelrheinischer Bautechniker*, in, "Deutsche Bauzeitung," III, 1869, n° 27, (pp. 318–322), pp. 321–322.

Röhrer, *Die Eisenbahnbrücke über die Donau bei Ingolstadt*, in, "Zeitschrift des bayerischen Architekten- und Ingenieur-Vereins," I, 1869, (pp. 9–11, 37–38, 47–49), pp. 9–11.

Martin Carl Philipp Gropius, *Die Provinzial-Irren-Anstalt zu Neustadt-Eberswalde*, in, "Zeitschrift für Bauwesen," XIX, 1869, coll. 147–194.

Stahl, *Die Construction der Taucherglocken zur Fundirung der Rheinbrücke bei Hamm*, in, "Zeitschrift des Vereins deutscher Ingenieure," vol. XIII, 1869, coll. 185–188.

Georg Gustav Erbkam, *Die Königliche National-Galerie*, in, "Zeitschrift für Bauwesen," XIX, 1869, coll. 265–282.

Joseph von Schlierholz, *Ueber Beton-Verwendung zu Hochbauzwecken, hauptsächlich zu ganzen Gebäuden*, in, "Allgemeine Bauzeitung," XXXV, 1870, pp. 260–265.

Dollinger, *Wärterhaus aus Béton*, in, "Deutsche Bauzeitung," IV, 1870, pp. 44–45.

C. Hense, *Kalk-Pisé-Steine*, in, "Deutsche Bauzeitung," IV, 1870, p. 91.

F., *Ueber einige Arten der Verwendung von Zement*, in, "Deutsche Bauzeitung," V, 1870, pp. 235–238.

Friedrich Wiebe, *Bauwerke aus Concret*, in, "Deutsche Industriezeitung," 1870, n° 1, pp. 7–8; and in, "Zeitschrift für Bauhandwerker," XIV, 1870, pp.61–62.

Brückwald, *Das Stadt-Theater in Leipzig von Langhans*, in, "Zeitschrift für Bauwesen," XX, 1870, coll. 291–298.

Anon., *Der Umbau des Bahnhofes zu Görlitz*, in, "Zeitschrift für Bauwesen," XX, 1870, coll. 471–482.

Lämmerhirt, *Albrecht Türrschmiedt*, in, "Deutsche Bauzeitung," V, 1871, pp. 330–331.

P. Ernst, *Ueber Cementröhren und deren Anfertigung*, in, "Zeitschrift für Praktische Baukunst," XXXI, 1871, coll. 235–242.

Anon., *Mittheilungen aus Vereinen*, in, "Deutsche Bauzeitung," VI, 1872, p. 385.

Anon., *Das Cementbau*, in, "Zeitschrift für Bauhandwerker," XVI, 1872, pp. 187–188.

Lehmann, *Bericht über einen Versuch mit Konkretbau*, in, "Verhandlungen des Vereins zur Beförderung des Gewerbefleisses in Preussen," LI, 1872, pp. 37–38.

Pichier, *Die Verbindungseisenbahn zwischen Düsseldorf und Neuss mit Ueberbrückung des Rheinstromes oberhalb Düsseldorf*, in, "Zeitschrift für Bauwesen," XXII, 1872, coll. 237–250, 369–384.

Anon., *Versuch mit Concretbau*, in, "Zeitschrift für Bauhandwerker," XVII, 1873, p. 143.

C. Gebhardt, *Der Aschenstampfbau*, in, "Zeitschrift für Bauhandwerker," XVII, 1873, pp. 159–160, 174–175.

Anon., *Ueber die Anwendung des Beton zur Herstellung von Wohnhäusers*, in, "Zeitschrift für Bauhandwerker," XVII, 1873, pp. 15, 28–31.

Franz Gustav Assmann, *Die Wohnungnoth in Berlin*, in, "Zeitschrift für Bauwesen," XXIII, 1873, coll. 111–130.

Freudenberg, *Die Canalisierung der Stadt Witten*, in, "Zeitschrift für Bauwesen," XXIII, 1873, coll. 135–152.

Julius Ludwig Quassowsky, *Ueber Fundirungen mit Senkbrunnen nebst Beschreibung einiger Fälle aus der Praxis*, in, "Zeitschrift für Bauwesen," XXIV, 1874, coll. 297–312.

Friedrich Eduard Salomon Wiebe, *Die Bauten der Breslau-Schweidnitz-Freiburger Eisenbahnim Oderthal bei Stettin*, in, "Deutsche Bauzeitung," IX, 1875, (n° 71, pp. 353–355, n° 73, pp. 363–365, n° 75, pp. 373–375), n° 73, pp. 363–365.

Friedrich Knapp, *Ueber Cemente*, in, "Polytechnisches Journal," CCXXII, 1876, pp. 147–153.

Anon., *Die Cementfrage*, in, "Thonindustrie-Zeitung," I, 1876, pp. 2–3, 37–38.

Anon., *Eisenbahnbrücke über die Weichsel bei Thorn*, in, "Zeitschrift für Bauwesen," XXVI, 1876, coll. 35–54, 197–218.

Bernhard Liebold, *Wohnhaus des Hrn. Kaufmann F. Planck in Holzminden*, in, "Deutsche Bauzeitung," XI, 1877, pp. 458–459.

Anon., *Zur Cementfrage*, in, "Thonindustrie-Zeitung," I, 1877, pp. 2–3.

Wilhelm von Haselberg, *Die Brücken der Berlin-Stettiner Eisenbahn im Oderthal bei Stettin*, in, "Zeitschrift für Bauwesen," XXIV, 1879, coll. 359–378.

Georg Christoph Mehrtens, *Zur Baugeschichte der alten Eisenbahnbrücken bei Dirschau und Marienburg*, in, "Zeitschrift für Bauwesen," XLIII, 1893, coll. 97–122.

Anon., *Ein Beitrag zum Betonieren von Hochbauten*, in, "Oeberösterreiche Bauzeitung," II, 1897, n° 9, pp. 69–70.

Otto Leube, *Zum 100. Geburtstag von Gustav Leube sen.*, in, "Tonindustrie–Zeitung," XXXII, 1908, n° 65, pp. 899–901.

Neumann, *Beiträge zur Geschichte der Mörtelbereitung*, in, "Zement," V, 1916, n° 6, p. 32, n° 7, pp. 37–38, n° 8, pp. 46–47, n° 9, pp. 53–55.

Kroher, *Staudacher Zementdachsteine. Ein Jubiläum*, in, "Tonindustrie–Zeitung," XLIII, 1919, pp. 443–444.

Anon., *E. Schwenk, Ulm. Zum 75jährigen Geschäftsjubiläum*, in, "Tonindustrie–Zeitung," XLVI, 1922, n° 138, p. 942.

Karl Goslich, *Einhundert Jahre Portlandzement*, in, "Zeitschrift für angewandte Chemie," XXXVII, 1924, pp. 265–267, 297–298, 504–505.

Hermann Bleibtreu, *Der erste Zementbrand der Stettiner Portland-Cementfabrik im Dezember 1853*, in, "Zement," XIV, 1925, n° 10, pp. 212–213.

Wilhelm Petry, *Zum 50 jährigen Bestehen des Vereins Deutscher Portlandzement-Fabrikanten. 75 Jahre deutscher Portlandzement. – Ein Rückblick auf die Anfänge der deutschen Portlandzementindustrie und auf das Schaffen von Hermann Bleibtreu*, in, "Der Bauingenieur," VIII, 1927, pp. 643–645.

Hans Joachim Straube, *Chr. P. Wilhelm Beuth*, in, "Abhandlungen und Berichte des Deutschen Museums," II, 1930, pp. 117–152.

K. Würzner, *Aus den ersten Tagen des Portlandzements in Deutschland*, in, "Zement-Kalk-Gips," V, 1952, pp. 133–137.

Hans Heinrich Quicke, *Peter Christian Wilhelm Beuth*, in, "Mitteilungen des Vereins für die Geschichte Berlins," LXIV, 1968, n° 14, pp. 187–190.

Ernst Meyer Hermann, *"Leicht wie ein Gedanke". Erste Große deutsche Kettenbrücke: 1836/39 bei Hameln erbaut*, in, "Die Weser," XLIV, 1970, n. 1, pp. 48–49.

Lars U. Scholl, *Georg Dietrich Wendelstadt und Johann Ludwig Böhlke, die Erbauer der Langbrücke Northeim*, in, "Notheimer Heimatblätter," 1977, vol. XLII, (n° 3, pp. 88–93, n° 4, pp. 134–139), n° 3, p. 92.

Ernst Kanow, *"Colonie Victoriastadt." Eine Berliner Wohnsiedlung mit mehr als hundert Jahre alten Wohnhäusern aus Beton*, in, "Architektur der DDR," XXX, 1981, pp. 50–53.

Julius Fekete, *Die Villa Merkel in Esslingen. Ein Fabrikantenwohnhaus der Gründerzeit*, in, "Esslinger Studien," 1982, vol. XXI, pp. 119–138.

Anon., *Friedrich Neuhaus – Ein vergessener Eisenbahnpioneer*, in, "Mitteilungen des Vereins für die Geschichte Berlins," LXXVIX, 1983, pp. 109–113.

Achim Hubel, *Die beiden Restaurationen des Bamberger Domes. Zur Geschichte der Denkmalpflege im frühen 19. Jahrhundert*, in, "Berichte des Historischen Vereins Bamberg," vol. CXXI, 1985, (pp. 45–90), pp. 78–81.

Sigfried Lotze, *Beiträge zu einer Geschichte hessischer Unternehmer im frühen 19. Jahrhundert. Die Familienbeziehungen niederhessischer Textil- und Farbenfabrikanten*, in, "Zeitschrift des Vereins für hessische Geschichte und Landeskunde," vol. XCVI, 1991, pp. 233–254.

Jos Tomlow, *Die Kuppel des Gießhauses der Firma Henschel in Kassel. Eine frühe Anwendung des Entwurfsverfahrens mit Hängemodellen*, in, "Architectura," vol. XXIII, 1993, pp. 151–171.

Michael Mende, *'Not only a Matter of Taste but … of the Laws of Mechanics': The Adoption of British Models in Nineteenth-Century Continental Suspension Bridge Design*, in, "Journal of Design History," vol. VI, 1993, (pp. 77–95), pp. 84–93.

Gerhard Weise, *Die Entwicklung der Zementherstellung in Thüringen bis zum Beginn des 20. Jahrhunderts*, in, "Thesis: Wissenschaftliche Zeitschrift der Bauhaus Universität Weimar," XLII, 1996, pp. 173–181.

Rotraut Trapp, *Die Brücken des Ludwig-Donau-Main-Kanals*, in, "Frankenland. Zeitschrift für fränkische Landeskunde und Kulturpflege," 2003, n° 4, pp. 287–306.

Berndt Weber, *Georgstrasse. Nach wem wurde die Georgstrasse in Berlin-Mitte bennant ?*, in, "Mitteilungen des Vereins für die Geschichte Berlins," CI, 2005, n°1.

Katja Laude, *Das Oberbuadepartement und die Untertanenbauten Möglichkeiten und Grenzen obrigkeitlicher Einflussnahme*, in, "Brandemburgische Denkmalpflege," vol. XVIII, 2009, n° 1, pp. 61–75.

Hermann Trautwein, *Dr. Ernst Gustav Leube, ein Pionier der deutschen Zementindustrie*, in, "Schwäbische Heimat," vol. IV, 2009, pp. 470–475.

Michael Plata, *Die Zementfabrik in Moorrege*, in, "Demokratische Geschichte," vol. XXII, 2011, pp. 125–142.

Karin Kraus, *Hydraulische Bindemittel im 19. Jahrhundert auf dem Gebiet der heutigen Bundesländer Hessen, Rheinland-Pfalz, Saarland und Thüringen*, in, "IFS-Berichte," Institut für Steinkonservierung e. V., Bericht n° 43, 2012.

Anon., *Kalksand-Mauerstein – vom Arzt entwickelt*, in, "Die Baustelle," 2014, n° 2, pp. 26–27.

Books and essays in books

Johann Joachim Winckelmann, *Anmerkungen über die Baukunst der Alten*, Leipzig, Johann Gottfried Dyck, 1762, p. 14.

Lukas Voch, *Allgemeines Baulexicon, oder Erklärung der deutschen und französischen Kunstwörter, in der bürgerlichen, Kriegs- und Schiffbaukunst, wie auch der Hydrotechnik und Hydraulik*, Augsburg, Leipzig, Matthäus Riechers sel. Söhnen, 1781, pp. 40–41, 285.

Johann Georg Krünitz, *Oekonomische Enzyclopädie, oder allgemeines System der Staats-, Stadt-, Haus- u. Landwirthschaft in alphabethischer Ordnung*, vol. XXXII, Berlin, Joachim Pauli, 1784, pp. 610–792.

Johann Karl Gottfried Jacobsson, Otto Ludwig Hartwig, Gottfried Erich Rosenthal, Johann Karl Gottfried, *Jacobssons technologisches Wörterbuch oder alphabetische Erklärung aller nützlichen mechanischen Künste, Manufakturen, Fabriken und Handwerker,*

wie auch aller dabey vorkommenden Arbeiten, Instrumente, Werkzeuge und Kunstwörter, nach ihrer Beschaffenheit und wahrem Gebrauche, Berlin, Stettin, Friedrich Nicolai, vol. IV, 1784, p. 279.

David Gilly, *Praktische Abhandlung aus der Landbaukunst, betreffend den Bau der sogenannten Lehm- oder Wellerwände wie man dieselben dauerhaft mit wenigen Kosten und einer wahren Holzersparung aufführen könnte*, Berlin, Friedrich Maurer, 1787.

David Gilly, *Beschreibung einer vortheilhaften Bauart mit getrockneten Lehmziegeln*, Berlin, [Lange], 1790.

Friedrich Ludwig Aster, *Gesammelte Nachrichten von dem Cemente aus Trasse uns Wasserdichten Mauerwerke der Holländer*, Dresden, Leipzig, Joh. Samuel Gerlach, (1773[1]), 1791[3].

Reinhard Woltman, *Beyträge zur hydraulische Architektur*, Göttingen, Johann Christian Dieterich, (vol. I, 1791, Vol. II, 1792, vol. III, 1794, vol. IV, 1799) vol. IV, pp. 371–416.

Friedrich August Alexander Eversmann, *Technologische Bemerkungen auf einer Reise durch Holland*, Freyberg, Annaberg, Crazische Buchhandlung, 1792.

Christian Ludwig Albrecht Patje, *Kurzer Abriß des Fabriken-, Gewerbe-, und Handlungszustandes in den Chur Braunschweig-Lüneburgischen Landen*, Göttingen, Vandenhoek und Ruprecht, 1796, p. 373.

Marcus Vitruvius Pollio, *Des Marcus Vitruvius Pollio Baukunst*, ed. August von Rode, Leipzig, Georg Joachim Göschen, voll. I , II, 1796.

David Gilly, *Handbuch der Land-Bau-Kunst vorzüglich in Rücksicht auf die Konstruktion der Wohn- und Wirtschaftsgebäuse für angehende Kameral-Baumeister und Oekonomen*, vol. I, Braunschweig, Friedrich Vieweg, 1797.

Carl Friedrich Wiebeking, *Allgemeine auf Geschichte und Erfahrung gegründete theoretisch-practische Wasserbaukunst*, Darmstadt, Johann Franz Peter Stahl, vol. I, 1798, vol. II, 1799, vol. III, 1801, vol. IV, 1805, vol. V, 1807.

Johann Beckmann, *Anleitung zur Technologie oder zur kenntniß der Handwerke, Fabriken und Manufacturen, vornehmlich derer, mit der Landwirtschaft, Polizey und Cameralwissenschaft in nächster Verbindung stehen*, (Göttingen, Wittwe Vandenhoeck, 1777[1], 1780[2], 1787[3], Göttingen, Wandenhoeck und Ruprecht, 1796[4], 1802[5], 1809[6]), Göttingen, Wandenhoeck und Ruprecht, 1802[7], pp. 308–318.

David Gilly, Johann Albert Eytelwein, *Praktische Anweisung zur Wasserbaukunst, welche eine Anleitung zum Entwerfen, Veranschlagen und Ausführen der am gewöhnlichsten vorkommenden Wasserbaue enthält*, vol. I, Berlin, auf Kosten der Verfasser, 1802; vol. II, Berlin, Realschulbuchhandlung, 1803.

David Gilly, *Ueber die Gründung der Gebäude auf ausgemauerte Brunnen*, Berlin, Realschulbuchhandlung, 1804.

Heinrich Gustav Flörcke, *Dr. Johann Georg Krünitz's Oekonomische Enzyclopädie, oder allgemeines System der Staats-, Stadt-, Haus- u. Landwirthschaft in alphabethischer Ordnung*, vol. XCIV, Berlin, Joachim Pauli, 1784, pp. 168–376.

Johann Albert Eytelwein, *Praktische Anweisung zur Wasserbaukunst, welche eine Anleitung zum Entwerfen, Veranschlagen und Ausführen der am gewöhnlichsten vorkommenden Wasserbaue enthält*, Berlin, Realschulbuchhandlung, vol. III, 1805, vol. IV, 1808.

Jean-Nicolas-Louis Durand, ed. Carl Friedrich Anton von Conta, *Grundlinien der bürgerlichen Baukunst*, Halle, Neue Societas- Buch und Kunsthandlung, 1806, pp. 18–23.

Carl Friedrich Wiebeking, *Beyträge zur Wasser- Brücken- und Straßenbaukunde oder wissenschaftliche Darstellung der in den neusten Zeiten ausgeführten oder in der Anlage begriffenen Bauwerke und Beschreibung der vorzüglichsten Maschinen*, Darmstadt, Johann Franz Peter Stahl, 1808.

Freidrich J. C. Schulz, *Versuch einiger Beiträge zur hydraulischen Architektur*, Königsberg, Friedrich Nicolovius, 1808.

Johann Friedrich John, *Chemisches Laboratorium. Oder Anweisung zur chemischen Analysen der Naturalien. Nebst Darstellung der nötigsten Reagenzien*, Berlin, Friedrich Maurer, 1808, pp. IX–XII.

Alois Hirt, *Die Baukunst nach den Grundsätzen der Alten*, Berlin, Realschulbuchhandlung, 1809.

Carl Friedrich Wiebeking, *Beyträge zur Brückenbaukunde: worin auch die neue Bauconstruction wohlfeiler und dauerhafter Bogenbrücken, nach welcher mehrere große Brücken vom Verfasser angegeben und ausgeführt sind, dargestellt ist; welche als eine Fortsetzung des Perronet'schen Werkes betrachtet werden können*, München, auf Kosten des Verfassers, 1809.

Carl Friedrich Wiebeking, *Beyträge zur Hafen-, Seeufer- und Flußbaukunde, oder: Abhandlung über die Verbesserung der Häfen von Venedig, und über die Erhaltung der Inseln vor Venedig; über die Verbesserung des Laufs der Brenta, des Bacchiglione, und der Schiffahrts-, so wie der Entwässerungs-Kanäle zwischen Venedig, Padua, Verona und der Etsch: nebst einem Vorschlage zur Anlage eines Seehafens vor Triest, auf genaue vom Verfasser angestellten Lokaluntersuchungen gegründet; und eine Beschreibung des Hafens*

Cronstadt, München, auf Kosten des Verfassers, Gedrückt bey Franz Hübschmann, 1810, (pp. 37–42), p. 42.

Johann Friedrich John, *Chemische Untersuchungen mineralischer, vegetabilischer und animalischer Substanzen. Fortsetzung des chemischen Laboratoriums*, Berlin, Friedrich Mauerer, 1810, (pp. IX–XX), p. IX.

Johann Friedrich John, *Chemische Untersuchungen mineralischer, vegetabilischer und animalischer Substanzen. Fortsetzung des chemischen Laboratoriums*, Berlin, J. E. Hitzig, 1811, pp. 22–24, 88–93.

Carl Friedrich Wiebeking, *Theoretisch-Practische Wasserbaukunst*, München, Joseph Zängl auf Kosten des Verfassers, vol. I, 1811, vol. II, 1812, vol. III, 1814, vol. IV, 1817.

Gabriel Christoph Benjamin Busch, *Handbuch der Erfindungen*, vol. IX, Eisenach, Wittekindt, 1817, pp. 333–334; vol. XII, Eisenach, Johann Friedrich Bärecke, 1822, p. 110.

Hamburgisches Address – Buch für das Jahr 1817, Hamburg, Hermann am Fischmarkt, 1817, p. 70.

Hamburgisches Address – Buch für das Jahr 1818, Hamburg, Hermann am Fischmarkt, 1818, p. 74.

Christian Friedrich von Bucholz, *Theorie und Praxis der pharmaceutisch - chemischen Arbeiten, oder Darstellung der Bereitungsmethoden der wichtigsten pharmaceutisch-chemischen Präparate: nach den neuesten Erfahrungen und rücksichtlich ihrer Brauchbarkeit und Vorzüglichkeit geprüft; nebst den theoretischen Erklärungen der dabey vorkommenden Erscheinungen, Beyfügung der vorzüglichsten Eigenschaften der abgehandelten Präparate, des zweckmässigsten Prüfungsverfahrens auf die Aechtheit oder Verfälschtheit der letztern und der nöthigen Literatur und des Geschichtlichen*, vol. I, Leipzig, Heinrich August Rottmann, Wien, Carl Gerold, 1818, pp. 208–217, 501–507.

Hamburgisches Address – Buch für das Jahr 1819, Hamburg, Hermann am Fischmarkt, 1819, p. 76.

John Friedrich John, *Ueber Kalk und Mörtel im Allgemeinen im Allgemeinen und den Unterschied zwischen Muschelschalen- und Kalksteinmörtel insbesondere; nebst Theorie des Mörtels*, Berlin, Duncker und Humblot, 1819.

Johann August Donndorff, *Supplemente zur Geschichte der Erfindungen in allen Theilen der Wissenschaften und Künste von der ältesten bis auf die gegenwärtige Zeit*, Quedlinburg, Leipzig, Gottfried Basse, vol. II, 1821, pp. 64–65, 226–227.

August Hermann Niemeyer, *Beobachtungen auf einer Deportationsreise nach Frankreich im Jahr 1807*, Halle, Buchhandlung des Waisenhauses, vol. I, 1824, vol. II, 1826.

André Étienne Justin Pascal Joseph François d'Audebert de Férussac, éd., *Bullettin des sciences technologiques*, vol. II, Paris, Au bureau du Bullettin (etc.), 1824, p. 106.

Plusieurs Professeurs du Jardin du Roi, et des principales Écoles de Paris, eds., *Dictionnaire des Sciences Naturelles es d'après l'état actuel de nos connoissances, dans lequel on traite méthodiquement des différens êtres de la nature, considérés soit en eux-mêmes,soit relativement a l'utilité qu'en peuvent retirer la médecine, l'agriculture, le commerce et les arts*, vol. XXXIII, Strasbourg, Paris, F. G. Levrault, Paris, Le Normant, 1824, p. 32.

Jean-Henry Hassenfratz, *Traité théorique et pratique de l'art de calciner la pierre calcaire et de fabriquer toutes sortes de mortiers, cimens, bétons, etc., soit à bras d'hommes, soit à l'aide de machines*, Paris, Carilian-Goeury, Mme Huzard, 1825, p. IX.

Moritz Karl Ernst von Prittwitz, ed., *Neue Versuche über den Kalk und Mörtel von L. J. Vicat und Andern*, Berlin, Posen, Ernst Sigfried Mittler, 1825.

Salomo Sachs, *Anleitung zur Erd-Bau-Kunst, (Pisé-Bau); mit Anwendung auf alle Arten von Stadt- und Land-Bauten, nebst einer vollständigen Lehre von der Konstruktion der Tonnen- Kappen- und Kreuzgewölbe in reinem Lehm und von der Anfertigung feuersicherer Dächer ohne alles Holzwerk, auch einer Anweisung, die Fundamente bis auf den Baugrund in bloßen Lehm anzufertigen*, Berlin, Carl Friedrich Amelang, 1825.

Herman Jan van der Wyck, *Uebersicht der Rheinschen und Eifeler erloschenen Vulkane und der Erhebungs - Gebilde, welche damit in geognostischer Verbindung stehen, nebst Bemerkungen über den technischen Gebrauch ihrer Produkte*, Bonn, Eduard Weber, 1826, pp. 87–110.

Gotthilf Hagen, *Beschreibung neuerer Wasserbauwerke in Deutschland, Frankreich und der Schweiz*, Königsberg, Gebrüder Bornträger, 1826.

Johann Christian Carl Rommerdt, *Handbuch der Land- und Wasserbaukunst für angehende und ausübende Forstmänner, Kameralisten und Oekonomen*, Erfurt, Gotha, Hennigs, 1827, pp. 47–57.

Louis Jacques Thénard, *Traité de chimie élémentaire, théorique et pratique*, vol. I, Bruxelles, Canongette et compagnie, 1829[7], p. 347.

Clement-Louis Treussart, *Mémoire sur les mortiers hydrauliques et sur les mortiers ordinaires*, Paris, Chez Carillan-Goeury, Anselin, Mahler et Compagnie, 1829, pp. 6–7, 49, 79–80, 142–148, 182, 214.

Johann Nepomuk Fuchs, *Ueber Kalk und Mörtel*, Leipzig, Barth, 1829.

Friedrich Panzer, *Abhandlung über die Aufbewahrung des Getreides in Silo's*, Würzburg, Carl Strecker, 1830.

Friedrich Fyck, *Die Verwaltung des Straße- und Brückenbaues mit Rücksicht auf möglichste Kosten-Ersparniß, Wohlthätigkeit für die ärmeren Volksklassen, uns Aufhebung der Frohdienste*, Cassel, Johann Christian Krieger, 1831, (pp. 57–67), p. 65.

Moritz Ferdinand Gaetzschmann, *Anleitung zur Grubenmauerung*, Schneeberg, Carl Schumann, 1831, pp. 20–47.

Ernst Ludwig Schubarth, *Elemente der technische Chemie, zum Gebrauch beim Unterricht königl. Gewerbinstitut und den Provizial-Gewerbeschulen*, vol. I, Berlin, August Rücker, 1831[1], pp. 402–16, 1833[2], pp. 392–402.

Friedrich Fick, *Die Verwaltung des Strassen- und Brückenbaues mit Rücksicht auf Möglichste Kosten-Ersparnis, Wohltätigkeit füt die ärmeren Volksklassen, und Aufhebung der Frohdienst*, Cassel, Johann Christian Krieger, pp. 23–67.

Ernst Ludwig Schubarth, *Elemente der technischen Chemie: zum Gebrauch beim Unterricht im Königl. Gewerbinstitut und den Provinzial-Gewerbschulen*, vol. I, Berlin, August Rücker, 1831[1], pp. 402–416; 1835[2], pp. 392–409.

Alexander von Lengerke, *Darstellung der Landwirthschaft in den Großherzogthümern Mecklenburg*, vol. I, Königsberg, Gebrüder Bornträger, 1831, pp. 122–126.

Friedrich Panzer, *Anleitung über die Bereitung des Mörtels aus hydraulischem Kalke in Beziehung auf die Auffindung des vorzüglich hierzu sich eignenden Mergels, so wie über das Verfahren, welches bei der Anwendung des hydraulischen Mörtels, sowohl bei Landgebäuden, als auch bei Bauten im Wasser zu beobachten ist*, Würzburg, Becker, 1831[1], München, Joseph Lindauer, 1832[2].

Heinrich Joseph Alois Freiherr von Pechmann, *Entwurf für den Kanal zur Verbindung der Donau mit dem Main*, München, s.n., 1832.

Heinrich Joseph Alois von Pechmann, *Atlas zu dem Entwurfe für den Kanal von Donau an den Main*, s.l., s.n., s.d.

Joseph-Mathieu Sganzin, *Lehrbuch des Straßen-, Brücken-, Kanal- und Hafen-Baukunde*, ed. H. F. Lehritter, G. H. Strauß, Regensburg, Friedrich Pustet, 1832, vol. I, pp. 116–151, pp. 160–162, vol. II, pp. 52–56, 206–207 n. 22; vol. II, pp. 52–56, 206–207, n. 22, 225–227.

Franz Joseph Ritter von Gerstner, *Handbuch der Mechanik*, vol. I, *Mechanik fester Körper*, Johann Spurny, Prag, 1833[2], p. 241.

Ludwig Friedrich Wolfram, *Lehre on den künstlichen Bausteinen und Verbindungsstoffen, oder Lehre von getrockneten Lehmsteinen, von der Zubereitung, vom Brennen und von der Anwendung aller Ziegeleierzeugnisse vom Brennen des Kalkes und Gipses, vom Baue der Ziegel-, Kalk- und Gipsöfen, von den Mörteln aller Art, von den Kitten u. s. w., mit Rücksicht auf bedeutende Bau- und Kunstwerke der Alten und Neuern*, Stuttgart, Carl Hoffmann, Wien, Carl Gerold, 1833, pp. 53–56, 78–116.

Carl Theodor von Kleinschrod, *Die Kanal-Verbindung des Rheins mit der Donau*, München, G. Franz, 1834.

Karl Friedrich Schinkel, *Grundlage der praktischen Baukunst*, vol. I, *Vorlegeblätter für Mauerer*, Berlin, Schenk & Gerstärcker, 1834, pp. 2, 9–10, pl. IV, XVIII.

Georg Christoph Hamberger, Johann Georg Meusel, Joahnn Wilhelm Sigismund Lundner, *Das gelehrte Teutschland, oder Lexikon der jetzt lebenden teutschen Schriftsteller*, vol. XXIII, Lemgo, Meyetsche Hof-Buchhandlung, 1834, pp. 106–107.

Carl Theodor von Kleinschrod, *Die Kanal-Verbindung des Rheins mit der Donau: unter Benützung amtlicher Quellen beleuchtet*, München, G. Frank'sche Buchhandlung, 1834.

Mark Bald Kittel, *Der Hydraulische oder Cäment-Kalk aus der Umgegend Aschaffenburgs und dessen Benutzung*, Aschaffenburg, M. I. Wailnad's Wittib und Sohn, 1836.

Friedrich Panzer, *Ueber das Vorkommen des hydraulischen Kalkes in der Keuperformation, über die Dauerhaftigkeit absolute und rückwirkende Festigkeit desselben und der Steine, als Beitrag zu dem Bau des Kanales zur Verbindung der Donau mit dem Main*, München, Joseph Lindauer, 1836.

Actiengesellschaft für den Ludwig-Donau-Main-Kanal, *Auszüge aus den Protokollen der zweiten Generalversammlung der Aektiengesellschaft für den Ludwig-Donau-Main-Kanal*, voll. I–XII, 1836–1848, esp.: vol. II, 1837, pp. 31–49, vol. III, 1838, p. 20; vol. IV, 1839, pp. 47–49; vol. V, 1840, p. 24.

François-Martin Lebrun, *Der Steinmörtel oder praktische Anweisung den Steinmörtel bei Gebäuden im Allgemeinen, besonders aber bei Bauten an und unter dem Wasser, bei Gewölben etc. statt jeder andern Art von Mauerarbeit Vortheil zu benutzen*, Ulm, Ernst Nübling, 1837.

Georg A. Lobe, *Wanderungen durch Cassel und die Umgegend. Eine Skizze für Einheimische und Fremde*, Cassel, Krieger, 1837, pp. 223–224, 227–228, n. 3.

Ernst Friedrich Martin Koch, *Gesammelte Erfahrungen über die Verarbeitung u. die verschiedenen Anwendungen des Cemente aus den Cementfabriken von Ernst Koch in Hessen-Kassel und Hanau*, Kassel, Leipizig, I. C. Kriegersche Buchhandlung, Theodor Fischer, 1838[1], Kassel, Theodor Fischer, 1851[2].

Gustav Leube, *Geognostische Beschreibung der Umgegend von Ulm*, Ulm, Ebner, 1839.

Friedrich Wilhelm Otto Ludwig Freiherr von Reden, *Das Königreich Hannover statistisch beschrieben, zunächst in Beziehung auf Landwirtschaft, Gewerbe und Handel*, vol. I, Hannover, Hahn'sche Hofbuchhandlung, 1839, pp. 462–465.

Gotthilf Heinrich Ludwig Hagen, *Handbuch der Wasserbaukunst*, vol. I, *Die Quellen*, Königsberg, Gebrüder Bornträger, 1841.

Johann Gottlieb Prochnow, *Anleitung zur Kunst, Wohnungen und Wirthschafts - Gebäude in sehr kurzer Zeit wohlfeil, Feuer- und Wetterfest zu erbauen aus reinem Sande und sehr wenigem Kalk*, Schwedt, J. C. W. Jantzen, 1842.

Johann Georg Ludwig Sievers, *Bauten für den Prinzen Kalr von Preussen*, Berlin, Deutscher Kunstverlag, 1842, p. 72.

Wilhelm and Gustav Leube, *Untersuchungen über das mineralische Material der Umgegend von Ulm in betreff seiner Verwendbarkeit für Bauzwecke und insbesondere seiner Bedeutung für den Festungsbau*, Ulm, Ernst Nübling, 1843.

Carl von Decker, *Algerien und die dortige Kriegführung*, Berlin, Friedrich August Herbig, 1844, pp. 384–389.

Heinrich Joseph Alois Freiherr von Pechmann, *Der Ludwig-Canal. Eine kurze Beschreibung deises Canal's und die Ausführung desselben*, München, Joseph Lindauer, 1846.

Heinrich Joseph Alois von Pechmann, *Der Ludwig-Kanal. Eine kurze Beschreibung dieses Kanal's und die Ausführung desselben*, München, Joseph Lindauer, 1846.

Johann Carl Wilhelm Lentze, *Bemerkungen über die grösseren Brücken-Bauwerke in Frankreich, England und Irland, aud einer Reise im Winter 1844–45 gesammelt*, Berlin, I. Petsch, 1846.

Friedrich Wilhelm Otto Ludwig Freiherr von Reden, *Die Eisenbahnen Deutschlands, Statistisch-geschichtliche Darstellung ihrer Entstehung, ihres Verhältnisses zu der Staatsgewalt, so wie ihrer Verwaltungs- und Betriebs-Einrichtungen*, Berlin, Posen, Bromberg, Ernst Siegfried Mittler, 1846, pp. 315–321.

Friedrich Schultheis, *Der Ludwig-Kanal- Seine Entstehung und Bedeutung als Handels-Strasse*, Nürnberg, Hofmann, 1847.

Louis-Joseph Vicat, *Praktische Anweisung den hydraulischen Kalk (Cement oder künstliche Puzzolane) zu bereiten und denselben bei Wasserbauten aller Art anzuwenden*, Quedlinburg, Leipzig, Gottfried Basse, 1847.

Johann Carl Leuchs, *Verbesserte Darstellung der Kitte, Mörtel, künstliche Steine, künstliche Massen, des Siegellaks, der Oblaten, der Löthe, und Anwendung der ersteren zum wohlfeilsten Bau von Häusern, Mauern, Kellern, Fußboden, Dächern, Rauchfängen, Wasserbecken, Wasserröhren, Brücken und Strassen*, Nürnberg, C. Leuchs & Co., 1848.

Julius von der Orbach, *Tufstein, Trass und hydraulischer Mörtel. Gesammelte Bemerkungen*, Coblenz, Karl Bädecker, 1849[1], Jacob Hölscher, 1862[2].

Theodor Bülau, *Das Haus der patriotischen Gesellschaft in Hamburg oder: Einiges über einen Neubau*, Hamburg, I. I. Nobiling, 1849.

Johann Christian Wedeke, *Handbuch der bürgerlichen Baukunst: Allgemein fasslich für Maurer- und Zimmermeister und die es werden wollen*, Quedlinburg, Leipzig, Gottfried Basse, vol. II, 1849, pp. 50–62.

Gustav Adolph Breymann, *Allgemeine Bau-Constructions-Lehre, mit besonderer Beziehung auf Hochbauwesen*, Stuttgart, Hoffmann, vol. I, 1849, pp. 25–26, 82–83; 1860[3], pp. 30–31, 33–38, 93–95, 121–127.

Johann Christian Wedeke, *Der Bau mit gestampftem Mörtel*, Quedlinburg, Leipzig, Gottfried Basse, 1850.

Willliam Löbe, *Encyclopädie der gesammten Landwirthschaft, der Staats-, Haus- und Forthwirthschaft und der in die Landwirthschaft einschlagenden technischen Gewerbe und Hülfswissenschaften*, vol. I, Leipzig, Otto Wigand, 1850, pp. 204–205.

Heinrich Müller, *Die Brückenbaukunde in ihrem ganzen Umfange. Ein Handbuch für Ingenieure und Baumeister*, Leipzig, Romberg, vol. I, 1850, pp. 33–43, 84–87, 225–238; vol. III, 1852, pp. 66–84.

August Wilhelm Hertel, *Die Lehre vom Kalk und Gyps, in ihrem ganzen Umfange begreifend: die Rohstoffe, das Brennen, die Brennmaterialien, die Oefen, die Theorie und das Verhalten des Aetzkalks und der hydraulischen Kalke, Bereitung der Mörtel, Cemente, des Stukko u.a.*, Weimar, Friedrich Voigt, 1851[1], 1860[3].

Johann Daniel Friedrich Engel, *Der Kalk-Sand-Pisébau*, Wriezen, Roeder, 1851.

[Friedrich Conrad Theodor] Krause, *Anleitung zur Kalk-Sand-Bau-Kunst*, Glogau, Flemming, 1851.

Johann Christian Wedeke, *Handbuch der Landbaukunst und der Landwirthschaftlicher Gewerbe für Baumeister, Landwirthe und Cameralisten*, vol. I, *Die Beschreibung der Baumaterialienlehre*, Leipzig, Romberg, 1851.

Gotthilf Heinrich Ludwig Hagen, *Handbuch der Wasserbaukunst*, vol. II, *Die Ströme und Kanäle*, Königsberg, Gebrüder Bornträger, 1852.

Anon., *Stenographische Berichte über die Verhandlungen der durch die Allerhöchste Verordnung vom 4. November 1851 einberufenen Kammern*, Zweite Kammer, vol. III, Berlin, Deckersche Geheime Ober-Hofbuchdruckerei, 1852, n° 162, pp. 1195–1197.

[Wilhelm Andreas] Becker, *Erfahrungen über den Portland-Cement*, Berlin, Gebauersche Buchhandlung, I. Petsch, 1853.

Heinrich Joseph Alois von Pechmann, *Der Ludwigskanal. Kurze Geschichte seines Baues und seiner noch bestehenden Mängel, sowie die Mittel, sie zu entfernen und zu verbessern und den Kanal zu seiner Vollkommenheit zu erheben*, Nürnberg, Friedrich Korn, 1854.

August Kopisch, *Die Königlichen Schlösser und Gärten zu Potsdam. Von der Zeit ihrer Gründung bis zum Jahre MDCCCLII*, Berlin, Ernst & Korn, 1854, p. 217.

Max Becker, *Der Brückenbau in seinem ganzen Umfange, und mit besonderer Rücksicht auf die neuesten Constructionen*, Stuttgart, Carl Mäcken, 1854, pp. 87, 97, 100, 131, 221.

J. Cohn, *Ueber die Wichtigkeit der Cemente: in Beziehung auf gesunde Wohnungen, dauerhafte Wasser- und Prachtbauten, Kunststein, Anlegung von Silo's, so wie auf Fabrik-Industrie; nebst Vorschlägen zur Gründung fester Steinufer als Bedämmung gegen Stromgefahr, so wie zur Nachforschung auf inländische Puzzolane oder Trass*, Breslau, Druck der Stadt- und Universität-Buchdruckerei von Grass, Barth, & Comp., 1855.

Carl Lentze, *Die im Bau begriffenen Brücken über die Weichsel bei Dirschau und über die Nogat bei Marienburg entworfen und in der Ausführung geleitet von Lentze*, Berlin, Ernst & Korn, 1855.

Leo Bergmann, *Baulexicon oder Realencyclopedie des gesammten Bauwesens*, vol. I, Leipzig, Ernst Schäfer, 1855, pp. 224–225.

Franz Graf Pocci, *Erinnerung an Friedrich Panzer, k. Oberbaurath*, München, C. Wolf & Sohn, 1855.

Max Becker, *Der Wasserbau in seinem ganzen Umfange*, Stuttgart, Carl Mäcken, 1856, pp. 15, 136, 191, 267, 280.

[Wilhelm Andreas] Becker, *Der feuerfeste Treppenbau von natürlichen und künstlichen Steinen. Nach den neuesten Erfindungen und Ausführungen mit besonderer Berücksichtigung der Constructionen zum praktischen Gebrauche bearbeitet*, Berlin, Ernst & Korn, 1857.

Georg Feichtinger, *Über die chemischen Eigenschaften mehrerer bayerischer hydraulischer Kalke im Verhältniss zu Portland-Cement: Zwei Abhandlungen. Über eine Auswitterung an einem Neubau*, München, J. G. Cotta'sche Buchhandlung, 1858.

Johann Christian Wedeke, Johann Andreas Romberg, *Handbuch der Landbaukunst und der Landwirthschaftlichen Gewerbe für Baumeister, Landwirte und Cameralisten*, vol. III, *Die Zimmerwerksarbeiten*, Leipzig, Romberg, 1858

Johann von Mihálik, *Praktische Anleitung zum Béton-Bau für alle Zweige des Bauwesens*, Wien, Joseph Stöckholzer v. Hirschfeld, 1858[1], 1859[2], Berlin, Theobald Grieben, 1860[2].

[Wilhelm Andreas] Becker, *Praktische Anleitung zur Anwendung der Cemente zu baulichen, gewerblichen, landwirthschaftlichen und Kunst-Gegenständen*, Berlin, *Nicolaische Verlagsbuchhandlung*, 1860[1], 1869[2].

Julius Manger, *Der Stettiner Portland-Cement in Versuche und Erfahrungen dargestellt und beleuchtet*, Stettin, F. Heffenland, 1860.

Emile Vuigner, Emile Fleur de Saint-Denis, *Pont sur le Rhin à Kehl. Détails pratiques sur les dispositions générales et d'exécution de cet ouvrage d'art*, Paris, Dunod, 1861.

Gotthilf Heinrich Ludwig Hagen, *Ueber Form und Stärke gewölbter Bogen*, Berlin, Ernst & Korn, 1862, p. 25.

Wilhelm Leonhardt, *Die Kitt-, Leim-, Cement- u. Mörtelfabrikation mit Einschluss der Kalk- und Gypsbrennerei*, Leipzig, Otto Spamer, 1863.

Johann Daniel Friedrich Engel, *Sammlung landwirthschaftlicher und ländlicher Bauausführungen*, Berlin, Ernst & Korn, 1864.

Johann Daniel Friedrich Engel, *Der Kalk-Sand-Pisébau und die Kalksand-Ziegelfabrikation*, Leipzig, C. H. Seemann, 1865.

Albert Julius Licht, *Der Portland-Cement und der hydraulischer Kalk der Cement-Fabrik zu Dirschau*, Danzig, Schroth, 1867.

Edmund Heusinger von Waldegg, *Die Kalk-, Ziegel- und Röhrenbrennerei*, vol. I, *Die Kalk- und Cementfabrikation einschließlich der Mörtel- und Bétonbereitung, Fabrikation der Kalksandziegel und anderer künstlicher Bausteine, sowie der Anfertigung der verschiedensten Kitte*, Leipzig, Theodor Thomas, 1867.

Johannes Christian Wilhelm Dalmann, *Der Hafen von Hamburg-Altona, insbesondere die Anlage des Sandthor-Quais in Hamburg*, Berlin, Ernst & Korn, 1868.

Friedrich Julius Alexander Lipowitz, *Die Portland-Cement-Fabrikation nach eigenen praktischen Erfahrungen*, Berlin, Julius Springer, 1868.

Wilhelm Michaelis, *Die hydraulischen Mörtel insbesondere der Portland-Cement in chemisch-technischer Beziehung für Fabrikanten, Bautechniker, Ingenieure und Chemiker*, Leipzig, Quant & Händel, 1869.

Gotthilf Heinrich Ludwig Hagen, *Brunnen, Wasserleitung und Fundirungen*, vol. II, Berlin, Ernst & Korn, 1870.

Theodor Stein, *Erweiterungsbauten der Berlin-Stettiner Eisenbahn ausgeführt 1864–1869*, Berlin, Ernst & Korn, 1870.

Hermann Zwick, ed., *Jahrbuch über die Leistungen und Fortschritte auf dem Gebiet der practischen Baugewerbe*, vol. I, Leipzig, Carl Scholtze, 1870, pp. 189–232, 253–257, 324–327.

Emil Böhmer, Friedrich Neumann, eds., *Handbuch für Anlage und Betrieb von Kalkwerken, Gypsmühlen und Cementfabriken; mit Rücksicht auf die Anwendung dieser Materialien in der Praxis, nebst Literatur-Nachweis und Tabellen*, Weimar, Friedrich Voigt, 1870 (4th ed. of Hertel's *Kalk, Gyps und Cement*).

Paul Loeff, *Gründliche Anleitung zum Bau von Kalk-, Zement-, Gyps- u. Ziegelöfen sowie zum Betrieb von Kalk-, Cement-, Gyps- u. Ziegelbrennereien*, Berlin, Peisener, 1870.

Max Becker, *Handbuch der Ingenieur-Wissenschaft*, vol. V, *Ausgeführte Constructionen des Ingenieurs*, n° VI, *Die Kabelschifffahrt und die Eisenbahnbrücke über den Neckar bei Jartfeld*, Stuttgart, Carl Mäcken, 1870, pp. 36–38.

Hermann Zwick, ed., *Jahrbuch über die Leistungen und Fortschritte auf dem Gebiet der practischen Baugewerbe*, vol. II, Leipzig, Carl Scholtze, 1871, pp. 246–250, 268–287, 340–342, 351–354, 358.

Heinrich Klose, *Der Portland-Cement und seine Fabrikation*, Wiesbaden, Kreisel, 1873.

Bernhard Liebold, *Zement*, Halle, Hendel, 1874.

Wilhelm Michaelis, *Zur Beurtheilung des Cementes*, Berlin, A. Seidel, 1876.

Adolph Kroher, *Cement und Cementwaaren-Fabrik in Staudach*, Augsburg, Rackl & Lochner, 1878.

Ferdinand Stöter, *Geschichte und Beschreibung des St. Nikolai Kirchenbaues in Hamburg*, Hamburg, Boysen, 1883.

Heinrich Wolffram, *Mittheilungen über Tuffstein, Trass und einige andere Baumaterialien der Vulkanische Eifel*, Diez an der Lahn, Franz Schickel, 1885.

Teichgräber, *Die Brücken im Oderthale bei Stettin*, in, Vv. Aa. eds, *Bericht über eine Bauwissenschaftliche Studienreise nach der Pommerschen Künste*, Berlin, Julius Springer, 1887, pp. 7–13.

Gustav Lang, *Zur Entwicklungsgeschichte des Spannwerke des Bauwesens*, Riga, N. Kymmel, 1890, p. 37.

Verein Deutscher Portland-Cement-Fabrikanten, ed., *Der Portland-Cement und seine Anwendungen im Bauwesen*, Berlin, Ernst Toeche, 1892[1], 1899[2], 1905[3].

Eduard Schmitt, Josef Durm, Hermann Ende, *Handbuch der Architektur - Die Hochbau-Konstruktionen - Raumbegrenzende Konstruktionen - Wände und Wandöffnungen*, (Darmstadt, Arnold Bergsträsser, 1891), Stuttgart, Arnold Bergsträsser, 1900, pp. 107–131.

Carl von Voit, *Max von Pettenkofer zum Gedächtniss*, München, Verlag de k. b. Akademie in Commission des G. Franz'schen Verlags, 1902, pp. 26–27.

Anton Hambloch, *Der rheinische Trass als hydraulischer Zuschlag in seiner Bedeutung für das Baugewerbe*, Andernach am Rhein, Selbstverlag des Verfassers, 1903.

Karl Goslich, *Stettiner Portland–Cement–Fabrick 1855–1905*, Stettin, H. Susenbeth, 1905

Feodor Ast, *Der Beton und seine Anwendung*, Berlin, Tonindustrie–Zeitung, 1907, pp. 141–157.

Friedrich Ritzmann, *Untersuchungen über Trass-, Kalk-, Sandmörtel*, doctoral dissertation, Technische Hochschule Karlsruhe, 1906, published in Karlsruhe, by Universitäts-Buchdruckerei von J. Hörning, 1907.

Anton Hambloch, *Der Trass, seine Entstehung, Gewinnung und Bedeutung im Dienste der Technik*, Berlin, Julius Springer, 1909.

Paul May, *Die bayrische Zementindustrie*, Leipzig, Deichert, 1909, pp. 1–39.

Friedrich Quietmeyer, *Zur Geschichte der Erfindung des Portlandzements*, doctoral dissertation, Königliche Technische Hochschule Hannover, 1911, published in Berlin, by Tonindustrie–Zeitung, 1911.

Königlich Preuzischen Ministerium der Oeffentlichen Arbeiten, ed., *Archiv für Eisenbahnwesen*, Berlin, Julius Springer, 1911, pp. 105, 911.

Conrad Matschoss, *Geschichte der Königlich Preußischen Technischen Deputation für Gewerbe. Zur Erinnerung an das 100 jährige Bestehen. 1811–1911*, in, Id., ed., *Beiträge zur Geschichte der Technik und Industrie. Jahrbuch des Vereins Deutscher Ingenieure*, Berlin, Julius Springer, 1911, pp. 239–275.

Wilhelm Petry, *Betonwerkstein*, München, Meissenbach Rissarth & Co, 1913.

Wilhelm Petry, *Betonwerststein und künstlerische Behandlung des Betons. Entwicklung von den ersten Anfängen der deutschen Kunststein-Industrie bis zur werksteinmäßigen Bearbeitung des Betons*, München, Deutscher Beton-Verein, Meisenbach Riffarth, 1913.

Ernst Madelung, *Die Entwicklung der Deutschen Portland-Zement-Industrie*, doctoral dissertation, Kgl. Bayr. Ludwig-Maximilians-Universität zu München, 1913, pp. 1–36.

Conrad Matschoss, *Preussens Gewerbeförderung und ihre grossen Männer*, Berlin, Verlags des Vereins Deutscher Ingenieure, 1921.

Otto Neustätter, *Max Pettenkofer*, Wien, Julius Springer, 1925, p. 23.

Julius Faulwasser, *Die St. Nikolai-Kirche in Hamburg*, Hamburg, Boysen & Maasch, 1926.

Friedrich Quietmeyer, *Die Mörtelkunde von ihren ersten Anfängen bis zur zielbewussten Herstellung des Portlandzementes*, in, Peter Hans Riepert, ed., *Die Deutsche Zementindustrie*, Charlottenburg, Zementverlag G.m.b.H, 1927, pp. 1–87.

Friedrich Schott, *Die Entwicklung der Fabrikation in Deutschland*, in, Peter Hans Riepert, ed., *Die Deutsche Zementindustrie*, Charlottenburg, Zementverlag G.m.b.H, 1927, (pp. 88–197), pp. 88–112.

Gustav Haegermann, *Die Entwicklung des Prüfungswesen*, in, Peter Hans Riepert, ed., *Die Deutsche Zementindustrie*, Charlottenburg, Zementverlag G.m.b.H, 1927, (pp. 265–396), pp. 265–305.

Günther Bugge, *Über die Anfänge der deutschen Zementfabrikation*, in, Julius Ruska, *Studien zur Geschichte der Chemie*, Berlin, Julius Springer, 1827, pp. 187–194.

Hans Joachim Straube, *Die Gewerbeförderung Preussens in der ersten Hälfte des 19. Jahrhunderts*, doctoral dissertation, Technische Hochschule Berlin, 1933.

Trassforschungsinstitut der Tuba, ed., *Der Rheinische Trass. Seine Eigenschaften und Anwendung im Bauwesen*, Andernach, Firma Tuffstein-Basaltlavawerke A.G., 1934.

Curt Piorkowski, *Dyckerhoff Portland-Zementwerke A.G.: Mainz-Amöneburg*, Leipzig, Arnd, 1937, pp. 5–13.

Adolf Brzesky, *Vom Trass und seiner Verwendung*, conference, Bauwesen des Bundes Deutscher Technik, Haus der Technik, Wien, March 3, 1939.

Fritz Schlabbach, *Paul Ludwig Simon 1771-1815*, doctoral dissertation, Technische Hochschule Berlin, 1938, published in Karlsruhe, Berlin, by Doering, 1939.

Verein Deutscher Portland- und Hüttenzementwerke, ed., *Deutscher Zement. 1852–1952*, Wiesbaden, Verein Deutscher Portland- und Hüttenzementwerke, 1952, pp. 9–48.

Historische Kommission bei der Bayerischen Akademie der Wissenschaften, Otto Stolberg-Wernigerode, ed., *Neue Deutsche Biographie*, vol. II, Berlin, Dunker & Humblot, 1955, pp. 297–298.

Historische Kommission bei der Bayerischen Akademie der Wissenschaften, Otto Stolberg-Wernigerode, ed., *Neue Deutsche Biographie*, vol. II, Berlin, Dunker & Humblot, 1955, pp. 200–201.

Bonner Portland-Zementwerk A.G., ed., *Ein Jahrhundert Bonner Zement*, Düsseldorf, Flieger, 1956, pp. 7–51.

Karl Wimmer, *Dr. Gustav Leube. Apotheker und Zementfabrikant 1808 – 1881*, in, Max Miller, Robert Uhland, *Schwäbische Lebensbilder*, Stuttgart, W. Kohlhammer, pp. 325–336.

Charles George Dobson, *The history of the concrete roofing tile. Its origin and development in Germany*, London, Batsford, 1959 (Id., *Geschichte des Betondachsteins - Seine Herkunft und Entwicklung in Deutschland*, Hildesheim, Hagemann, 1959).

Harm-Hinrich Brandt, *Die Industrie- und Handelskammer Kassel und ihre Vorlaeufer, 1763–1963*, Kassel, Industrie- und Handelskammer Kassel, 1963, pp. 11–34, 78–79.

Gustav Haegermann, Günter Huberti, *Vom Cœment zum Zement – Die erneuerte Bauwesen*, Wiesbaden, Berlin, Bauverlag, 1964.

Adolf Leonhardt, *Von der Cementware zum Konstruktiven Stahlbetonfertigteil*, Wiesbaden, Berlin, Bauverlag, 1964.

Robert Oedl, Werner Leube, Gustav Schefold, *125 Jahre Gebrüder Leube in Ulm und 100 Jahre Gartenau*, Gartenau, Zementwerk Gebrüder Leube, 1964.

Gert v. Klass, *Weit spannt sich der Bogen: 1865-1965: die Geschichte der Bauunternehmung Dyckerhoff & Widmann*, Wiesbaden, Bartels, (1955[1]), 1965[2], pp. 7–53.

Rolf Fuhlrott, *Deutschsprachige Architektur-Zeitschriften. Entstehung und Entwicklung der Fachzeitschriften für Architektur in der Zeit 1789–1918*, München, Verlag Dokumentation, 1975.

Hans Scheller, *Der Nordkanal zwischen Neuss und Venlo*, Neuss, Stadtarchive, 1980.

Otmar Schäuffelen, *Die Bundesfestung Ulm und ihre Geschichte: Europas grösste Festungsanlage*, Ulm, Vaas, 1980, p. 43.

Harold Breyer, *Max von Pettenkofer Arzt im Vorfeld der Krankheit*, Leipzig, Hirzel, Teubner, 1980, pp. 44–45.

Gert Kollmer, *Die Industrieentwicklung einer württembergischen Amtsstadt am Beispiel von Blaubeeren*, in, Hansmarzin Decker-Hauff, Imme Eberl, eds., *Blaubeuren. Die Entwicklung einer Siedlung in Südwestdeutschland*, Sigmaringen, Jan Thorbecke, 1986, (pp. 627–698), pp. 635–637.

Stephan Rolph, *Johann Hermann Maack: 1809 bis 1868: Bauinspektor der Brücken und Schleusen in Hamburg von 1841 bis 1868: sein Lebensweg und seine Bauwerke*, Hamburg, C. W. Dingwort, 1987.

Karl Wieninger, *Max von Pettenkofer. Das Leben eines Wohltäters*, München, Hugendubel, 1987, pp. 117–121.

Robert Oedl, Werner Leube, *Gustav Schefold, 150 Jahre Leube (1838-1938)*, Salzburg, Druckhaus Nonntal, 1988.

Reinhard Wegner, Margarete Kühn, ed., *Karl Friedrich Schinkel. Die Reise nach Frankreich und England: Das Tagebuch*, München, Deutscher Kunstverlag, 1990, pp. 46, 125.

Armin Niemeyer, *Ein Vorläufer des Betonbaues am Rande Berlins*, in, Ulrich Großmann, Klaus Freckmann, Fred Kaspar, Ulrich Klein, eds., *Aus den Forschungen des Arbeitskreises für Haus- und Siedlungsforschung*, Marburg, Jonas, 1991, pp. 97–108.

Helmut Albrecht, *Vom Caementum zu Zement*, in, Id., ed., *Kalk und Zement in Württemberg: Industriegeschichte am Südrand der Schwabischen Alb*, Ubstadt-Weiher, Landesmuseum für Technik und Arbeit in Mann, Verlag Regionalkultur, 1991, pp. 117–130.

W. Kasig, B. Weiskorn, *Zur Geschichte der Deutschen Kalkindustrie und ihrer Organisationen*, Köln, Bundesverband der Deutschen Kalkindustrie 1992, pp. 54–79.

Heike Linderkamp, *Die Industrialisierung in Buxtehude. Anfänge und Entwicklung bis zum Beginn des 20. Jahrhunderts*, Buxtehude, Buxtehude Stadtarchiv, 1992.

Helmut Reihlen, *Christian Peter Wilhelm Beuth. Eine Betrachtung zur preußischen Politik der Gewerbeförderung in der ersten Hälfte des 19. Jahrhunderts und zu den Drakeschen Beuth-Reliefs*, Berlin, Köln, Beuth Verlag, 1992.

Max Liebold, *Beitrag zur Geschichte der Firma B. Liebold & Co. AG Holzenden und der Firma Haberland & Guckes-Liebold AG.*, ed. Matthias Seeliger, in, Heimat- und Geschichtsverein Holzenden e. V., ed., *Jahrbuch für den Landkreis Holzminden*, vol. X-XI, Holzminden, 1994.

Mohamed Scharabi, *Architekturgeschichte des 19. Jahrhundert*, Tübingen, Wasmuth, 1993, p. 161.

Fritz Scheidegger, *Kleine Geschichte der Bindemittel*, in, Id., *Aus der Geschichte der Bautechnik*, vol. I, Basel, Boston, Berlin, Birkhäuser, pp. 75–86.

Reinhart Strecke, *Schinkel und der Leuchtturm auf Kap Arkona*, in, Stiftung Preußischer Kulturbesitz, ed., *Jahrbuch Preußischer Kulturbesitz*, vol. XXXII, Berlin, Gebr. Mann Verlag, 1996, pp. 297–319.

Christina Czymay, *"Colonie Victoriastadt" in Berlin-Lichtenberg*, in, Landesdenkmalamt Berlin, ed., *Beiträge zur Denkmalpflege in Berlin*, n° 12, *Großstadtdenkmalpflege. Erfahrungen und Perspektiven*, Berlin, Schelzky und Jeep, 1997, pp. 60–63

Jochen Stark, Bernd Wicht, *Geschichte der Baustoffe*, Wiesbaden, Berlin, Bauverlag, 1998, pp. 53–93.

Birgit Lang, *Gestaltung mit Beton im Garten. Kreativer Umgang mit dem Werkstoff im 19. und 20. Jahrhundert*, in, Stiftung Preußische Schlösser und Gärten Berlin-Brandenburg, ed., *Jahrbuch Stiftung Preußische Schlösser und Gärten Berlin-Brandenburg*, vol. III, Berlin, Akademie-Verlag, 2002, pp. 32–41.

Walter Buschmann, *Die Bonner Zementfabrik. Beitrag zur Entstehungs- und Entwicklungsgeschichte der Zementproduktion in Deutschland*, in, Hartwig Schmidt, ed., *Zur Geschichte des Stahlbetonbaus. Die Anfänge in Deutschland. 1850 bis 1910*, Berlin, Ernst & Sohn, 1999, pp. 2–10.

Christina Czymay, *"Colonie Victoriastadt" in Berlin-Rummelsburg*, in, Hartwig Schmidt, ed., *Zur Geschichte des Stahlbetonbaus. Die Anfänge in Deutschland. 1850 bis 1910*, Berlin, Ernst & Sohn, 1999, pp. 11–15.

Andreas Kahlow, *Stampfbeton. Frühe Anwendungsbeispiele im Hochbau*, in, Hartwig Schmidt, ed., *Zur Geschichte des Stahlbetonbaus. Die Anfänge in Deutschland. 1850 bis 1910*, Berlin, Ernst & Sohn, 1999, pp. 16–26.

Hartwig Schmidt, *Vom Zementkunststein zum Betonwerkstein*, in, Hartwig Schmidt, ed., *Zur Geschichte des Stahlbetonbaus. Die Anfänge in Deutschland. 1850 bis 1910*, Berlin, Ernst & Sohn, 1999, pp. 27–35.

Reinhart Strecke, *Anfänge und Innovation der preußischen Bauverwaltung. Von David Gilly zu Karl Friedrich Schinkel*, Köln, Weimar, Wien, Böhlau, 2000, pp. 86–116.

Heike Hinz, "Berlin-Lichtenberg, Türrschmidtstrasse 17. Frühe Betonhäuser in der Victoriastadt ab 1871. Zur Entstehung und Sanierung des Wohnhauses in der Türrschmidtstrasse 17", final report for "Aufbaustudiengang Denkmalpflege," Technische Universität Berlin, 2002.

Hort Knobe, *Hamelner Wasserbauwerke an der Weser. Die Geschichte der Schleusen und Wehre, der Münsterbrücke und des Hafens*, Bielefeld, Verlag für Regionalgeschichte, 2003, pp. 117–121.

Werner Lorenz, *Von mancherlei Schwierigkeit, einer historischen Konstruktion gerecht zu werden. Anmerkungen zu neueren Äußerungen über die*

Schutzkuppel von St. Nikolai, in, Stiftungsrat vom Generaldirektor der Stiftung Preussische Schlösser und Gärten Berlin-Brandenburg, ed., *Jahrbuch Stiftung Preußische Schlösser und Gärten Berlin-Brandenburg*, vol. V, 2003, (pp. 127–149), p. 140.

Wieland Ramm, *History and construction od the old Vistula Bridges in Tczew*, in, Santiago Huerta, ed., *Proceedings of the First International Congress on Construction History*, Madrid, January 20–24, 2003, Madrid, Instituto Juan de Herrera, Escuela Técnica Superior de Arquitectura, pp. 1699–1707.

Wieland Ramm, ed., *Zeugin der Geschichte: die alte Weichselbrücke in Dirschau*, Kaiserslautern, Technische Universität Kaiserslautern, Fachgebiet Massivbau und Baukonstruktion, 2004.

Uta Hassler, ed., *Häuser aus Beton. Vom Stampfbeton zum Grosstafelbau*, Tübingen, Berlin, Ernst Wasmuth, 2004.

Reinhart Strecke, *Inventar zur Geschichte der preußischen Bauverwaltung 1723-1848*, vol. I–II, Berlin, Geheimen Staatsarchivs Preußischer Kulturbesitz, 2005.

Hans-Werner Hahn, *Die industrielle Revolution in Deutschland*, München, Oldenbourg, 2005.

Sergej Fedorov, *Carl Friedrich von Wiebeking und das Bauwesen in Russland. Zur Geschichte deutsch-russischer Architekturbeziehungen 1800-1840*, München, Berlin, Deutsche Kunstverlag, 2005.

Reimar F. Lacher ed., *Kunstler(auto)biographien*, Berlin, Berlin-Brandenburgische Akademie der Wissenschaften, 2005, (pp. 65–72), pp. 71–72, n. 154.

Thomas Brunsch, *Die historische Verwendung zementgebundener Kunststeine im Außenraum im 19. und frühen 20. Jahrhundert unter besonderer Berücksichtigung Berlins und Brandenburgs*, doctoral dissertation, Technische Universität von Berlin, 2006.

Margarete Siebeck, Gabriele Grassegger, Helmut Kollmann, Gabriele Patitz, Michael Walter, Wolfgang Franz, *Villa Merkel (1873) Untersuchungen und Restaurierungen einer reichverzierten Betonfassade*, in, Gabriele Grassegger, Gabriele Patitz, eds., *Natursteinsanierung Stuttgart 2007. Neue Natursteinrestaurierungsergebnisse und messtechnische Erfassungen sowie Sanierungsbeispiele*, Conference Proceedings, Stuttgart, March 16, 2007, Karlsruhe, Fraunhofer IRB Verlag, pp. 95–121.

Rudolf Vierhaus ed., *Deutsche Biographische Enzyklopädie*, vol. VIII, München, K. G. Sauer, 2007, p. 130.

Wieland Ramm, *Über die faszinierende Geschichte des Betonbaues vom Beginn bis zur Zeit dem 2. Weltkrieg*, in, Deutscher Ausschuss für Stahlbeton im DIN Deutsches Institut für Normung e.V., ed., *Gebaute Visionen: 100 Jahre Deutscher Ausschuss für Stahlbeton, 1907 – 2007*, Berlin, Wien, Zürich, Beuth Verlag, 2007, pp. 27–48.

"Beiträge zur Genealogie und Geschichte der Familie Quistorp. Der Stettiner Zweig", in, www.quistorp.de, 2008, pp. 111–112.

Silke Schulte, *Geschichte der Stadt Hameln*, in, Niedersächsisches Ministerium für Inneres, ed., *Niedersachsenbuch 2009. Rattenfängerstadt Hameln*, Hameln, C. W. Niemeyer, 2009, pp. 24–32.

Staatliche Museen zu Berlin – Stiftung Preußischer Kulturbesitz, ed., *The Neues Museum Berlin*, Leipzig, Seemann, 2009, pp. 124, 176–182.

Albrecht Hoffmann, *Ernst Koch (1786-1860). Ein Pionier der hessischen Zementindustrie*, in, Landkreiskassel, ed., *Jahrbuch 2010*, Kassel, 2010, pp. 69-74.

Petra Egloffstein, *Entwicklung von Kalk-Trass-Mörteln im preußischen Festungsbau am Beispiel der Festung Ehrenbreitstein, Koblenz*, in, Institut für Steinkonservierung e. V., ed., *Hydraulische Bindmittel im 19. Jahrhudert auf dem Gebient der heutigen Bundesländer Hessen, Rheinland-Pfalz, Saarland und Thüringen*, Mainz, Institut für Steinkonservierung e. V., 2012, pp. 15–20.

Roberto Gargiani, *Concrete from archaeology to invention 1700-1769*, Lausanne, EPFL Press, 2013.

Stefan M. Holzer, *Statische Beurteilung historischer Tragwerke*, Berlin, Ernst & Sohn, 2013, p. 267–268.

Edgar Breuss, Wilhelm Günther, Kurt Kaindl, Rudolf Zrost, *175 Jahre Leube. Chronik eines Familienunternehmens 1838 bis 2013*, St. Leonhard, Leube, 2013.

Knut Stegmann, *Das Bauunternehmen Dyckerhoff Et Widmann. Zu den Anfängen des Betonbaus in Deutschland 1865–1918*, Tübingen, Berlin, Ernst Wasmuth, 2014.

Archival sources

Geheimes Staatsarchiv Preußischer Kulturbesitz in Berlin (GStA)

"Versuche sowie Anfertigung künstlicher Steine," 1810–1826, GstA PK I.HA Rep. 93B, Ministerium für öffentliche Arbeiten, n° 1394; esp.: Königliche Ober-Bau-Deputation, letter to the Departement für die Gewerbe und den Handel im Ministerio des Innern, December 25, 1810, n. fol.; Königliche Ober-Bau-Deputation, letter to Friedrich von Schuckmann, October 26, 1811, n. fol.; Paul Ludwig Simon, "Der Geheime-Ober-Baurath Simon berühket über die Versuche mit der Zubereitung künsltícher Steine", October 20, 1811, n. fol.; Schwarz, "Uebersicht von den im Monat Juni 1812 in dem Hause des verstorbenen Geheimen Ober-Bauraths Herrn Simon angefertigten Mörtelsteinen", March 18, 1815, n. fol.; Königliche Ober-Bau-Deputation, letter to Heinrich von Bülow, April 30, 1815, n. fol.

"Verfertigung von künstlichen Bausteinen Patente," 1812–1878, GstA PK I.HA Rep. 120 D XII 1 n° 12, vol. I; esp.: Salomo Sachs, letter to the Ober-Bau-Deputation, October 17, 1825, foll. 22v–23r/v.

"Die Bereitung des Kalks und Gips, die Anfertigung von Mörtel und die Patenterteilung," 1818–1903, GStA PK I.HA Rep. 120, D XII 1 n° 3, vol. I; esp.: [Johann Heinrich] Krahmer, letter to the Ministerium für den Handel das Gewerbe und das Bau–Wesen, August 28, 1821, n. fol.; Id., "Roman cement," foll. 1, 1r–9v.

"Vorschläge und Schriften über Verbesserungen in der Baukunst," 1835–1866, GstA PK I.HA Rep. 93B, Ministerium für öffentliche Arbeiten, n° 990; esp.: foll. 80–124 r/v, about Prochnow's building technique with lean conglomerate.

"Die von dem Ackerbürger Prochnow zu Bahn neu eingeführte Sand-Kalk- Pisée-Baumethode," 1847–1863, GStA PK, I. HA Rep. 93 D, Technische Oberbaudeputation, n° 45/2; esp.: foll. 2–20 r/v, about Prochnow's building technique with lean conglomerate.

Hermann Bleibtreu, "Beschreibung eines neuen Verfahrens zur Darstellung hydraulischer Cemente," 1852, GStA, I HA Rep. 120 TD.D, n° 206, foll. 1–4 r/v.

"Cement, Mörtelbereitung," 1822–1877, GStA, I HA Rep. 120 TD C, n° 62.

GStA PK, XI. HA AKS, Allgemeine Kartensammlung, G, n° 51833.

General Inspection des Ingenierus–Corps und der Festungen, "Vorläufige Bestimmungen über die mit Rücksichtigkeit auf die Wirkung des gezogenen 28 cm Mörsers den Bombensicheren Eisendecken fortan zu gebende Construction," 1875, GStA IV. HA Preußische Armee, Rep. 16, Militärvorschriften, n° 667.

Staatliche Museen zu Berlin - Preußischer Kulturbesitz Zentralarchiv (SMB–ZA)

K. F. Schinkel, diary of the journey to France, England and Scotland, 1826, SMB-ZA, IV/NL Schinkel 94, fol. XXIX.

Brandenburgische Landeshauptarchiv (BLHA)

"Acta des Stadtbaurath Gersterbergk zu Berlinbetreffend Offerten über Lieferungen von Portland-Cement, Kalk, Gyps," 1858–1873, BLAB, A Rep. 010–01–01, n° 321; esp.: Carl Görcke, statement with attachments, April 1, 1858, foll. 1–2 r/v.; foll. 3–34, about the cement production and trade in and around Berlin.

"Acta des Stadtbaurath Gerstenberg zu Berlin betreffend Lieferung von Kunststeine, Granit, Platten, Marmorgliederung", BLHB, A, Rep. 10.01.01, n° 336, foll. 1–15.

"Acta der Königlichen Polizei-Präsidii zu Berlin betreffend Kanalisation," 1837–1859, BLHA, A Pr. Br. Rep. 030, n° 20112; esp.: "Annalen der Landwirthschaft in den Königlich Preußischen Staaten", XXI, 1863, pp. 182–189.

Stadt Archiv Soest (SAS)

Carl Lentze, «Verzeichnis meiner Mappen und ihres Inhalts», n.d., SAS, P. 1.382.

Königlichen Ministerieum für Handel, Gewerbe und Oeffentliche Arbeiten, ed., *Die im Bau befindlichen Brücken über die Weichsel bei Dirschau und über die Nogat bei Marienburg. Entworfen und in der Ausführung geleitet von Lentze*, Berlin, Ernst und Korn, 1855, SAS, P. 1.528.

Johann Carl Wilhelm Lentze, *Bemerkungen über die grösseren Brücken-Bauwerke in Frankreich, England und Irland, aud einer Reise im Winter 1844–45 gesammelt*, Berlin, I. Petsch, 1846, SAS, P. 1.813.

[Ascribed to] Johann Carl Wilhelm Lentze, notebook, SAS, P. 1.436, fol. 25.

Johann Carl Wilhelm Lentze, collection of drawings and drafts, SAS, P. 1.590.

Staatsarchiv Hamburg (SAH)

"Angebote von Zement durch in- und ausländische Firmen" 1834–1889, SAH, 326-2 I, Strom und Hafenbau I, 160.

"Bau und Unterhaltung der Landungsanlagen auf dem Grasbrook" 1838–1863, SAH, 326-2 I, Strom und Hafenbau I, n° 713, n. fol.

"Ausbau und Erweiterung des Sandtorhafens bis zum verlegten Brooktor" 1862–1869, SAH, 326-2 I, Strom und Hafenbau I, 576.

"Ausbau der Kaimauer im Sandtorhafen von Abendroths Mühle bis zum Brooktor" 1862–1865, SAH, 326-2 I, Strom und Hafenbau I, 641.

"Hagen-Dalmannscher Grasbrook-Hafenplan", SAH, 326-2 I, Strom und Hafenbau I, 574.

"Begutachtung von hydraulischem Kalk und Beton," 1845–1862," SAH, 321–2, Baudeputation, B451; esp.: John William Thomas Giles, "Bericht an eine hochlöbliche Bau-Deputation", April 8, 1845, n. fol.; Franz Gustav Joachim Forsmann, Johann Hermann Maack, "Bericht den Vorschlag des Ingenieurs Giles über die Anwendung des hydraulischen Kalkes beim Bau der Hammerbrook-Schleuse u. s. w. betreffend", April 14, 1845, n. fol.; Giles, "Sr. Wohlweisheit Herrn Senator Ienisch Präses hochlöblicher Bau-Deputation," April 22, 1845, n. fol.

SAH – Kartensammlung

"Sandthorhafen," SAH, 720–1, 126–1, Häfen und Flüsse – Hafenpläne 5, pl. 186–01; 186–75 III; 187–01.

"Grasbrokhafen," SAH, 720–1, 126–1, Häfen und Flüsse – Hafenpläne 5, pl. 187–12, 187–13a, 187–14; Hafenpläne 6, pl. 186–92–III, 187–31 III, 187–31 IV; Hafenpläne 7, pl. 187–51.

"Nikolaikirche," SAH, 720-1 361–2, pl. III 35a; XII 35; XII 36b; XII 36c; XII 48; XII 49.

Stadtarchiv Buxtehude

"Brunkhorst und Westphalen", Rep. HG. 28, 18; esp.: "Königlisches Ministerium des Innern. Resolution", August 9, 1835,Rep. HG. 29,18, n. fol.

"Brunkhorst und Westphalen", Rep. HG. 28, 28; esp.: "Fragen, welche von Seiten der auswärtigen Mitglieder des Gewerbe-Vereins den einzelnen Besitzern der Fabriken und von Gewerbe-Anstalten zur Beantwortung vorgelegen sind", 1834, n. fol.

Heike Linderkamp, "Die Industrialisierung in Buxtehude. Anfänge und Entwicklung bis zum Beginn des 20. Jahrhunderts," Buxtehude Stadtarchiv, Buxtehude, 1992.

Niedersächsisches Landesarchiv – Standort Hannover (NLA)

"Zusammenstellung wichtiger Erscheinungen und Ergebnisse in Beziehung auf die bei dem Hamelner Kettenbrückenbau gemachten Erfahrungen," n.d., NLA, Hann. 109, n° 1069, esp. foll. 18–43 r/v.

"Anlegung einer Zementfabrik zu Hannover durch den Bankier A. Meyer und den Ingenieur-Kapitän Wendelstadt," 1832–1847, NLA, Hann. 80, Hannover n° 15903; esp.: "Königligles Privilegium zu Cementbereitung," 1833, n. fol.

"Beeinträchtigung des dem Bankier Meyer in Hannover erteilten Privilegium zur Anlage von Zementfabriken," 1833, NLA, Hann. 74, Hameln, n° 3102; esp.: Adolph Meyer, letter to the Königliche Großbritannische Hannoversche Landdrost zu Hannover, Juny 21, 1833

"Direktion der Indirekten Angaben und Zölle zu Göttingen – Die Etablissements zur Verarbeitung des rohen Traß, sowie die Zementfabriken im hiesigen Direktionsbezirk," 1840, NLA, Hann. 85a, n° 642.

"Anlage eines Zement-Steinbruchs in den Forsten der Ämter Hameln und Grohnde und der Betrieb desselben durch den Ingenieur Kapitän Wegebaumeister Wendelstedt und den Bankier Adolph Meyer hieselbst," 1833–1842, NLA, Hann. 76a, n° 828.

NLA Hannover – Plankammer

"Mannheim, die Kettenbrücke über den Neckar in XVI Zeichnungen", n.d., NLA, Karten Deutschland n° 73 SD/1 pg, pl. 3, 12, 13.

"Kettenbrücke über die Weser bei Hameln," 1833, NLA, Kartensammlung, n° 13 b, Hameln, pl. 1k, 2k, 3k, 4pg.

Stadtarchiv Holzminden

"Zeichnungen; Werbeschriften der Fa. Vorwohler-Cementbau-Gesellschaft B. Liebold & Co.," 1877–1902, E.4 n° 3.

Name index